and Partial Differential Equations

Springer
Berlin
Heidelberg
New York
Barcelona
Hong Kong
London
Milan
Paris
Singapore
Tokyo

L. Ambrosio · N. Dancer

Calculus of Variations and Partial Differential Equations

Topics on Geometrical Evolution Problems and Degree Theory

Edited by G. Buttazzo, A. Marino,
M. K. V. Murthy

Springer

Authors and Editors

Luigi Ambrosio
Scuola Normale Superiore
Piazza Cavalieri 7
56100 Pisa, Italy

Norman Dancer
University of Sydney
School of Mathematics
NSW 2006 Sydney, Australia

Giuseppe Buttazzo
Antonio Marino
M. K. V. Murthy
Università di Pisa
Dipartimento di Matematica
Via Buonarroti 2
56127 Pisa, Italy

Mathematics Subject Classification (1991): 60J15, 30D05, 35Q15, 30F10, 60K25

Library of Congress Cataloging-in-Publication Data

Calculus of variations and partial differential equations : topics on geometrical evolution
problems and degree theory / edited by G. Buttazzo, A. Marino, M.K.V. Murthy.
   p. cm.
   Includes bibliographical references and index.
   ISBN 3540648038 (softcover : alk. paper)
   1. Calculus of variations. 2. Topological degree. 3. Differential equations, Partial. 4.
Mathematical physics. I. Buttazzo, Giuseppe. II. Marino, A. III. Murthy, M. K. V. (M.
K. Venkatesha) IV. Summer School on "Calculus of Variations and Partial Differential
Equations" (1996 : Pisa, Italy)

QC20.7.C3 C35 1999
515'.64--dc21
                                                                           98-054672

ISBN 3-540-64803-8 Springer-Verlag Berlin Heidelberg New York

© Springer-Verlag Berlin Heidelberg 2000
Printed in Germany

Cover design: *design & production* GmbH, Heidelberg

SPIN: 10688575   41/3143 - 5 4 3 2 1 0 - Printed on acid-free paper

Preface

The project of editing this book originated in a two-week Summer School on *"Calculus of Variations and Partial Differential Equations"* which was held in Pisa in September 1996.

The School was attended by more than 150 participants, mostly Ph. D. students, but also by several more experienced scientists. The success achieved encouraged us to collect into a book the course material of the two main lecturers and the complementary articles of the eight other speakers on related subjects.

Special importance was attached to the self-contained presentation of these rather advanced research topics, and we hope that our work will contribute to filling the gap between the level of basic courses and more advanced research.

Pisa, March 3, 1998

G. Buttazzo   A. Marino   M.K.V. Murthy

Table of Contents

I Geometric Evolution Problems

II Degree Theory on Convex Sets and Applications to Bifurcation

---

List of Authors

Giovanni Alberti
Dipartimento di Matematica
Via Buonarroti 2
Università di Pisa
56127 Pisa (Italy)
e-mail: alberti@dm.unipi.it

Luigi Ambrosio
Scuola Normale Superiore
Classe di Scienze
Piazza dei Cavalieri, 7
56126 Pisa (Italy)
e-mail: ambrosio@sns.it

Giovanni Bellettini
Università di Roma "Tor Vergata"
Via della Ricerca Scientifica
00133 Roma (Italy)
e-mail: bellettini@axp.mat.uniroma2.it

Vieri Benci
Dipartimento di Matematica
Applicata "U. Dini"
Università di Pisa, Via Bonanno 25/b
56126 Pisa (Italy)
e-mail: benci@dma.unipi.it

Andrea Braides
SISSA
Via Beirut 4
34014 Trieste (Italy)
e-mail: braides@sissa.it

Giovanna Cerami
Dipartimento di Matematica e Applicazioni
Università di Palermo
Via Archirafi, 34
90123 Palermo (Italy)
e-mail: cerami@ipamat.math.unipa.it

E. N. Dancer
School of Mathematics and Statistics
University of Sydney
NSW 2006 (Australia)
e-mail: dancer@maths.su.oz.au

Donato Fortunato
Dipartimento di Matematica
Università di Bari, Via Orabona 4
70125 Bari (Italy)
e-mail: fortunat@pascal.dm.uniba.it

Antonio Leaci
Dipartimento di Matematica
Università di Lecce
Via Arnesano
73100 Lecce (Italy)
e-mail: leaci@ingle01.unile.it

Matteo Novaga
Dipartimento di Matematica
Università di Pisa
Via Buonarroti, 2
56127 Pisa (Italy)
e-mail: novaga@dm.unipi.it

Donato Passaseo
Dipartimento di Matematica
Università di Lecce
Via Arnesano
73100 Lecce (Italy)
e-mail: passaseo@ilenic.unile.it

Part I

Geometric Evolution Problems

Introduction to Part I

In many problems arising from Mathematical Physics and from Applied Mathematics the objects involved are surfaces, or more generally lower dimensional sets. Some nowadays well-known examples that could be quoted in this respect as model cases are the phase transitions problem, the Mumford and Shah image segmentation problem in computer vision, and the geometric evolution problems such as for instance the motion by mean curvature.

In some phase transitions problems one has to deal with the minimization of a double well potential with a volume constraint:

$$\min\left\{\int_\Omega W(u)\,dx \ : \ \int_\Omega u\,dx = \text{constant}\right\}. \qquad (\wp)$$

Since problem $(\wp)$ above has too many nonphysical solutions, the singular perturbation

$$\min\left\{\int_\Omega [\varepsilon|Du|^2 + W(u)]\,dx \ : \ \int_\Omega u\,dx = \text{const}\right\}. \qquad (\wp_\varepsilon)$$

was proposed by Cahn and Hilliard in 1958 to select, by taking the limit as $\varepsilon \to 0^+$, physically reasonable solutions.

It was first conjectured by De Giorgi in 1975 and then proved by Modica and Mortola in 1977 that the solutions selected in this way jump between two zeros of the potential W across a surface of least area.

The Mumford and Shah image segmentation problem can be described by the following minimization problem:

$$\min\left\{\alpha \int_{\Omega\setminus K} |\nabla u|^2\,dx + \beta\mathcal{H}^{n-1}(K\cap\Omega) + \gamma\int_{\Omega\setminus K} |u-g|^2\,dx\right\}, \qquad (MS)$$

where Ω represents the domain of a digitalized image g, and α, β, γ are three given positive constants. By minimizing the functional in (MS) among all closed subsets K of $\overline{\Omega}$ and all functions $u \in C^1(\Omega\setminus K)$ one looks for a piecewise smooth approximation u of the input g, filtering noise effects. The feature of this model is that diffusion across sharp discontinuities of g is prevented, thus leading to a good reconstruction of the edges of the image.

The motion of a surface according to its mean curvature has been widely studied in the last two decades and is, among geometric evolution problems, the best known. Since singularities may appear during the evolution, several weak formulations have been proposed to describe the long time behaviour of surfaces. One of the possibilities is to represent the evolving surface as the level set of an auxiliary function u solving (in the weak sense of the theory of viscosity solutions) the partial differential equation

$$\frac{\partial u}{\partial t} = |Du|\,\text{div}\left(\frac{Du}{|Du|}\right). \qquad (MC)$$

The formal interpretation of this equation is that any level set of $u(t,\cdot)$ flows by mean curvature. In this connection we mention that Huisken and Ilmanen recently, using the inverse mean curvature flow in a Riemannian manifold and a PDE analogous to (MC), proved the Riemannian Penrose inequality in general relativity.

The lecture notes by Luigi Ambrosio, organized in a self-contained way, are devoted to a presentation of the level set approach and the theory of viscosity solutions for the mean curvature flow problem. Related topics are the fine properties of the distance functions, the differentiability properties of semiconvex and semiconcave functions and the parabolic version of the Modica and Mortola approximation theorem, first proposed by Allen and Cahn in 1979.

A new and very general version of the convergence of problems (p_ε) is contained in the contribution by G. Alberti, where the Dirichlet integral is replaced by a nonlocal functional. The strong connections between the level set–viscosity approach to the mean curvature flow problem and the theory of barriers, proposed by De Giorgi in 1994, are investigated in the contribution by G. Bellettini. The contribution by A. Braides deals with the approximation of the Mumford and Shah functional by a family of nonlocal problems. Finally, the contribution by A. Leaci is devoted to the existence of optimal pairs (u, K) for problem (MS) and to their regularity properties.

Geometric evolution problems, distance function and viscosity solutions

L. Ambrosio

1 Introduction

The mean curvature flow is a geometric initial value problem. Starting from a smooth initial surface Γ_0 in $\mathbf{R}^n$, the surfaces Γ_t evolve in time with normal velocity equal to their mean curvature vector. By parametric methods of differential geometry many results have been obtained for convex surfaces, graphs or planar curves (see for instance Altschuler & Grayson [AG92], Ecker & Huisken [EH89], Gage & Hamilton [GH86], Grayson [Gra87], and Huisken [Hui84]). However, for $n \geq 3$, initially smooth surfaces may develop singularities. For example, a "dumbbell" region in $\mathbf{R}^3$ splits into two pieces in finite time (cf. [Gra89a]) or a "fat" enough torus closes its interior hole in finite time (cf. [SS93]). Also it can be seen that smooth curves in $\mathbf{R}^3$ may self intersect in finite time.

These examples show that the parametric method may fail, and therefore several weak solutions of the mean curvature flow, defined even after the appearence of singularities, have been proposed. Basically, these weak solutions can be divided in two groups, the *measure theoretic* ones and the *set theoretic* ones.

In his pioneering work, Brakke [Bra78] used geometric measure theory to construct a varifold solution. In Brakke's setting a surface (with multiplicities) is identified with a measure and the mean curvature vector is identified by a suitable integration by parts formula (as the gradient in Sobolev spaces) related to the divergence theorem (cf. (7)). Even though Brakke's solutions are not unique (roughly speaking, they are measure theoretic subsolutions) Brakke's work had a great influence on later developments, also because of his partial regularity result. A different approach, still measure theoretic and based on time discretization and on a recursive minimization problem has been proposed by Almgren, Taylor & Wang [ATW93] (see also [TCH93, LS95, Amb95a]).

For codimension one surfaces, a completely different approach (initially suggested in the physics literature by Ohta, Jasnaw & Kawasaki [JKO82] and for numerical calculations by Sethian [Set85] and Osher & Sethian [OS88]) represents the evolving surface as the level set of an auxiliary function solving an appropriate nonlinear parabolic equation. This "level set" approach has been developed by Chen, Giga & Goto [CGG91] and, independently, by Evans & Spruck [ES91]. Given an initial hypersurface Γ_0, one selects a uniformly continuous function $u_0 : \mathbf{R}^n \to \mathbf{R}$ such that

$$\Gamma_0 = \{x \in \mathbf{R}^n \,|\, u_0(x) = 0\}.$$

Then, the idea is to flow *any* level set of $u(t, \cdot)$ by mean curvature, in order to get a parabolic PDE satisfied by $u(t, x)$, with the initial condition $u(0, \cdot) = u_0$. Assuming that u is smooth and that its spatial gradient ∇u does not vanish, we can orient the level sets $\{u = \tau\} = \partial\{u < \tau\}$ by $\nu := \nabla u/|\nabla u|$ to get

$$u_t = |\nabla u| \operatorname{div}\left(\frac{\nabla u}{|\nabla u|}\right) = \Delta u - \frac{\langle \nabla^2 u \nabla u, \nabla u \rangle}{|\nabla u|^2} \tag{1}$$

because $-u_t/|\nabla u|$ is the normal velocity along ν and $-\operatorname{div}(\nu)$ is the (scalar) mean curvature along ν (cf. (11)).

The equation (1) is nonlinear, degenerate along ∇u, and it is not well defined when ∇u is zero. Evans & Spruck and Chen, Giga & Goto overcame these difficulties by using the theory of viscosity solutions ([CL83, CEL84, CIL92, FS93]). In particular, in [CGG91, ES91] it is proved that there exists a unique viscosity solution u of (1) with the initial condition $u(0, \cdot) = u_0$ and that

$$\Gamma_t := \{u(t, \cdot) = 0\}$$

depends (of course) on Γ_0, but not on the auxiliary function u_0. Therefore (Γ_t) is a well defined evolution of Γ_0 satisfying the semigroup property. Other interesting properties of (Γ_t), including Hausdorff dimension estimates, local time existence of classical solutions are obtained in [ES92a, ES92b, ES95]. Also, in [CGG91] it has been proved that the level set approach for hypersurfaces, is general enough to deal with equations more general than the mean curvature flow, including anisotropic flows with driving forces. Finally, in [AS96] the level set approach has been extended to higher codimension mean curvature flow problems.

More intrinsic and set theoretic definitions related to level set solutions have also been studied: in [Son93b], Soner builds weak solutions by using the signed distance function $\bar{d}(x, U_t)$ from time dependent domains U_t (see also [BSS93]). The basic idea is to use the inequalities (in the viscosity sense)

$$\bar{d}_t \geq \Delta\bar{d} \quad \text{in } \{\bar{d} > 0\}, \qquad \bar{d}_t \leq \Delta\bar{d} \quad \text{in } \{\bar{d} < 0\}$$

to characterize the flow of ∂U_t by mean curvature.

In [Ilm93b], Ilmanen used smooth classical solutions as test functions to define set theoretic subsolutions. These subsolutions, based on the *avoidance* principle, have been later used in [Ilm94] to prove a connection between the varifold solutions of Brakke and the level set solutions (see also [AS96]). In [Gio94b] De Giorgi introduced, extending Ilmanen's approach, a general notion of barrier based on the *inclusion* principle. Barriers are very closely related to the level set solutions (see [BP95b, BP95c, BN98]).

Finally, the singular limit as $\epsilon \downarrow 0$ of a reaction–diffusion equation with a cubic nonlinearity

$$u_t = \Delta u + \frac{1}{\epsilon^2}u(1 - u^2) \tag{2}$$

also provides an approximation and a possibly different definition for mean curvature flow of boundaries (see §14 and [AC79, BK91, Gio90a, CI90, MS90, ESS92,

BSS93, NPV94]). The equations (2) are, up to a time scaling, the gradient flows of the Modica & Mortola functionals

$$F_\epsilon(u) := \frac{1}{2} \int_{\mathbf{R}^n} \left(\epsilon |\nabla u|^2 + \frac{(1 - u^2)^2}{2\epsilon} \right) dx$$

which Γ^- converge to a constant multiple of the area functional as $\epsilon \downarrow 0$ (cf. Theorem 28). This provides an heuristic justification of why, with a suitable time scale, motion by mean curvature should be expected in the limit. The limiting behaviour of solutions of (2) can be studied either with measure theoretic tools (see [Ilm94, Son93a]) or with set theoretic ones (see [ESS92, BSS93]). For this reason, this approach provides an important link between measure theoretic and set theoretic solutions.

These notes are mostly devoted to the set theoretic approach in codimension 1 and higher than 1; since I aim to a (as much as possible) self contained presentation of this subject and of the relevant techniques involved, a lot of basic definitions (second fundamental form, smooth flows) and of preliminary results (properties of distance function, viscosity solutions of equations of first and second order, semiconcave and semiconvex functions) are needed before entering in the heart of the matter, in §10. The goal of the notes is also to give a systematic account of the main properties of distance function, particularly in connection with geometric evolution problems.

In the last sections, after proving consistency of the level set approach with classical solutions, the level set approach will be compared with the barrier approach of De Giorgi, the distance approach of Soner and with the approximations provided by (2). A short introduction to the theory of semilinear parabolic equations, needed to have existence and bounds for the solutions of (2), is provided.

I had the occasion to write these notes in September 1996, when I was asked by Marino and Buttazzo to give a series of lectures in Pisa on this topic. I would like to dedicate this work to the memory of Ennio De Giorgi, with admiration for him as a scientist and as a man. I also would like to thank G. Bellettini, C. Mantegazza, M. Novaga and H.M. Soner for many stimulating conversations.

2   Main notations

– $B_r(x)$ denotes the open ball of radius r centered at x in $\mathbf{R}^n$, $B_r = B_r(0)$, $B = B_1$ and $\mathbf{S}^{n-1} = \partial B$;
– $\mathcal{L}^n$ denotes the Lebesgue measure in $\mathbf{R}^n$, $\omega_n = \mathcal{L}^n(B)$ and Sym^n denotes the space of $n \times n$ symmetric matrices;
– given a function u, u_* and u^* respectively denote the lower semicontinuous and the upper semicontinuous envelope of u, defined on the closure of the domain of u;
– ∇u and $\nabla^2 u$ stand for the gradient and the Hessian of u, while Du and $D^2 u$ stand for first order and the second order distributional derivative;
– $\mathrm{dist}(x, E)$, $\bar{d}(x, E)$, $\eta(x, E)$ respectively stand for the distance function, the signed distance function and the squared distance function from a set $E \subset \mathbf{R}^n$, defined in §4.

3   Second fundamental form and mean curvature vector

In this section we will introduce the main geometric concepts we will deal with, namely, the second fundamental form and the mean curvature vector of a smooth (embedded) manifold $\Gamma \subset \mathbf{R}^n$.

Definition 1 (tangential gradient). Let $\Gamma \subset \mathbf{R}^n$ be a smooth k-dimensional manifold, let $x \in \Gamma$ and let ϕ be a function defined in a neighbourhood of x and differentiable at x; we denote by

$$\nabla^\Gamma \phi(x) = \left(\nabla_1^\Gamma \phi(x), \ldots, \nabla_n^\Gamma \phi(x)\right)$$

the *tangential gradient* of ϕ at x, i.e., the projection of $\nabla \phi(x)$ on $T_x \Gamma$.

It is easy to check that $\nabla^\Gamma \phi$ has an intrinsic meaning: if g and g' are equal in $U \cap \Gamma$ for some neighbourhood U of x, then $\nabla^\Gamma g(x) = \nabla^\Gamma g'(x)$. Now we define the second fundamental form of Γ, a 3-tensor which contains all the informations on the curvature of Γ.

Definition 2 (second fundamental form and principal curvatures). Let $\Gamma \subset \mathbf{R}^n$ be a smooth k-dimensional manifold and let $x \in \Gamma$; we will denote by $\mathbf{B}_x : T_x \Gamma \times T_x \Gamma \to N_x \Gamma$ the *second fundamental form* of Γ at x, defined by

$$\mathbf{B}_x(\xi, \eta) := -\sum_{\alpha=1}^{n-k} \langle \xi, d_\eta \nu^\alpha(x) \rangle \nu^\alpha(x). \tag{3}$$

In (3), $\nu^1, \ldots, \nu^{n-k}$ are smooth functions defined in a neighbourhood U of x, such that

$$\left\{ \nu^1(y), \ldots, \nu^{n-k}(y) \right\}$$

is an orthonormal basis of $N_y \Gamma$ for any $y \in U \cap \Gamma$ and $d_\eta \nu^\alpha$ is the derivative of ν^α along the direction η. Clearly, $\mathbf{B}_x$ does not depend on the orientation of Γ and it can be proved that $\mathbf{B}_x$ is bilinear and symmetric (see for instance [Sim84]). Given any unit vector $p \in N_x \Gamma$, the *principal curvatures* along the direction p are the eigenvalues $\kappa_1(x), \ldots, \kappa_k(x)$ of the symmetric bilinear form on $T_x \Gamma$

$$B_x^p(\xi, \eta) := -\langle \mathbf{B}_x(\xi, \eta), p \rangle.$$

Remark 1. In the codimension 1 case $k = (n-1)$, (3) becomes much simpler because there is only one normal direction. The principal curvatures, unlike $\mathbf{B}_x$, depend on the orientation. If $\Gamma = \partial U$ we will canonically use the orientation on Γ induced by the outer normal to ∂U. With this convention in force, boundaries of convex sets have nonnegative principal curvatures. Also the computation of tangential gradients is simple in codimension 1:

$$\nabla^\Gamma \phi = \nabla \phi - \langle \nabla \phi, \nu \rangle \nu = P_\nu \nabla \phi, \tag{4}$$

where

$$P_\nu := I - \nu \otimes \nu. \tag{5}$$

Definition 3 (mean curvature vector). Let $\Gamma \subset \mathbf{R}^n$ be as in Definition 2 and let $x \in \Gamma$. The *mean curvature vector* $\mathbf{H}(x)$ of Γ at x is the trace of $\mathbf{B}_x$, i.e.:

$$\mathbf{H}(x) = \sum_{i=1}^{k} \mathbf{B}_x(e_i, e_i) \tag{6}$$

where $\{e_1, \ldots, e_k\}$ is any orthonormal basis of $T_x\Gamma$ (it is easy to check that the definition is well posed, because $\mathbf{B}_x$ is symmetric).

The mean curvature vector is normal to Γ and does not depend on the orientation. The importance of this vector lies in the *divergence formula*

$$\int_\Gamma \mathrm{div}^\Gamma X \, d\mathcal{H}^k = -\int_\Gamma \langle \mathbf{H}, X \rangle \, d\mathcal{H}^k \qquad \forall X \in C_c^1(\mathbf{R}^n; \mathbf{R}^n) \tag{7}$$

valid if Γ has no boundary (without this assumption there is an extra term in (7), as in Gauss–Green's formula). We recall that, by definition

$$\int_\Gamma g(x) \, d\mathcal{H}^k(x) := \int_U g(\Phi(y)) J_k\Phi(y) \, dy \tag{8}$$

where the k *dimensional Jacobian* $J_k\Phi$ is given by

$$J_k\Phi := \left| \det \langle \frac{\partial\Phi}{\partial y_i}, \frac{\partial\Phi}{\partial y_j} \rangle \right| . \tag{9}$$

The equation (8) makes sense if $\mathrm{supp}\,(g)$ is contained in the image of the local parametrization $\Phi : V \subset \mathbf{R}^k \to \Gamma$; this assumption on the support of g can be removed by using partitions of unity, and it can be proved that the integral is independent of the parametrizations and of the partitions used (see for instance [Car76]). In (7), $\mathrm{div}^\Gamma X$ is the *tangential divergence* of X, i.e.,

$$\mathrm{div}^\Gamma X := \sum_{i=1}^{n} \nabla_i^\Gamma X_i.$$

The divergence formula can be proved using the identity (see for instance [Car76])

$$\int_\Gamma \mathrm{div}^\Gamma Y \, d\mathcal{H}^k = 0 \tag{10}$$

for any C_c^1 tangential vectorfield Y. In fact, to prove (7), splitting the field X in tangential part $Y = X'$ and normal part

$$X'' = \sum_{\alpha=1}^{n-k} \langle X, \nu^\alpha \rangle \nu^\alpha = \sum_{\alpha=1}^{n-k} X^\alpha \nu^\alpha$$

we can assume by (10) and the perpendicularity of $\mathbf{H}$ that X is normal, hence

$$-\int_\Gamma \langle X, \mathbf{H} \rangle \, d\mathcal{H}^k = \sum_{\alpha=1}^{n-k} \int_\Gamma X^\alpha \mathrm{div}^\Gamma \nu^\alpha \, d\mathcal{H}^k = \sum_{\alpha=1}^{n-k} \int_\Gamma \mathrm{div}^\Gamma (X^\alpha \nu^\alpha) \, d\mathcal{H}^k$$

$$-\sum_{\alpha=1}^{n-k} \int_\Gamma \langle \nabla^\Gamma X^\alpha, \nu^\alpha \rangle \, d\mathcal{H}^k = \int_\Gamma \mathrm{div}^\Gamma X \, d\mathcal{H}^k .$$

Remark 2. In the codimension 1 case $k = (n - 1)$, (6) becomes

$$\mathbf{H} = -\sum_{i=1}^{n-1} \langle e_i, d_{e_i} \nu \rangle \nu = -\sum_{i=1}^{n-1} \langle e_i, d_{e_i} \nu \rangle \nu - \langle \nu, d_\nu \nu \rangle \nu$$

$$= -(\mathrm{div}^\Gamma \nu)\nu = -(\mathrm{div}\, \nu)\nu + \sum_i \langle \nabla\nu_i, \nu \rangle \nu_i = -(\mathrm{div}\, \nu)\nu \qquad (11)$$

because (by differentiation of the identity $|\nu|^2 = 1$) $\sum_1^n \nu_i d\nu_i = 0$.
In addition, if $\kappa_1, \dots, \kappa_{n-1}$ are the principal curvatures induced by ν, (6) gives

$$\mathbf{H} = -(\kappa_1 + \dots + \kappa_{n-1})\nu. \qquad (12)$$

Recalling our convention about the sign of principal curvatures of convex sets, we see from (12) that $\mathbf{H}$ points "inside" when Γ is the boundary of a convex set. More generally, the geometric meaning of $\mathbf{H}$ (in any codimension) is the direction where area "decreases most". To make this statement precise, let us assume for simplicity that Γ is compact and without boundary, and let us consider a smooth 1-parameter family of smooth functions

$$\Phi_t : \mathbf{R}^n \to \mathbf{R}^n \qquad t \in J \subset \mathbf{R} \text{ interval}$$

such that $0 \in J$ and $\Phi_0(x) \equiv x$ (the typical example of this situation is $\Phi_t(x) = x + tg(x)$ with $g \in C_c^\infty(\mathbf{R}^n)$ and $J = (-\epsilon, \epsilon)$ for small ϵ). Then, setting

$$g(x) := \frac{d}{dt}\Phi_t(x)\Big|_{t=0}$$

using the area formula to compute $\mathcal{H}^k(\Gamma_t)$ (see [Sim84]) it can be proved that

$$\frac{d}{dt}\mathcal{H}^k(\Gamma_t)\Big|_{t=0} = \frac{d}{dt}\int_\Gamma J_k \Phi_t \, d\mathcal{H}^k\Big|_{t=0} = \int_\Gamma \mathrm{div}^\Gamma g \, d\mathcal{H}^k = -\int_\Gamma \langle \mathbf{H}, g \rangle \, d\mathcal{H}^k \qquad (13)$$

the last equality being justified by the divergence theorem.

We conclude this section showing how $\mathbf{B}_x$ and $\mathbf{H}$ can be computed for codimension 1 surfaces using the standard representation as level sets of functions. Hence, assume that Γ concides in a neighbourhood U of x with $\{u = \tau\} = \partial\{u < \tau\}$ for some $u \in C^2(U)$ such that $\nabla u \neq 0$ in U. Then, setting $\nu := \nabla u(x)/|\nabla u(x)|$, it is easy to show that

$$B_x^\nu = \frac{P_\nu \nabla^2 u P_\nu}{|\nabla u|} \qquad \text{on } T_x\Gamma \times T_x\Gamma \qquad (14)$$

and therefore

$$\langle \mathbf{H}, \nu \rangle = -\mathrm{trace}\left(\frac{P_\nu \nabla^2 u P_\nu}{|\nabla u|}\right) = -\Delta u + \frac{\langle \nabla^2 u \nabla u, \nabla u \rangle}{|\nabla u|^2}. \qquad (15)$$

In the higher codimension case Γ is the intersection of $(n - k)$ level sets (i.e., the set of solutions of a system) and it seems quite difficult to get a manageable formula extending (14) and (15) to this case. We will see in the next section that the distance function provides a representation of $\mathbf{B}_x$ and $\mathbf{H}$ which works in any codimension and which is very useful in codimension 1 as well.

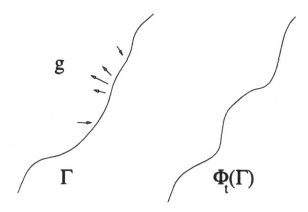

Fig. 1. First variation of area along a field g

4   Distance function

Let $E \subset \mathbf{R}^n$ be a set; we will denote by

$$d(x, E) := \inf_{y \in E} |x - y|$$

the *distance function* from E, and we will drop E when no ambiguity arises. The *signed distance function* is defined by

$$\bar{d}(x, E) := d(x, E) - d(x, \mathbf{R}^n \setminus E).$$

We notice that, by definition, $\bar{d}(x, E)$ is equal to $-d(x, \partial E)$ in E, and equal to $d(x, \partial E)$ in the complement of E. Moreover

$$\bar{d}(x, \mathbf{R}^n \setminus E) = -\bar{d}(x, E), \qquad E \subset F \implies \bar{d}(x, E) \geq \bar{d}(x, F). \quad (16)$$

Finally, the *squared distance function* is defined by

$$\eta(x, E) = \frac{1}{2} d^2(x, E).$$

The factor $1/2$ is introduced just for convenience, to simplify several identities below.

It is easy to check that the Lipschitz constants of d and of $\bar{d}$ are less than 1; in particular, by Rademacher's theorem, these functions are almost everywhere differentiable in $\mathbf{R}^n$ and $|\nabla d| \leq 1$, $|\nabla \bar{d}| \leq 1$. A more precise result is stated below:

Theorem 1 (differentiability of distance). *Let $E \subset \mathbf{R}^n$ be a nonempty closed set and let $x \in \mathbf{R}^n \setminus E$. Then, d is differentiable at x if and only if there exists a unique $y \in E$ such that $d(x) = |x - y|$. In this case*

$$\nabla d(x) = \frac{x - y}{|x - y|} = \frac{x - y}{d(x)}.$$

In particular, $|\nabla d| = 1$ at any differentiability point.

Proof. Assume that d is differentiable at x and let $y \in E$ be any least distance point; then

$$d^2(x + \epsilon z) \geq d^2(x) + 2\epsilon d(x)\langle \nabla d(x), z \rangle + o(\epsilon)$$

$$d^2(x + \epsilon z) \leq |x + \epsilon z - y|^2 = d^2(x) + 2\epsilon\langle x - y, z \rangle + o(\epsilon)$$

for any $z \in \mathbf{R}^n$. Comparing the two expressions we obtain $d(x)\nabla d(x) = x - y$, because z is arbitrary. In particular, y is uniquely determined by d.

The proof of the opposite implication depends on the local semiconcavity of d (see Exercise 7 and Proposition 15). In particular, we will use the fact that if d is not differentiable at x then the first order super jet $J^{1,+}d(x)$ is not a singleton. If $J^{1,+}d(x)$ is not a singleton, we can find sequences (x_h^1), (x_h^2), both converging to x, such that d is differentiable at x_h^i and

$$p^1 := \lim_{h \to \infty} \nabla d(x_h^1) \neq p^2 := \lim_{h \to \infty} \nabla d(x_h^2).$$

Denoting by $y_h^i = x_h^i - d(x_h^i)\nabla d(x_h^i)$ the corresponding least distance points on E, it follows by the previous implication that $x - d(x)p^1$ and $x - d(x)p^2$ are distinct least distance points from x.   □

Remark 3. (1) An useful consequence of Theorem 1 is the following: if d is differentiable at $x \notin E$, then $y = x - d(x)\nabla d(x)$ is the nearest point on E and d is differentiable at any point $z \in S \setminus \{y\}$ with $\nabla d(z) = \nabla d(x)$, where S is the segment joining x to y; indeed, a simple application of the triangle inequality shows that y is the unique nearest point to any $z \in S$. We will often use this fact in the following.

(2) A similar result applies to the $\bar{d}(x, E)$ on the complement of ∂E. The squared distance function is differentiable on E and satisfies the identity $|\nabla \eta|^2 = 2\eta$ at any differentiability point.

Of course, the smoothness of d, $\bar{d}$ and the smoothness of ∂E are related. We now state the following local regularity result, whose proof essentially depends on the implicit function theorem (see also [Fed59, Fu85, Foo84, DZ94]):

Theorem 2 (regularity of distance).

(i) *Let $\Omega \subset \mathbf{R}^n$ be a bounded open set with smooth boundary; then, $\bar{d}(x, \Omega)$ is smooth in a tubular neighbourhood U of $\partial\Omega$. Moreover, the outer normal to Ω is given by $\nabla\bar{d}$.*

(ii) *Let $\Gamma \subset \mathbf{R}^n$ be a compact, smooth k-dimensional manifold without boundary; then, $\eta(x, \Gamma)$ is smooth in a tubular neighbourhood U of Γ and*

$$\eta(x + p) = \frac{1}{2}|p|^2 \tag{17}$$

for any $x \in \Gamma$ and any vector $p \in N_x\Gamma$ such that $x + p \in U$. In particular, for any $x \in \Gamma$ the matrix $\nabla^2\eta(x)$ is the orthogonal projection on $N_x\Gamma$.

Proof. We will only prove (ii), the proof of the first statement being analogous. Fix $x_0 \in \Gamma$. By the smoothness of Γ, there exist a constant $s > 0$ and a smooth orthonormal vector field

$$(\nu^1, \ldots, \nu^{n-k}) : B_s(x_0) \cap \Gamma \to \mathbf{R}^{n(n-k)}$$

spanning the normal space to Γ. Set

$$\Phi(x, \alpha) := x + \sum_{i=1}^{n-k} \alpha_i \nu^i(x) \qquad x \in B_s(x_0) \cap \Gamma, \ \alpha \in \mathbf{R}^{n-k}.$$

Using local coordinates, it is easy to see that the Jacobian $J_n \Phi(x_0, 0)$ (see (9)) is equal to 1. Hence, by the implicit function theorem, there exists $r \in (0, s)$ satisfying:
(1) in $(B_r(x_0) \cap \Gamma) \times B_r^{n-k}(0)$, Φ is one to one;
(2) $V = \Phi\big((B_r(x_0) \cap \Gamma) \times B_r^{n-k}(0)\big)$ is an open set containing x_0.
  For $y \in V$, let

$$\Psi(y) := (x(y), \alpha(y)) \in (B_r(x_0) \cap \Gamma) \times B_r^{n-k}(0)$$

be the smooth inverse of Φ. Choose $\sigma \in (0, r/2)$ such that $B_\sigma(x_0) \subset V$. We will relate the functions $x(y)$, $\alpha(y)$ to the distance function. Indeed, for $y \in B_\sigma(x_0)$, let $x \in \Gamma$ be any minimizer of the distance, i.e., $d(y) = |x - y|$. Then, since

$$|x - y| \leq |x_0 - y| < \sigma, \qquad |y - x_0| < \sigma$$

it follows that $x \in B_r(x_0) \cap \Gamma$, hence $x = x(y)$. Moreover, $d(y) = |\alpha(y)|$ and consequently

$$2\eta(y) = |\alpha(y)|^2 = \sum_{i=1}^{n-k} \alpha_i^2(y) \qquad \forall y \in B_\sigma(x_0).$$

Hence η is smooth and (17) holds in $B_\sigma(x_0)$ by construction. Since Γ is compact, we use a covering argument to extend these properties to a tubular neighbourhood $\{\eta < \sigma^2/2\}$.
  Since $\eta(x + p) = o(|p|^2)$ for $p \in T_x\Gamma$, $\nabla^2\eta$ vanishes on $T_x\Gamma$, so that (17) implies that $\nabla^2\eta(x)$ is the orthogonal projection on $N_x\Gamma$. $\square$

  The compactness assumption in Theorem 2 can be weakened, getting regularity only in a neighbourhood of the manifold. By differentiation of the identities $|\nabla \bar{d}|^2 = 1$ and $|\nabla \eta|^2 = 2\eta$ on U we obtain (here the subscript stands for differentiation and the summation convention is understood)

$$\bar{d}_{ij}\bar{d}_i = 0, \qquad \bar{d}_{ijk}\bar{d}_i + \bar{d}_{ij}\bar{d}_{ik} = 0 \qquad \text{on } U \qquad (18)$$

$$\eta_{ij}\eta_i = \eta_j, \qquad \eta_{ijk}\eta_i + \eta_{ij}\eta_{ik} = \eta_{jk} \qquad \text{on } U. \qquad (19)$$

Using (11), (14) and the identity $|\nabla \bar{d}|^2 = 1$ we can easily compute in the codimension one case $\mathbf{B}_x^{\nu}$ (choosing $\nu = \nabla \bar{d}$) and $\mathbf{H}$

$$B_x^{\nu} = \nabla^2 \bar{d} \quad \text{on } T_x \partial \Omega, \qquad \mathbf{H} = -\Delta \bar{d} \cdot \nu \qquad (20)$$

because, by (18), $\nabla^2 \bar{d} \nabla \bar{d} = 0$.

The characterization of geometric evolution problems by PDE satisfied by the distance function in the ambient space $\mathbf{R}^n$ requires the knowledge of $\nabla^2 \bar{d}$ not only on the manifold, but also in a neighbourhood of it. To this aim, we state two separate results, the first one concerning $\bar{d}$ and the second one concerning η.

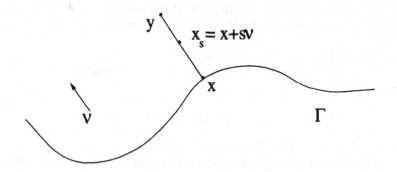

Fig. 2. Propagation of eigenvalues on normal lines

Theorem 3 (eigenvalues of $\nabla^2 \bar{d}$). *Let Ω, U be as in Theorem 2(i) and let $y \in U$, $x = y - \bar{d}(y)\nabla \bar{d}(y) \in \partial \Omega$, $\nu = \nabla \bar{d}(y)$. Then, denoting by $\mu_1, \ldots, \mu_{n-1}$ the eigenvalues of $\nabla^2 \bar{d}(y)$ corresponding to directions orthogonal to ν, it holds*

$$\mu_i = \frac{\kappa_i}{1 + \bar{d}(y)\kappa_i} \qquad i = 1, \ldots, (n-1) \qquad (21)$$

where $\kappa_1, \ldots, \kappa_{n-1}$ are the principal curvatures of $\partial \Omega$ at x along $\nu = \nabla \bar{d}(x)$.

Proof. By (18) and (20) we can choose an orthonormal basis $\{z_1, \ldots, z_n\}$ of $\mathbf{R}^n$ in which $\nabla^2 \bar{d}(x)$ is diagonal, such that $z_n = \nu$ corresponds to the zero eigenvalue κ_n and

$$\nabla^2 \bar{d}(x) z_i = \kappa_i z_i \qquad i = 1, \ldots, n.$$

Let $B(s) = \nabla^2 \bar{d}(x + s\nu)$ for any $s \in \mathbf{R}$ such that $x + s\nu \in U$; by (18) we infer

$$B_{ij}'(s) = \bar{d}_{ijk}\nu_k = \bar{d}_{ijk}\bar{d}_k = -\bar{d}_{ik}\bar{d}_{kj} = -B_{ij}^2(s)$$

where all derivatives of $\bar{d}$ are evaluated at $x + s\nu$. The solution of the (matrix valued) ODE $B' = -B^2$ is

$$B(s) := \sum_{i=1}^{n} \beta_i(s) z_i \otimes z_i$$

where
$$\beta_i'(s) = -\beta_i^2(s), \quad \beta_i(0) = \kappa_i \qquad i = 1, \ldots, n.$$

Solving the ODE's above, we find $\beta_i(s) = \kappa_i/(1 + s\kappa_i)$. $\qquad\qquad\square$

Theorem 4 (eigenvalues of $\nabla^2\eta$). *Let Γ, U be as in Theorem 2(ii) and let $y \in U$, $x = y - \nabla\eta(y) \in \Gamma$, $k = \dim\Gamma$. Then, denoting by $\lambda_1, \ldots, \lambda_n$ the eigenvalues of $\nabla^2\eta(y)$, it holds*

$$\lambda_i = \begin{cases} \dfrac{d(y)\kappa_i}{1 + d(y)\kappa_i} & \text{if } 1 \leq i \leq k; \\[2mm] 1 & \text{if } k < i \leq n \end{cases} \tag{22}$$

where $\kappa_1, \ldots, \kappa_k$ are the principal curvatures of Γ at x along $\nabla d(y) \in N_x\Gamma$.

Proof. We will only prove the existence of real numbers κ_i satisfying (22); their characterization as principal curvatures along $p := \nabla d(y)$ is based on Theorem 3 and proved in Theorem 3.5 of [AM96].

Let us set $B(s) = \nabla^2\eta(x + sp)$ for any $s \geq 0$ such that $x + sp \in U$; arguing as in Theorem 3 we find that the last identities in (19) imply that B fulfills the (matrix valued) ODE

$$B'(s) = \frac{B(s) - B^2(s)}{s}.$$

for $s > 0$. As in Theorem 3, it turns out that

$$B(s) = \sum_{i=1}^{n} \lambda_i(s) z_i \otimes z_i$$

where $\{z_1, \ldots, z_n\}$ is a diagonal basis for $B(0)$ (recall that B is continuous at 0) and the eigenvalues $\lambda_i(s)$ satisfy

$$\lambda_i'(s) = \frac{\lambda_i(s)(1 - \lambda_i(s))}{s}. \tag{23}$$

Since $B(0)$ has k eigenvalues equal to 0 and $(n-k)$ ones equal to 1, the smallest k eigenvalues of $B(s)$ converge to 0 as $s \downarrow 0$ and the remaining $(n-k)$ ones converge to 1. Since $\beta_i(s) := \lambda_i(s)/s$ solve $\beta_i'(s) = -\beta_i^2(s)$, whose only unbounded solution near 0 is $1/s$, it turns out that the largest k eigenvalues are equal to 1; the smallest $(n - k)$ ones have linear growth in 0, so that, setting

$$\kappa_i := \lim_{s \downarrow 0} \frac{\lambda_i(s)}{s} \qquad i = 1, \ldots, k$$

from the ODE $\beta_i' = -\beta_i^2$ with initial condition κ_i we infer $\beta_i(s) = \kappa_i/(1 + s\kappa_i)$, i.e., (22). $\qquad\qquad\square$

Remark 4. Theorem 3 can be reformulated by saying that the principal curvatures μ_i of $\{\bar{d} = \tau\}$ and the principal curvatures κ_i of $\{\bar{d} = 0\}$ are related by $\mu_i = \kappa_i/(1 + \tau\kappa_i)$. Analogously, Theorem 4 means that the level sets $\{d = \tau\}$ have $(n - k - 1)$ principal curvatures equal to $1/\tau$ and the remaining k ones equal to $\kappa_i/(1 + \tau\kappa_i)$. Notice also that $\kappa \mapsto \kappa/(1 + \kappa\tau)$ is nondecreasing in its domain for any $\tau \in \mathbf{R}$.

We have seen in (20) that $\bar{d}$ provides a representation of the second fundamental form and of the mean curvature vector in the codimension 1 case. These representations can be extended to higher codimension using the squared distance function:

Theorem 5 (representation of the second fundamental form). *Let Γ be a smooth manifold in $\mathbf{R}^n$ without boundary and let $x \in \Gamma$. Denoting by $\pi : \mathbf{R}^n \to T_x\Gamma$ the orthogonal projection, and by $e_1, \ldots, e_n$ the canonical basis of $\mathbf{R}^n$, the following identities hold*

$$\mathbf{B}_{ij}^k := \left\langle \mathbf{B}_x\left(\pi(e_i), \pi(e_j)\right), e_k \right\rangle = \left(\eta_{im}\eta_{mj} - \eta_{ij}\right)_k(x)$$

$$\eta_{ijk}(x) = -\mathbf{B}_{ij}^k - \mathbf{B}_{jk}^i - \mathbf{B}_{ki}^j.$$

Theorem 5, proved in [AM96], is mainly based on Theorem 4. Since this result will not be used in the following, we omit the proof of it. Indeed, we are more interested in the representation of the trace $\mathbf{H}(x)$ of $\mathbf{B}_x$:

Theorem 6 (representation of the mean curvature vector). *Let Γ be a compact, smooth manifold in $\mathbf{R}^n$ without boundary. Then*

$$\mathbf{H}(x) = -\nabla\Delta\eta(x) \qquad \forall x \in \Gamma. \tag{24}$$

Proof. Let U be the tubular neighbourhood given by Theorem 2 and let $k = \dim\Gamma$. We begin with the proof of the identity

$$\mathbf{H}_i = d_j^\Gamma P_{ij} \qquad i = 1, \ldots, n \tag{25}$$

where P_{ij} is the matrix of orthogonal projection on $T_x\Gamma$. To prove (25) we use the divergence formula (10). Let $\phi \in C_c^\infty(\mathbf{R}^n, \mathbf{R}^n)$ be a smooth vectorfield and let $X = P\phi$ its tangential component; it holds

$$0 = \int_\Gamma \operatorname{div}^\Gamma X \, d\mathcal{H}^k = \int_\Gamma d_j^\Gamma\left(P_{ji}\phi_i\right) d\mathcal{H}^k$$

$$= \int_\Gamma P_{ij}d_j^\Gamma\phi_i + \phi_i d_j^\Gamma P_{ji} \, d\mathcal{H}^k = \int_\Gamma d_i^\Gamma\phi_i + \phi_i d_j^\Gamma P_{ji} \, d\mathcal{H}^k.$$

Comparing this expression with (7) we obtain (25), because ϕ is arbitrary. Since $P + \nabla^2\eta = I$ on Γ, it holds

$$\mathbf{H}_i = d_j^\Gamma P_{ij} = -d_j^\Gamma \eta_{ij} = -P_{jk}\eta_{ijk} = -\eta_{ijj} + \eta_{jk}\eta_{ijk}.$$

Now we claim that $2\eta_{jk}\eta_{ijk}$ vanishes on Γ for any i; indeed, this quantity is the i-th derivative of the square norm $\sum_{j,k} \eta_{jk}^2$ of $\nabla^2\eta$, i.e., the i-th derivative of the sum of the squares of the eigenvalues. Since, by Theorem 4, this sum is equal to $n - k + o(d)$ near Γ, it follows that its gradient vanishes on Γ. $\qquad\square$

5   Smooth flows

To define a smooth flow of manifolds $(\Gamma_t)_{t \in [0,T]}$ we initially follow a parametric (Lagrangian) viewpoint, i.e., we see each Γ_t as a deformation of the initial manifold Γ_0.

Definition 4 (smooth flow). Let $\Gamma \subset \mathbf{R}^n$ be a compact, connected k-dimensional manifold in $\mathbf{R}^n$ without boundary and let $(\Gamma_t)_{t \in [0,T]}$ be subsets of $\mathbf{R}^n$. We say that (Γ_t) is a *smooth flow* starting from Γ if there exists a smooth parametrization map $\Phi : [0,T] \times \Gamma \to \mathbf{R}^n$ such that $\Phi(0,y) = y$ for any $y \in \Gamma$ and

(i) $\Phi(t,\cdot) : \Gamma \to \Gamma_t$ is a bijection for any $t \in [0,T]$;
(ii) for any $t \in [0,T]$ the (tangential) k dimensional Jacobian $J_k\Phi(t,\cdot)$ (cf. (9)) is nowhere equal to 0 on Γ.

We call any such Φ *parametrization* of the flow; in addition the orthogonal projection $\mathbf{v}(t,y)$ of $\Phi_t(t,y)$ on $N_x\Gamma_t$ will be called *normal velocity* of $x := \Phi(t,y)$. We say that the parametrization is *normal* if $\Phi_t(t,y)$ is perpendicular to Γ_t at $\Phi(t,y)$ whenever $t \in [0,T] \times \Gamma$.

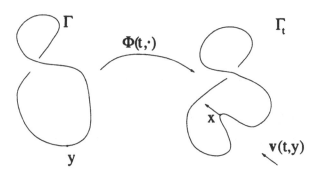

Fig. 3. Smooth flow

Remark 5. (a) Conditions (i), (ii) ensure that Γ_t is a compact, connected, smooth manifold without boundary for any $t \in [0,T]$.
(b) Normal parametrizations of the flow depend only on $(\Gamma_t)_{t \in [0,T]}$; indeed, by Theorem 7(ii) below, any such parametrization Φ solves the Cauchy problem

$$\Phi_t(t,y) = -\nabla \eta_t(t,\Phi(t,y)) \quad (t,y) \in [0,T] \times \Gamma, \qquad \Phi(0,y) = y \qquad (26)$$

where $\eta(t,\cdot)$ is the squared distance function from Γ_t. Because of tangential components in the velocity, a generic parametrization does not depend only on Γ_t.

Arguing as in Theorem 2 it can be proved that the squared distance function

$$\eta(t,x) := \frac{1}{2}\mathrm{dist}^2(x,\Gamma_t) \qquad (t,x) \in [0,T] \times \mathbf{R}^n$$

from a smooth flow $(\Gamma_t)_{t\in[0,T]}$ is smooth in $\{\eta < \sigma^2/2\}$ for σ small enough, and a corresponding property holds for the signed distance function

$$\bar{d}(t,x) := \bar{d}(x,U_t) \qquad (t,x) \in [0,T] \times \mathbf{R}^n$$

from a codimension 1 flow. Therefore, it is natural to use these functions to characterize the normal velocity of the flow.

Theorem 7 (characterization of normal velocity with distance). *Let* $(\Gamma_t)_{t\in[0,T]}$ *be a smooth flow. Then*

(i) *if* $\Gamma_t = \partial U_t$ *and* $\bar{d}(t,\cdot)$ *is the signed distance function from* U_t *oriented by the outer normal* $\nu_t(x) = \nabla\bar{d}(t,x)$, *then the normal velocity of* $x \in \Gamma_t$ *is equal to* $-\bar{d}_t(t,x)\nu_t(x)$;

(ii) *if* $\eta(t,\cdot)$ *is the squared distance function from* Γ_t, *then the normal velocity of* $x \in \Gamma_t$ *is* $-\nabla\eta_t(t,x)$.

Proof. (i) Let us fix $y \in \Gamma$; by differentiating in t the identity

$$\bar{d}\big(t,\Phi(t,y)\big) = 0$$

we get

$$\bar{d}_t\big(t,\Phi(t,y)\big) + \langle\nabla\bar{d}\big(t,\Phi(t,y)\big), \Phi_t(t,y)\rangle = 0.$$

Hence

$$\mathbf{v}(t,y) = \langle\nu_t(x), \Phi_t(t,y)\rangle\nu_t(x) = -\bar{d}_t(t,x)\nu_t(x).$$

(ii) Let us fix an index $k \in \{1,\ldots,n\}$ and $y \in \Gamma$; by differentiating in t the identity

$$\eta_k\big(t,\Phi(t,y)\big) = 0$$

we get

$$\eta_{kt}\big(t,\Phi(t,y)\big) + \eta_{ki}\big(t,\Phi(t,y)\big)\Phi_t^i(t,y) = 0.$$

Recalling that $\nabla^2\eta$ gives the projection on normal space we get

$$\mathbf{v}(t,y) = -\nabla\eta_t\big(t,\Phi(t,y)\big).$$

$\square$

We now prove that any smooth flow has a normal parametrization; this parametrization is unique by Remark 5(b).

Theorem 8 (existence of normal parametrizations). *Any smooth flow* $(\Gamma_t)_{t\in[0,T]}$ *admits a unique normal parametrization.*

Proof. Let us consider the ODE

$$\Psi_t(t,y) = -\nabla \eta_t(t,\Psi(t,y)), \quad \Psi(0,y) = y \qquad (t,y) \in [0,T] \times \Gamma. \quad (27)$$

By a continuity argument, we can find $\tau > 0$ (depending only on η) such that the solution of the ODE above satisfies

$$\bigcup_{t\in[0,\tau]} \{t\} \times \Psi(t,\cdot)(\Gamma) \subset \{\eta < \sigma^2/2\}, \qquad J_k\Psi(t,y) \neq 0 \quad \forall(t,y) \in [0,\tau] \times \Gamma. \quad (28)$$

We will prove that Ψ provides the normal parametrization of (Γ_t) in $[0,\tau]$. From differentiation in time of the identity $\eta_{ik}\eta_k = \eta_i$ we get

$$\eta_{tik}\eta_k + \eta_{ik}\eta_{tk} = \eta_{ti}. \quad (29)$$

Now, let us fix $y \in \Gamma$ and set

$$x_t := \Psi(t,y), \qquad g(t) := \eta(t,x_t).$$

Using (29) we find

$$\begin{aligned} 2g'(t) &= \sum_k \eta_k^2(t,x_t)' = 2\eta_k(t,x_t)\big[\eta_{kt}(t,x_t) + \eta_{ki}(t,x_y)\Psi_t^i(t,y)\big] \\ &= 2\eta_k(t,x_t)\big[\eta_{kt}(t,x_t) - \eta_{ki}(t,x_t)\eta_{it}(t,x_t)\big] \\ &= 2\eta_k(t,x_t)\eta_i(t,x_t)\eta_{kit}(t,x_t) \leq Cg(t). \end{aligned}$$

Since $g(0) = 0$, g is identically 0 in $[0,\tau]$; this proves that $\eta(t,\Psi(t,y)) \equiv 0$, so that $\Psi(t,y) \in \Gamma_t$ whenever $(t,y) \in [0,\tau] \times \Gamma$; now, $\Psi(t,\cdot)(\Gamma)$ is a compact subset of Γ_t, relatively open by (28), therefore $\Psi(t,\cdot)$ is onto for any $t \in [0,\tau]$.

(ii) To show that $\Psi(t,\cdot)$ is one to one for any $t \in [0,\tau]$, it suffices to prove that

$$I := \big\{t \in [0,\tau] \,|\, \Psi(t,\cdot) \text{ is not 1-1}\big\}$$

is open and closed in $[0,\tau]$. Indeed, if $t \in I$ and $\Psi(t,x) = \Psi(t,y)$ for $x \neq y$, then we can fix $r \in (0, |x-y|/2)$ to find (because of (28)) $\sigma, \delta > 0$ such that

$$\Psi\big(s, B_r(x)\big) \supset B_\sigma\big(\Psi(s,x)\big), \qquad \Psi\big(s, B_r(y)\big) \supset B_\sigma\big(\Psi(s,y)\big)$$

for any $s \in B_\delta(t) \cap [0,\tau]$. Possibly reducing δ, we can assume that

$$\Psi(t,x) \in B_\sigma\big(\Psi(s,x)\big), \quad \Psi(t,y) \in B_\sigma\big(\Psi(s,y)\big) \qquad \forall s \in B_\delta(t) \cap [0,\tau].$$

In particular, since $B_r(x)$ and $B_r(y)$ are disjoint, $\Phi(s,\cdot)$ is not one to one for any $s \in B_\delta(t) \cap [0,\tau]$. To check the closure property, we first notice that (28) and the smoothness of Ψ imply the existence, given $(t,x) \in [0,\tau] \times \Gamma$, of $r > 0$ such that

$$\Phi(s,\cdot) \text{ is 1-1 in } B_r(x) \text{ for any } s \in B_r(t) \cap [0,\tau]. \quad (30)$$

Given a sequence $(t_h) \subset I$ converging to t, let $y_h, y_h' \in \Gamma$ such that $\Phi(t_h, y_h) = \Phi(t_h, y_h')$ and $y_h \neq y_h'$. Assuming both sequences $(y_h), (y_h')$ converging, clearly

(30) prevents the possibility that their limits are the same; denoting by y, y' their limits we find $\Phi(t,y) = \Phi(t,y')$, therefore $t \in I$.

(iii) Choosing Γ_τ as new initial surface, the argument can be repeated in $[\tau, 2\tau]$, $[2\tau, 3\tau]$ and so on, getting a finite number of normal parametrizations Ψ^i such that $\Psi^0 = \Psi$ and

$$\Psi^i_t(t,y) = -\nabla\eta_t\big(t, \Psi^i(t,y)\big), \quad \Psi^i(i\tau, y) = y \qquad (t,y) \in [i\tau, (i+1)\tau] \times \Gamma_{i\tau}.$$

In the end, a unique parametrization defined in $[0,T] \times \Gamma$ can be recovered by setting

$$\Psi(t,y) = \Psi^0(t,y) \quad (t,y) \in [0,\tau] \times \Gamma,$$

$$\Psi(t,y) = \Psi^1\big(t, \Psi^0(\tau,y)\big) \quad (t,y) \in [\tau, 2\tau] \times \Gamma, \dots$$

It turns out that Ψ is the unique solution of (27) in $[0,T]$. $\qquad\qquad$ □

Remark 6. In some applications it is more convenient to keep tangential components in the velocity, to simplify the equations. This is for instance the case for the mean curvature flow for graphs $\Gamma_t = \{y = u(t,x)\}$: the PDE

$$u_t = \sqrt{1 + |\nabla u|^2}\, \mathrm{div}\left(\frac{\nabla u}{\sqrt{1 + |\nabla u|^2}}\right)$$

corresponds to the parametrization

$$\Phi(t,x) := \big(x, u(t,x)\big)$$

whose derivative $(0, u_t(t,x))$ is "vertical".

The following theorem provides a partial converse of Theorem 2(ii): any smooth function η satisfying $|\nabla\eta|^2 = 2\eta$ is locally the squared distance function from a smooth manifold. The proof, due to Novaga [Nov96], is based on a local version of Theorem 12.

Theorem 9. *Let $U \subset \mathbf{R}^n$ be a bounded open set and let $\eta \in C^\infty(\overline{U})$ be satisfying $|\nabla\eta|^2 = 2\eta$ in U. Then, if $\eta > 0$ on ∂U, each connected component of*

$$K := \{x \in U \mid \eta(x) = 0\}$$

is a smooth manifold in $\mathbf{R}^n$ without boundary and near to K the function η coincides with the squared distance function from K.

Proof. Since $\eta \geq 0$ has only interior minimizers, K is compact, nonempty and contained in U. Let $d = \sqrt{2\eta}$; this function satisfies the equation $|\nabla d|^2 = 1$ in $A := U \setminus K$. Let us prove that

$$d(x) = \inf\left\{d(y) + |x - y| \,\big|\, y \in \partial A,\ [x,y] \subset A\right\} \qquad \forall x \in A. \qquad (31)$$

Let $w(x)$ be the right hand side in (31); since $|\nabla d| = 1$ in A, $d(x) \leq d(y) + |x - y|$ for any $y \in \partial A$ such that the segment $[x, y)$ is contained in A, therefore $d(x) \leq w(x)$. To show the opposite inequality, fix $x_0 \in A$ and set $x_s := x_0 - s\nu$, with $\nu = \nabla d(x_0)$ and $s > 0$; denoting by y_0 the first point on ∂A on the halfline $x_0 - s\nu$, it holds

$$\nabla d(x_s)' = \nabla^2 d(x_s)\nu = \nabla^2 d(x_s)(\nabla d(x_s) - \nu) \qquad s \in [0, |y_0 - x_0|)$$

therefore the function $g(s) := |\nabla d(x_s) - \nu|^2$ satisfies

$$g'(s) \leq 2C_t g(s) \quad s \in [0, t] \qquad \text{with} \qquad C_t := \sup_{0 \leq s \leq t} \|\nabla^2 d(x_s)\|$$

for any $t \in [0, |y_0 - x_0|)$. Since $g(0) = 0$, this implies that $\nabla d(x_s) = \nu$ for any $s \in [0, |y_0 - x_0|)$, hence

$$d(x_0) = d(y_0) + |y_0 - x_0| \geq w(x_0).$$

In particular (31) implies that near to K the function η coincides with the squared distance function from K. Let $\sigma > 0$ be such that $\eta > \sigma$ on ∂U. Denoting by V a connected component of $\{\eta < \sigma\}$, we will prove that $K \cap V$ is a smooth connected manifold. Since $d(y_0) < d(x_0)$ and $y_0 \in \partial U \cup K$, $x_0 \mapsto y_0$ maps V onto $K \cap V$, hence $K \cap V$ is connected.

(i) We will first prove that $\nabla^2 \eta$ has constant rank on $K \cap V$; this will be achieved by showing that $\nabla^2 \eta$ has only eigenvalues in $\{0, 1\}$, so that the rank coincides with the trace. Since trace $(\nabla^2 \eta)$ is continuous and integer valued on K, this implies that the rank is constant in $K \cap V$.

Let $x \in K$; since $\eta(y) \leq |y - x|^2/2$, all the eigenvalues of $\nabla^2 \eta(x)$ do not exceed 1. Let E be the vector space generated by all vectors $v \in S^{n-1}$ which are limit of vectors $y - x/|y - x|$ such that $y \in K$ and y converges to x; since

$$0 = \eta(y) = \eta(x) + \langle \nabla^2 \eta(x)(y - x), (y - x) \rangle + o(|y - x|^2)$$

dividing both sides by $|y - x|^2$ and passing to the limit we obtain $\langle \nabla^2 \eta(x)v, v \rangle = 0$ for any generating vector of E, therefore $\nabla^2 \eta(x)$ is zero on E. Let $w \in S^{n-1}$ be an eigenvector of $\nabla^2 \eta(x)$ orthogonal to E, let λ be its eigenvalue and let $y_t \in K$ be the least distance point of $x + tw$ from K; since

$$\frac{\lambda}{2} t^2 + o(t^2) \geq \eta(x + tw) = \frac{1}{2}|y_t - x - tw|^2$$

we can assume that $(y_{t_h} - x)/t_h$ converges to some $v \in E$ for some infinitesimal sequence (t_h), hence dividing both sides by t_h^2 and passing to the limit as $h \to \infty$ we get

$$\lambda \geq |v|^2 + |w|^2.$$

Since $|w| = 1$ and $\lambda \leq 1$, this proves that $\lambda = 1$.

(ii) By step (i), there exists an integer $m \in [0, n]$ such that $\nabla^2 \eta$ has rank (and trace) m on $K \cap V$. We notice that $m > 0$, because the argument of Theorem 4 shows that

$$L := \{x \in V \mid \nabla^2 \eta(x) = 0\}$$

is open; since V is connected, either L is empty or $L = V$, i.e., η is linear in V. The equation $|\nabla\eta|^2 = 2\eta$ excludes the second case, hence $L = \emptyset$.

We will prove that for any $x \in K \cap V$ there exists $r > 0$ such that $K \cap B_r(x)$ is a smooth k dimensional manifold, with $k := n - m$. Let $\nu^1, \ldots, \nu^m$ be an orthonormal basis of $E^\perp$; since $\nabla^2\eta|_E = I$, we can choose $r > 0$ so small that $B_r(x) \subset V$ and the jacobian of the map

$$F(y) := \left(F^1(y), \ldots, F^m(y)\right) = \left(\langle\nabla\eta(y), \nu^1\rangle, \ldots, \langle\nabla\eta(y), \nu^m\rangle\right)$$

has rank m for any $y \in B_r(x) \cap K$. Since F vanishes on K, the implicit function theorem implies that $K \cap B_r(x)$ is contained in a k-dimensional manifold Γ, the zero set of F in $B_r(x)$. Possibly reducing r, we can assume that Γ is connected. Clearly, $K \cap B_r(x)$ is relatively closed in Γ; notice that the rank of the Jacobian of the map

$$V \ni x \mapsto \pi(x) := x - \nabla\eta(x) \in K \cap V \subset \Gamma$$

is identically equal to k on V. By applying the local invertibility theorem to suitable restrictions of π to k-dimensional manifolds $\Gamma' \subset V \setminus K$ it is not hard to see that $K \cap B_r(x)$ is relatively open in Γ, hence $K \cap B_r(x) = \Gamma$ and this concludes the proof of the theorem.   □

Theorem 9 allows an alternative formulation of smooth flow, without *any* reference to parametrizations: $(\Gamma_t)_{t\in[0,T]}$ is a smooth flow if

$$K := \bigcup_{t\in[0,T]} \{t\} \times \Gamma_t$$

is compact and there exist an open set A and a smooth function η in $[0,T] \times A$ such that

$$K = \{\eta = 0\} \subset [0,T] \times A, \qquad |\nabla\eta|^2 = 2\eta \quad \text{on } [0,T] \times A.$$

This is the definition of smooth flow adopted in [Gio94b].

Definition 5 (smooth mean curvature flow). We say that $(\Gamma_t)_{t\in[0,T]}$ is a *smooth mean curvature flow* if the normal velocity of Γ_t at x is $\mathbf{H}_t(x)$, where $\mathbf{H}_t$ is the mean curvature vector of Γ_t according to Definition 3.

From (20) and Theorem 7(i) we infer a characterization of mean curvature flows based on the distance function: a smooth flow $\Gamma_t = \partial U_t$ is a mean curvature flow if and only if

$$\bar{d}_t(t, x) = \Delta\bar{d}(t, x) \qquad \forall t \in [0, T], \ x \in \partial U_t. \tag{32}$$

Analogously, (24) and Theorem 7(ii) yield a characterization of mean curvature flows based on the squared distance function and true in any codimension:

$$\nabla\eta_t(t, x) = \nabla\Delta\eta(t, x) \qquad \forall t \in [0, T], \ x \in \Gamma_t. \tag{33}$$

In many applications (see for instance Theorem 24) it is convenient to characterize smooth flows by PDE's satisfied not only on $\bigcup_{t\in[0,T]}\{t\} \times \Gamma_t$, but also in a tubular neighbourhood of the flow.

Theorem 10 (distance function and smooth mean curvature flows).
Let $(\Gamma_t)_{t \in [0,T]}$ be a smooth flow. Then

(i) if $\Gamma_t = \partial U_t$ and $\bar{d}(t, \cdot)$ is the signed distance function from U_t, then Γ_t is a mean curvature flow if and only if

$$\bar{d}_t = \sum_\mu \frac{\mu}{1 - \bar{d}\mu} \qquad \text{(here } \mu \text{ varies among the eigenvalues of } \nabla^2 \bar{d}) \quad (34)$$

in a tubular neighbourhood $\{|\bar{d}| < \sigma\}$ of $\bigcup_{t \in [0,T]}\{t\} \times \Gamma_t$.
(ii) if $\eta(t, \cdot)$ is the squared distance function from Γ_t, then Γ_t is a mean curvature flow if and only if

$$\eta_t = \sum_{\lambda < 1} \frac{\lambda}{1 - \lambda} \qquad \text{(here } \lambda \text{ varies among the eigenvalues of } \nabla^2 \eta) \quad (35)$$

in a tubular neighbourhood $\{\eta < \sigma^2/2\}$ of $\bigcup_{t \in [0,T]}\{t\} \times \Gamma_t$.

Proof. (i) Let (t, y) be in the tubular neighbourhood, $\nu := \nabla \bar{d}(y)$ and let $x = y - \bar{d}(t, y)\nu$ be its projection on ∂U_t. Denoting by μ_i, κ_i respectively the eigenvalues of $\nabla^2 \bar{d}(t, y)$, $\nabla^2 \bar{d}(t, x)$, we will prove the identities

$$\kappa_i = \frac{\mu_i}{1 - \bar{d}(t, y)\mu_i}, \qquad \bar{d}_t(t, y) = \bar{d}_t(t, x). \quad (36)$$

In fact, the first equality in (36) has been proved in Theorem 3; to prove the second one we set $b(s) := \bar{d}_t(t, x + s\nu)$ and differentiate with respect to s:

$$b'(s) = \bar{d}_{ti}(t, x + s\nu)\nu_i = \bar{d}_{ti}(t, x + s\nu)\bar{d}_i(t, x + s\nu_i) = 0$$

because $\sum_i \bar{d}_i^2 \equiv 1$. If Γ_t is a mean curvature flow, from (32) and (36) we infer

$$\bar{d}_t(t, y) = \bar{d}_t(t, x) = \Delta \bar{d}(t, x) = \sum_{i=1}^n \kappa_i = \sum_{i=1}^n \frac{\mu_i}{1 - \bar{d}(t, y)\mu_i}$$

so that the PDE in the statement is satisfied. The proof of the opposite implication is analogous.
(ii) Assume that the equation is satisfied; let $t \in [0, T]$, $x \in \Gamma_t$ and, for a given normal direction p, let $x_s = x + sp$ and $B(s) = \eta_t(t, x_s) - \Sigma(s)$, with

$$\Sigma(s) := \sum_{\lambda < 1} \frac{\lambda(t, x_s)}{1 - \lambda(t, x_s)} \qquad s \in [0, \sigma).$$

Notice that $B(0) = 0$; a simple computation based on the fact that $s \mapsto \lambda(t, x_s)$ fulfils (23) shows that $s\Sigma'(s) = \Sigma(s)$, hence $\Sigma(s)/s$ is constant in $(0, \sigma)$. In addition,

$$\eta_t(t, x_s)' = \eta_{ti}(t, x_s)p_i = \frac{\eta_{ti}(t, x_s)\eta_i(t, x_s)}{\delta(t, x_s)} = \frac{\eta_t(t, x_s)}{s}$$

because (by differentiation of $\eta_i^2 = 2\eta$) $\eta_i\eta_{it} = \eta_t$; therefore $s \mapsto \eta_t(t, x_s)/s$ is constant too and so is $B(s)/s$. If the PDE is satisfied, then

$$\langle \nabla(\eta_t - \Delta\eta), p \rangle = \langle \nabla(\eta_t - \sum_{\lambda < 1} \lambda), p \rangle = \lim_{s \downarrow 0} \frac{1}{s} \sum_{\lambda < 1} \left(\frac{\lambda(t, x_s)}{1 - \lambda(t, x_s)} - \lambda(t, x_s) \right)$$

$$= \lim_{s \downarrow 0} \sum_{\lambda < 1} \frac{1}{s} \frac{\lambda^2(t, x_s)}{1 - \lambda(t, x_s)} = 0.$$

Since p is an arbitrary normal direction, (33) follows.

Conversely, if Γ_t is a mean curvature flow then the same identities above show that $B(s)/s$ tends to 0 as $s \downarrow 0$, hence $B(s)/s$ and $B(s)$ are identically 0 in $(0, \sigma)$. $\square$

Remark 7. Equation (34) has been used by Evans & Spruck to show short time existence of smooth codimension 1 mean curvature flows. Equation (35), first proved by Novaga in [Nov96], provides a similar formulation of smooth mean curvature flows which makes sense in any dimension and codimension.

6   Viscosity solutions

In this section we give a self contained introduction to the theory of viscosity solutions, first introduced for first order equations by Crandall, Evans and Lions (see [CL83, CEL84, Lio82]) and later extended to second order equations by Jensen, Ishii, Lions, Souganidis (see [Ish89, Jen88, JLS88]). A complete survey paper on this topic is [CIL92].

Definition 6 (viscosity solutions). Let $A \subset \mathbf{R}^n$ be a locally compact set, W a dense subset of $A \times \mathbf{R} \times \mathbf{R}^n \times \mathrm{Sym}^n$ and $E : W \to \overline{\mathbf{R}}$. We say that $u : A \to \mathbf{R}$ is a *viscosity subsolution* of

$$E(x, u, \nabla u, \nabla^2 u) = 0 \tag{37}$$

if $u^* < \infty$ in A and

$$E_*(x_0, u^*(x_0), \nabla\phi(x_0), \nabla^2\phi(x_0)) \leq 0$$

whenever $x_0 \in A$, $\phi \in C^\infty(U)$ in a neighbourhood U of x_0 and $u^* - \phi$ has a local maximum at x_0. Analogously, u is a *viscosity supersolution* of (37) if $u_* > -\infty$ in A and

$$E^*(x_0, u_*(x_0), \nabla\phi(x_0), \nabla^2\phi(x_0)) \geq 0$$

whenever $x_0 \in A$, $\phi \in C^\infty(U)$ in a neighbourhood U of x_0 and $u_* - \phi$ has a local minimum at x_0. The function u is called *viscosity solution* of (37) if it is both a viscosity subsolution and a viscosity supersolution.

Remark 8. (1) The local compactness of A is a natural assumption because local maximizations or minimizations are involved (see for instance Proposition 2 and Theorem 14). This assumption allows open and closed domains.

(2) We notice that u is a viscosity subsolution (supersolution) if and only if u^* (u_*) is a viscosity subsolution (supersolution) of the same equation. However, without continuity assumptions on u, very simple equations are plenty of viscosity solutions: for instance the Dirichlet function $\chi_{\mathbf{Q}}$ is a viscosity solution of the equation $u' = 0$.

(3) In general a viscosity solution of $E = 0$ *is not* a viscosity solution of $-E = 0$. For instance, it is not hard to see that

$$f(t) := (1 - t) \wedge (1 + t) \qquad A = (-1, 1)$$

is a viscosity solution of $|u'| - 1 = 0$ but it is not a viscosity solution of $1 - |u'| = 0$. On the other hand, if E is odd with respect to $(u, \nabla u, \nabla^2 u)$, then $u \mapsto -u$ maps subsolutions (supersolutions) of $E = 0$ into supersolutions (subsolutions) of $E = 0$. If E is even with respect to $(u, \nabla u, \nabla^2 u)$, then the same map transforms subsolutions (supersolutions) of $E = 0$ into supersolutions (subsolutions) of $-E = 0$.

(4) Possibly modifying the comparison function ϕ out of a neighbourhood of x_0 it is not restrictive to assume that U contains A and that the the relative maximum (minimum) is global. This can be achieved by a covering argument based on the local boundedness from above of u^*. Moreover, possibly replacing ϕ by the fourth order perturbation

$$\phi(x) + |x - x_0|^4 \qquad (\phi(x) - |x - x_0|^4)$$

which does not affect first and second order derivatives at x_0, we can make the maximum (minimum) strict. Finally, since only the derivatives of ϕ are involved, we can assume that the maximum (minimum) is 0. Summing up, it is not restrictive to assume that

$$u^*(x_0) - \phi(x_0) = 0, \qquad u^*(x) - \phi(x) \le -|x - x_0|^4 \quad \forall x \in A.$$

Graphically, this means that the graph of u touches the graph of ϕ from below (above) at x_0. In many applications to geometric evolution problems, in which one is interested to level sets, this means that $\{\phi \le \phi(x_0)\}$ is inside $\{u^* \le u^*(x_0)\}$ and the corresponding level sets touch only at x_0.

Exercise 1. Let $A \subset \mathbf{R}^n$ be a locally compact set, let $u : A \to \mathbf{R}$ be a viscosity subsolution of (37) and assume that $u^* - \phi$ has a relative maximum at $x_0 \in A$ for some C^2 function ϕ in a neighbourhood of x_0. Prove that

$$E_*\big(x_0, u^*(x_0), \nabla\phi(x_0), \nabla^2\phi(x_0)\big) \le 0.$$

Hint: assuming that the relative maximum is strict in a compact neighbourhood K of x_0, approximate uniformly ϕ in K by functions $\phi_h \in C^\infty(K)$ and prove that

$$\lim_{h \to \infty} \max_K (u^* - \phi_h) = \max_K (u^* - \phi) = u^*(x_0) - \phi(x_0)$$

with convergence of maximizers to x_0.

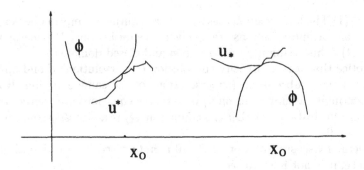

Fig. 4. Test functions for subsolutions and supersolutions

Using Exercise 1 it can be proved that if we use C^2 test functions ϕ we get an equivalent definition of viscosity subsolution and supersolution. A even weaker, but still equivalent, definition is based on sub and super jets:

Definition 7 (super and sub jets). Let $A \subset \mathbf{R}^n$ be a locally compact set and let $u : A \to \mathbf{R}$. The *second order* super jet $J_A^{2,+}u(x_0)$ of u at $x_0 \in A$ is the collection of all pairs $(p, X) \in \mathbf{R}^n \times \mathrm{Sym}^n$ such that

$$u(x) \leq u^*(x_0) + \langle p, x - x_0\rangle + \frac{1}{2}\langle X(x - x_0), (x - x_0)\rangle + o(|x - x_0|^2). \quad (38)$$

The *second order* sub jet $J_A^{2,-}u(x_0)$ of u at $x_0 \in A$ is defined in a similar way: (p, X) belongs to this set if and only if

$$u(x) \geq u_*(x_0) + \langle p, x - x_0\rangle + \frac{1}{2}\langle X(x - x_0), (x - x_0)\rangle + o(|x - x_0|^2). \quad (39)$$

Since (38) and (39) are local properties, we will simply write $J^{2,\pm}u(x_0)$ if x_0 belongs to the interior of A.

Remark 9. (1) The definition of viscosity subsolution can be reformulated in terms of super jets as follows: u is a viscosity subsolution in A if and only if

$$E_*(x_0, u^*(x_0), p, X) \leq 0 \qquad \forall x_0 \in A, \ (p, X) \in J_A^{2,+}u(x_0). \quad (40)$$

One implication is trivial, because a Taylor expansion of ϕ shows that

$$(\nabla\phi(x_0), \nabla^2\phi(x_0)) \in J_A^{2,+}u(x_0)$$

whenever $\phi \in C^2(U)$ and $u^* - \phi$ has a relative maximum at x_0. The opposite implication is based on Exercise 2. A similar characterization is true for viscosity supersolutions: u is a viscosity supersolution in A if and only if

$$E^*(x_0, u_*(x_0), p, X) \geq 0 \qquad \forall x_0 \in A, \ (p, X) \in J_A^{2,-}u(x_0). \quad (41)$$

(2) if E has a continuous extension to $A \times \mathbf{R} \times \mathbf{R}^n \times \mathrm{Sym}^n$ and $u \in C^2(V)$ in an open set V containing A, then

$$u \text{ viscosity solution} \quad \Longrightarrow \quad u \text{ classical solution.} \qquad (42)$$

Indeed, we need only to use (40), (41) and a second order Taylor expansion of u. The same implication separately holds for subsolutions and supersolutions.

Exercise 2. Let $A \subset \mathbf{R}^n$ be a locally compact set, $x_0 \in A$ and let $\psi : A \to \mathbf{R}$ be an upper semicontinuous function such that $\psi(x) \leq o(|x - x_0|^2)$. Then, there exist $\phi \in C^2(\mathbf{R}^n)$ and $r > 0$ such that $\psi \leq \phi$ in $A \cap B_r(x_0)$ and

$$\phi(x_0) = 0, \qquad \nabla\phi(x_0) = 0, \qquad \nabla^2\phi(x_0) = 0.$$

The opposite implication in (42) does not hold in general, as Exercise 3 shows. This motivates the following definition:

Definition 8 (degenerate ellipticity). We say that the function $G : A \times \mathbf{R} \times \mathbf{R}^n \times \mathrm{Sym}^n \to \overline{\mathbf{R}}$ is degenerate elliptic if

$$Y \leq X \quad \Longrightarrow \quad G(x, s, p, X) \leq G(x, s, p, Y)$$

(here $Y \leq X$ if and only if all eigenvalues of $X - Y$ are nonnegative). We say that the equation (37) is degenerate elliptic if both E_* and E^* are degenerate elliptic.

One of the fundamental ingredients in the construction of subsolutions (supersolutions) is the following proposition.

Proposition 1. *If A is open and E_* (E^*) is degenerate elliptic, then any classical subsolution (supersolution) of (37) is a viscosity subsolution (supersolution).*

Proof. Let x_0, ϕ be as in Definition 6. Then, the classical maximum principle implies

$$\nabla\phi(x_0) = \nabla u(x_0), \qquad \nabla^2\phi(x_0) \geq \nabla^2 u(x_0),$$

hence

$$\begin{aligned} E_*\big(x_0, u(x_0), \nabla\phi(x_0), \nabla^2\phi(x_0)\big) &= E_*\big(x_0, u(x_0), \nabla u(x_0), \nabla^2\phi(x_0)\big) \\ &\leq E_*\big(x_0, u(x_0), \nabla u(x_0), \nabla^2 u(x_0)\big) \leq 0. \end{aligned}$$

$\square$

Exercise 3. The function $u(t) := t^2$ is a classical solution of $u'' - 2 = 0$ but it is not a viscosity solution of the same equation.

Proposition 2. *Let $\mathcal{F}$ be a nonempty family of viscosity subsolutions of (37) in $A \subset \mathbf{R}^n$ and let*

$$u(x) := \sup\left\{ v(x) \,|\, v \in \mathcal{F} \right\} \qquad \forall x \in A.$$

Then, u is a viscosity subsolution of (37) in $A \cap \{u^* < \infty\}$.

Proof. Let $x_0 \in A \cap \{u^* < \infty\}$ and assume that $K := A \cap \overline{B}_r(x_0)$ is compact and

$$u^*(x_0) - \phi(x_0) = 0, \qquad u^*(x) - \phi(x) \le -|x - x_0|^4 \quad \forall x \in K$$

for some $\phi \in C^\infty(B_r(x_0))$. By the definition of u, we can find a sequence (x_h) converging to x_0 and a sequence $(v_h) \subset \mathcal{F}$ such that

$$u^*(x_0) = \lim_{h\to\infty} u(x_h) = \lim_{h\to\infty} v_h(x_h).$$

Let y_h be maximizers of $v_h^* - \phi$ on K and assume (extracting a subsequence, if necessary) that (y_h) converges to some $y \in K$; since $x_h \in K$ for h large enough, it holds

$$u^*(y_h) - \phi(y_h) \ge v_h^*(y_h) - \phi(y_h) \ge v_h^*(x_h) - \phi(x_h) \ge v_h(x_h) - \phi(x_h).$$

Passing to the limit as $h \to \infty$, we obtain that

$$u^*(y) - \phi(y) \ge \limsup_{h\to\infty} v_h^*(y_h) - \phi(y_h) \ge \liminf_{h\to\infty} v_h^*(y_h) - \phi(y_h) \ge 0$$

hence $y = x_0$ and $v_h^*(y_h) - \phi(y_h)$ converges to 0, i.e., $v_h^*(y_h)$ converges to $\phi(x_0) = u^*(x_0)$. Finally, passing to the limit as $h \to \infty$ in

$$E_*\big(y_h, v_h^*(y_h), \nabla\phi(y_h), \nabla^2\phi(y_h)\big) \le 0$$

we get $E_*\big(x_0, u^*(x_0), \nabla\phi(x_0), \nabla^2\phi(x_0)\big) \le 0$. $\qquad\qquad\square$

One of the basic existence theorems in the theory of viscosity solutions is based on Perron's method: under suitable assumptions a maximal subsolution is a solution.

Theorem 11 (Perron). *Let f, g respectively be a subsolution and a supersolution of (37) in an open set $A \subset \mathbf{R}^n$ such that*

$$f(x) \le g(x), \quad f_*(x) > -\infty, \quad g^*(x) < \infty \qquad \forall x \in A$$

and let us assume that the equation is degenerate elliptic. Then, there exists a viscosity solution u of (37) in A satisfying

$$f(x) \le u(x) \le g(x) \qquad \forall x \in A.$$

Proof. Let $\mathcal{F}$ be the nonempty collection of all viscosity subsolutions of (37) in A less than g and let u be as in Proposition 2. Then, since $u^* \leq g^* < \infty$, we know that u is a viscosity subsolution in A; to prove that u is viscosity supersolution we argue by contradiction.

Since $u \geq f$, we have $u_* \geq f_* > -\infty$ in A. Let $x_0 \in A$ and $\phi \in C^\infty(U)$ such that $u_* - \phi$ (see **Remark 8(4)**) satisfies

$$u_*(x_0) - \phi(x_0) = 0, \qquad u_*(x) - \phi(x) \geq |x - x_0|^4 \quad \forall x \in A$$

and assume by contradiction that

$$E^*\left(x_0, u_*(x_0), \nabla\phi(x_0), \nabla^2\phi(x_0)\right) < 0. \tag{43}$$

For $\delta > 0$ given, let $w = (\phi + \delta^4) \vee u$; since $u_*(x_0) = \phi(x_0)$, the set $\{w > u\}$ is not empty (indeed, x_0 is in the closure of this set). Therefore, a contradiction with the maximality of u will be achieved if we prove that for δ small enough w is a subsolution and $w \leq g$.

By the upper semicontinuity of E^*, for δ sufficiently small it holds

$$E^*\left(x, \phi(x) + \delta^4, \nabla\phi(x), \nabla^2\phi(x)\right) \leq 0 \qquad \forall x \in A \cap B_{2\delta}(x_0),$$

so that $\phi + \delta^4$ is a classical subsolution in $A \cap B_{2\delta}(x_0)$. By Proposition 2 and Proposition 1, w is a viscosity subsolution in the same set. On the other hand, in $A \setminus \overline{B}_\delta(x_0)$ it holds

$$u(x) \geq u_*(x) \geq \phi(x) + |x - x_0|^4 \geq \phi(x) + \delta^4$$

hence $w = u$ in $A \setminus \overline{B}_\delta(x_0)$ and w is a viscosity subsolution there. Since the open sets $A \cap B_{2\delta}(x_0)$, $A \setminus \overline{B}_\delta(x_0)$ cover A, w is a viscosity subsolution in A.

Finally, we prove that $w \leq g$ for δ sufficiently small. Indeed, we first notice that $u_*(x_0)$ is strictly less than $g_*(x_0)$: equality would imply that $g_* - \phi \geq u_* - \phi$ has a strict minimum at x_0, so that

$$E^*\left(x_0, g_*(x_0), \nabla\phi(x_0), \nabla^2\phi(x_0)\right) \geq 0,$$

in contradiction with (43). Since

$$\phi(x_0) = u_*(x_0) < g_*(x_0)$$

we obtain that for δ small enough $\phi + \delta^4$ is less than g_* in $A \cap B_\delta(x_0)$, hence w is less than g in A. □

7   First order equations and distance function

Let $A \subset \mathbf{R}^n$ be an open set and $u : A \to \mathbf{R}$. The definition of viscosity subsolution (supersolution) for a first order equation

$$H(x, u, \nabla u) = 0 \tag{44}$$

is similar to Definition 6. In this case, since H does not depend on second order derivatives, classical solutions are always viscosity solutions and an equivalent definition could be given with C^1 test functions ϕ. Moreover, it is clear that Proposition 2 and Theorem 11 are still valid, as a particular case, for first order equations.

Let us define, in analogy with (38) and (39), the first order *super and sub jets*

$$J_A^{1,+}u(x) := \left\{ p \in \mathbf{R}^n \mid u(y) \leq u^*(x) + \langle p, y - x \rangle + o(|y - x|) \right\} \qquad (45)$$

$$J_A^{1,-}u(x) := \left\{ p \in \mathbf{R}^n \mid u(y) \geq u_*(x) + \langle p, y - x \rangle + o(|y - x|) \right\}. \qquad (46)$$

These super and sub jets are closed and convex, possibly empty, and it turns out that u is a viscosity subsolution in A of (44) if and only if

$$H_*\left(x, u^*(x), p\right) \leq 0 \qquad \forall p \in J_A^{1,+}u(x), \ x \in A. \qquad (47)$$

Analogously, u is a viscosity supersolution in A of (44) if and only if

$$H^*\left(x, u_*(x), p\right) \geq 0 \qquad \forall p \in J_A^{1,-}u(x), \ x \in A. \qquad (48)$$

We will study a very special case of (44), namely the *eikonal equation* $|\nabla u|^2 - 1 = 0$. Let $C \subset \mathbf{R}^n$ be a nonempty closed set, let $A = \mathbf{R}^n \setminus C$ and let d be the distance function from C; since the Lipschitz constant of d is less than 1, it is easy to check (cf. Lemma 1) that d is a global viscosity subsolution of the equation; on the other hand, since

$$d(x) = \inf_{y \in C} |x - y|$$

and since for any $y \in C$ the function $x \mapsto |x - y|$ is a viscosity (classical) supersolution in A, by Proposition 2 we infer that d is a viscosity supersolution in A. Summing up, we have proved that d is a viscosity solution of

$$|\nabla u|^2 - 1 = 0 \qquad \text{in } A. \qquad (49)$$

We will prove now that this equation actually characterizes the distance function:

Theorem 12 (viscosity characterization of distance functions). *Let $C \subset \mathbf{R}^n$ be a closed set and let $A := \mathbf{R}^n \setminus C$. Let $u \in C(\overline{A})$ be a nonnegative viscosity solution of the equation (49) vanishing on ∂A. Then, C is not empty and*

$$u(x) = \text{dist}(x, C) \qquad \forall x \in \overline{A}.$$

Proof. Let $w(x) = \text{dist}(x, C)$ and extend u to all $\mathbf{R}^n$ setting $u = 0$ on C; we have to prove that $u \equiv w$. We will first assume that A is bounded. It is easy to check that $|\nabla u|^2 - 1 \leq 0$ in the whole $\mathbf{R}^n$, in the viscosity sense, so that Lemma 1 below yields that the Lipschitz constant of u is less than 1. In particular, for any $x \in \mathbf{R}^n$ and any $y \in C$ it holds

$$u(x) \leq u(y) + |x - y| = |x - y|.$$

As y is arbitrary, the inequality $u \leq w$ follows.

To prove the opposite inequality we argue by contradiction. Assume that $w(x_0) > u(x_0)$ for some $x_0 \in \mathbf{R}^n$; then, we can find $\lambda_0 > 0$ and $\gamma_0 > 0$ such that

$$\max_{x,\,y}\left[w(x) - (1 + \lambda_0)u(y) - \frac{1}{2\epsilon}|x - y|^2\right] \geq \gamma_0$$

for any $\epsilon > 0$. Indeed, if we evaluate the function above at (x_0, x_0), we find

$$w(x_0) - (1 + \lambda_0)u(x_0)$$

which is strictly positive for small λ_0. Choosing $\epsilon \in (0, 2\gamma_0)$, we will find a contradiction.

Let $(\bar{x}, \bar{y})$ be maximizers; since w is a subsolution, we get

$$\left|\frac{\bar{x} - \bar{y}}{\epsilon}\right| \leq 1. \tag{50}$$

Now we claim that (50) implies that $\bar{y} \in A$; indeed, if this were not true we would get

$$\gamma_0 \leq w(\bar{x}) - \frac{1}{2\epsilon}|\bar{x} - \bar{y}|^2 \leq \left(|\bar{x} - \bar{y}| - \frac{1}{2\epsilon}|\bar{x} - \bar{y}|^2\right)$$

$$= |\bar{x} - \bar{y}|\left(1 - \frac{|\bar{x} - \bar{y}|}{2\epsilon}\right) \leq \frac{|\bar{x} - \bar{y}|}{2},$$

hence $2\gamma_0 \leq |\bar{x} - \bar{y}| \leq \epsilon$, contradicting our choice of ϵ. Therefore $\bar{y} \in A$ and since u is a supersolution in A we get

$$\left|\frac{\bar{x} - \bar{y}}{\epsilon}\right| \geq 1 + \lambda$$

contradicting (50).

If A is unbounded, we notice that the function

$$u_R(x) := u(x) \wedge \operatorname{dist}(x, \mathbf{R}^n \setminus B_R)$$

is a viscosity supersolution in $A \cap B_R$ (by Proposition 2); since the Lipschitz constant of u_R is less than 1, it is also a subsolution. Therefore the previous argument shows that

$$u_R(x) = \operatorname{dist}(x, C \cup (\mathbf{R}^n \setminus B_R)) \qquad x \in \mathbf{R}^n, \ R > 0.$$

Letting $R \uparrow \infty$ we obtain that $u(x) = \operatorname{dist}(x, C)$. Since u is real valued, C is not empty. $\qquad \square$

Exercise 4. (global solutions) Show that the equation $|\nabla u|^2 - 1$ has no global, continuous viscosity solution bounded from below, and find a global continuous solution unbounded from below. Hint: given $m_\epsilon \downarrow \inf u$, apply Theorem 12 to $u - m_\epsilon$ in $\{u > m_\epsilon\}$ to find that u has a minimum point $\bar{x}$. The definition of viscosity supersolution is violated at $\bar{x}$.

Remark 10. (1) The argument adopted in Theorem 12 shows that (under suitable assumptions on H) if u is a viscosity subsolution of (44) in Ω and if v is a viscosity supersolution of (44) in Ω, then

$$u \leq v \quad \text{on } \partial\Omega \quad \Longrightarrow \quad u \leq v \quad \text{on } \Omega.$$

The introduction of two variables x, y in the maximization problem is explained by the necessity to compare u and v with smooth test functions.

(2) More generally, it can be proved (see for instance Theorem 5.3 of [Lio82] and [BD98a]) under suitable regularity and growth conditions on H, Ω, g that the unique viscosity solution of the equation (44) in Ω satisfying $u = g$ on $\partial\Omega$ is the *value function* of a variational problem:

$$u(x) = \inf\left\{ g(\gamma(T)) + \int_0^T L(\gamma(t), \gamma'(t))\, dt \,\middle|\, T > 0,\ \gamma \in \mathcal{A}_T(x) \right\} \qquad (51)$$

where

$$\mathcal{A}_T(x) := \{\gamma \in C^1([0,T], \overline{\Omega}) \,|\, \gamma(0) = x,\ \gamma(T) \in \partial\Omega,\ \gamma([0,T)) \subset \Omega\}.$$

For this reason, (44) is usually called Hamilton–Jacobi equation. In (51), the Lagrangian $L(x,p)$ is the conjugate function of H, i.e.

$$L(x,q) := \sup\left\{ \langle p, q\rangle - H(x,p) \,|\, p \in \mathbf{R}^n \right\}$$

and the crucial assumption on $H(x,p)$ for the validity of (51) is the convexity with respect to p.

(3) The name *viscosity* solutions is related to the approximation of u by the functions $(u^\epsilon)_{\epsilon > 0}$ solving

$$H(x, u, \nabla u) - \epsilon \Delta u = 0 \quad \text{in } \Omega$$

in the classical (or distributional) sense, with suitable boundary conditions. This approximation process can be used in place of Theorem 11 to get existence results. In fact, by Theorem 14 below, one needs only to know that (a subsequence) of u^ϵ is locally uniformly converging in Ω to some function u to get a viscosity solution of (44). Analogous approximation arguments can also be used for specific second order equations (see for instance [ES91] and Remark 19).

Lemma 1. *Let $u \in C(\overline{B}_R)$. Then, u satisfies $|\nabla u|^2 - 1 \leq 0$ in B_R in the viscosity sense if and only if the Lipschitz constant of u in $\overline{B}_R$ is less than 1.*

Proof. Assume that the Lipschitz constant of u is less than 1, and let $\phi \in C^{\infty}(B_R)$ such that $u - \phi$ has a relative maximum, equal to 0, at $x_0 \in B_R$. From the inequalities

$$\phi(x_0) + \langle \nabla\phi(x_0), z \rangle + o(|z|) \geq \phi(x_0 + z) \geq u(x_0 + z) \geq u(x_0) - |z|$$

we infer $\langle \nabla\phi(x_0), z \rangle + o(|z|) \geq -|z|$, hence $|\nabla\phi(x_0)| \leq 1$.

To show the opposite implication, let $r \in (0, R)$ and let u^{ϵ} be the sup convolutions of u (see (66)). Then by Proposition 4, for ϵ small enough u^{ϵ} is a viscosity subsolution in B_r (Remark 16); since u^{ϵ} is a Lipschitz function, the differential inequality is also satisfied almost everywhere in B_r, hence $\text{Lip}(u^{\epsilon}, B_r) \leq 1$. Passing to the limit first as $\epsilon \downarrow 0$ then as $r \uparrow R$, the conclusion follows.   □

8 Γ convergence and the stability theorem

Since the definitions of viscosity subsolution and supersolution involve local maximization or minimimization conditions, it is natural to investigate several properties of these functions in connection with the Γ convergence, a variational convergence introduced by De Giorgi and Franzoni in [GF75] (a general introduction to this topic is available in [Mas93]).

In fact, the Γ^- (respectively Γ^+) convergence induces convergence of minimizers to minimizers (maximizers) and the convergence of extremal values as well (see Theorem 13 below). This leads to a general stability theorem for supersolutions (subsolutions) which, besides Proposition 2, is very useful in many proofs. In particular, we will see that the property of being a viscosity subsolution (supersolution) is stable under uniform convergence.

Definition 9 (Γ limits). Let (X, d) be a metric space and let $f_h : X \to \overline{\mathbf{R}}$ be functions. We set

$$\Gamma^- \liminf_{h\to\infty} f_h(x) := \inf\left\{ \liminf_{h\to\infty} f_h(x_h) \,|\, x_h \to x \right\}$$

$$\Gamma^- \limsup_{h\to\infty} f_h(x) := \inf\left\{ \limsup_{h\to\infty} f_h(x_h) \,|\, x_h \to x \right\}$$

$$\Gamma^+ \liminf_{h\to\infty} f_h(x) := \sup\left\{ \liminf_{h\to\infty} f_h(x_h) \,|\, x_h \to x \right\}$$

$$\Gamma^+ \limsup_{h\to\infty} f_h(x) := \sup\left\{ \limsup_{h\to\infty} f_h(x_h) \,|\, x_h \to x \right\}.$$

We say that the sequence (f_h) Γ^- converges if

$$\Gamma^- \liminf_{h\to\infty} f_h(x) = \Gamma^- \limsup_{h\to\infty} f_h(x) \qquad\qquad \forall x \in X.$$

The common value of the two functions above is called Γ^- *limit* of (f_h) and denoted by

$$\Gamma^- \lim_{h\to\infty} f_h.$$

Analogously, we say that the sequence (f_h) Γ^+ converges if

$$\Gamma^+ \liminf_{h\to\infty} f_h(x) = \Gamma^+ \limsup_{h\to\infty} f_h(x) \qquad \forall x \in X.$$

The common value of the two functions above is called Γ^+ *limit* of (f_h) and denoted by

$$\Gamma^+ \lim_{h\to\infty} f_h.$$

Remark 11. (1) The sequence $\sin(hx)$ Γ^- converges to -1 and Γ^+ converges to 1. An example showing that in general the Γ limit differs from the pointwise limit is the sequence $f_h(x) := hxe^{-2h^2x^2}$, which is pointwise converging to 0 while

$$\Gamma^- \lim_{h\to\infty} f_h(x) = \begin{cases} -\dfrac{1}{2\sqrt{e}} & \text{if } x = 0; \\ 0 & \text{if } x \neq 0 \end{cases} \qquad \Gamma^+ \lim_{h\to\infty} f_h(x) = \begin{cases} \dfrac{1}{2\sqrt{e}} & \text{if } x = 0; \\ 0 & \text{if } x \neq 0. \end{cases}$$

(2) It is not hard to see by a diagonal argument that the "inf" and "sup" in the definitions above are attained; moreover,

$$x \mapsto \Gamma^- \liminf_{h\to\infty} f_h(x), \qquad x \mapsto \Gamma^- \limsup_{h\to\infty} f_h(x)$$

are lower semicontinuous, while

$$x \mapsto \Gamma^+ \liminf_{h\to\infty} f_h(x), \qquad x \mapsto \Gamma^+ \limsup_{h\to\infty} f_h(x)$$

are upper semicontinuous. Finally, the $\Gamma^\pm$ limits are unique and stable under subsequences.

The connection between Γ^+ and Γ^- limits is given by

$$\Gamma^+ \liminf_{h\to\infty} f_h(x) = -\Gamma^- \limsup_{h\to\infty} [-f_h](x)$$

$$\Gamma^+ \limsup_{h\to\infty} f_h(x) = -\Gamma^- \liminf_{h\to\infty} [-f_h](x). \tag{52}$$

In the following we will only make statements for Γ^- limits, the corresponding one for Γ^+ limits can be easily deduced from (52). We also notice that a sequence (f_h) Γ^- converges to X if and only if

$$\liminf_{h\to\infty} f_h(x_h) \geq f(x) \qquad \forall (x_h) \subset X, \ x_h \to x \tag{53}$$

and

$$\forall \epsilon > 0 \, \forall x \in X \, \exists (x_h) \subset X, \ x_h \to x \quad \text{s.t.} \quad \limsup_{h\to\infty} f_h(x_h) \leq f(x) + \epsilon. \tag{54}$$

Indeed, the first inequality is equivalent to $f \leq \Gamma^- \liminf_{h\to\infty} f$ and the second one is equivalent to $f \geq \Gamma^- \limsup_{h\to\infty} f$.

Proposition 3 (properties of Γ limits).

(i) If $f_h = f$ for any $h \in \mathbf{N}$, then

$$\Gamma^- \lim_{h \to \infty} f_h = f_*.$$

The same formula holds if f_h converges to f locally uniformly in X;
(ii) if $g : X \to \mathbf{R}$ is continuous, then

$$\Gamma^- \lim_{h \to \infty} (f_h + g) = \Gamma^- \lim_{h \to \infty} f_h + g;$$

(iii) if $f_h = (f_h)_* \uparrow f$, then

$$\Gamma^- \lim_{h \to \infty} f_h = f;$$

(iv) if X is separable, any sequence (f_h) admits a Γ^- converging subsequence.

Proof. (i) and (ii) are a straightforward consequence of the definitions.
(iii) For any $k \in \mathbf{N}$ it holds

$$\liminf_{h \to \infty} f_h(x_h) \geq \liminf_{h \to \infty} f_k(x_h) \geq f_k(x) \qquad \forall x_h \to x, \ x \in X.$$

Passing to the limit in k we get (53). Inequality (54) is trivial with $x_h = x$ for any h.
(iv) Let $\mathcal{U}$ be the collection of all balls with rational radius centered in a countable dense subset of X. By a diagonal argument, we can find a subsequence $f_{k(h)}$ such that $\inf_B f_{k(h)}$ converges as $h \to \infty$ to some $c(B) \in \overline{\mathbf{R}}$. Assuming for notational simplicity $k(h) = h$, let us set

$$f(x) := \sup\{c(B) \,|\, B \ni x, \ B \in \mathcal{U}\}.$$

We will prove that the sequence (f_h) Γ^- converges to f. Let $(x_h) \subset X$ be any sequence converging to $x \in X$ and let $B \in \mathcal{U}$ containing x. Since $x_h \in B$ for h large enough, it holds

$$\liminf_{h \to \infty} f_h(x_h) \geq \liminf_{h \to \infty} \inf_B f_h = c(B).$$

Since B is arbitrary, (53) follows.
    To prove (54) we choose a sequence of balls $B_k \ni x$ belonging to $\mathcal{U}$ and such that $c(B_k) \uparrow f(x)$. Possibly reducing B_k, we can assume that their radius tends to 0 as $k \to \infty$; it holds

$$\limsup_{h \to \infty} \inf_{B_k} f_h = c(B_k) \leq f(x) \qquad \forall k \in \mathbf{N}.$$

This implies the existence of a sequence $k(h)$ tending to ∞ such that

$$\limsup_{h \to \infty} \inf_{B_{k(h)}} f_h \leq f(x).$$

Choosing $x_h \in B_{k(h)}$ such that $f_h(x_h) \leq \inf_{B_{k(h)}} f_h + 2^{-h}$, (54) follows. $\square$

Exercise 5. Let X be a locally compact metric space and let $f_h, f : X \to \mathbf{R}$ be functions. Show that (f_h) converges to f locally uniformly if and only if

$$\Gamma^- \lim_{h \to \infty} f_h = f = \Gamma^+ \lim_{h \to \infty} f_h.$$

Theorem 13 (variational properties of Γ^- limits). *Let (f_h) be a sequence Γ^- converging to f, and assume that $f_h = f_{h*}$ and there exists a compact set $K \subset X$ such that*

$$\inf_X f_h = \inf_K f_h \qquad \forall h \in \mathbf{N}. \tag{55}$$

Then, if $x_h \in K$ are minimizing points for f_h, every limit point of (x_h) is a minimizer of f and

$$\min_X f = \lim_{h \to \infty} \min_X f_h. \tag{56}$$

In particular, if f has a unique minimizer x, then (x_h) converges to x.

Proof. Let x be the limit of any subsequence $x_{h(k)}$. Then, (53) yields

$$f(x) \le \liminf_{k \to \infty} f_{h(k)}(x_{h(k)}) = \liminf_{k \to \infty} \min_X f_{h(k)} \tag{57}$$

On the other hand, for any $x' \in X$ we can find a sequence (x'_h) converging to x' such that the upper limit of $f_h(x'_h)$ is less than $f(x')$, so that

$$f(x') \ge \limsup_{h \to \infty} f_h(x'_h) \ge \limsup_{k \to \infty} \min_X f_{h(k)}. \tag{58}$$

From (57) and (58) we infer that x minimizes f; moreover, (57) and (58) with $x' = x$ imply

$$\min_X f = \lim_{k \to \infty} \min_X f_{h(k)}.$$

Since the subsequence is arbitrary and (x_h) lies in a compact set, (56) follows.$\square$

Remark 12. A local variant of Theorem 13, which can be proved by a similar argument, is the following. Assume in place of (55) that X is locally compact and that x is a strict relative minimum of $f = \Gamma^- \lim_{h \to \infty} f_h$, i.e.,

$$f(y) > f(x) \qquad \forall y \in K \setminus \{x\}$$

for some compact neighbourhood K of x. Then, if x_h are minimizing points for the restriction of f_h to K, the sequence (x_h) converges to x and

$$\min_K f = f(x) = \lim_{h \to \infty} \min_K f_h.$$

If $f = \Gamma^- \liminf_{h \to \infty} f_h$ the same conclusions are true for a subsequence $h(k)$, the one satisfying

$$f(x) = \lim_{k \to \infty} f_{h(k)}(y_k)$$

for some sequence (y_k) converging to x.

By the convergence of local strict minimizers we infer the following very useful result (see [BP87, CGG91]), which enables to pass to the limit in u *and* in E in (37).

Theorem 14 (stability theorem). *Let $A \subset \mathbf{R}^n$ be a locally compact set, and let (u_h) be a sequence of upper semicontinuous functions defined in A, satisfying*

$$E_h\left(x, u_h, \nabla u_h, \nabla^2 u_h\right) \leq 0 \qquad in\ A$$

in the viscosity sense. Assume that $u := \Gamma^+ \limsup\limits_{h \to \infty} u_h$ is nowhere equal to $-\infty$ in A. Then,

$$E\left(x, u, \nabla u, \nabla^2 u\right) \leq 0 \qquad in\ A \cap \{u < \infty\}$$

in the viscosity sense, with $E := \Gamma^- \liminf\limits_{h \to \infty} E_{h*}$.

Proof. Let $x_0 \in A$, $r > 0$ and let $\phi \in C^\infty(B_r(x_0))$ such that $u - \phi$ has a strict relative maximum at x_0. Assuming that r is small enough to ensure the compactness of $K := A \cap \overline{B}_{r/2}(x_0)$ and the validity of the implication

$$x \in K \setminus \{x_0\} \qquad \Longrightarrow \qquad u(x) - \phi(x) < u^*(x_0) - \phi(x_0)$$

we denote by x_h any maximizer of u_h on K. Then, the Γ^+ convergence of (u_h) and Remark 12 imply the convergence of (a subsequence of) x_h to x and of $u_h(x_h) - \phi(x_h)$ to $u(x_0) - \phi(x_0)$. In particular, $u_h(x_h)$ converges to $u(x_0)$.

Since u_h are viscosity subsolutions in A, it holds

$$E_{h*}\left(x_h, u_h(x_h), \nabla\phi(x_h), \nabla^2\phi(x_h)\right) \leq 0$$

for any index h such that $x_h \in A \cap B_{r/2}(x_0)$. Passing to the limit as $h \to \infty$ we obtain

$$E\left(x, u(x_0), \nabla\phi(x_0), \nabla^2\phi(x_0)\right) \leq \liminf_{h \to \infty} E_{h*}\left(x_h, u_h(x_h), \nabla\phi(x_h), \nabla^2\phi(x_h)\right) \leq 0.$$

$$\square$$

9   Semiconvex and semiconcave functions

In this section we study the main properties of semiconvex (semiconcave) functions, i.e., functions whose second order distributional derivative can be estimated from below (above). This class of functions plays an important rôle in the proof of comparison theorems for viscosity solutions of second order equations because of the following two facts:

(a) any viscosity subsolution can be locally uniformly approximated from above by semiconvex subsolutions of the same equation (at least for equations $E(\nabla u, \nabla^2 u) = 0$);

(b) any semiconvex function satisfies an approximate version of the classical maximum principle.

Definition 10. Let $\Omega \subset \mathbf{R}^n$ be an open set and $u : \Omega \to \mathbf{R}$. We say that u is *semiconvex* in Ω, and we write $u \in S\Omega$, if there exists a constant $c \geq 0$ such that

$$x \mapsto u(x) + \frac{c}{2}|x|^2$$

is convex in any ball $B \subset \Omega$. The smallest constant c for which this property holds is called *semiconvexity constant* of u and denoted by $\mathrm{sc}(u, \Omega)$.

Remark 13. Obviously, any C^2 perturbation of a convex function is semiconvex, and the semiconvexity constant can be estimated with the C^2 norm of the perturbation. The definitions of semiconcave function and of semiconcavity constant can be given in a similar way.

An important example of semiconcave function is the squared distance function from *any* nonempty set E: indeed, the identities

$$\mathrm{dist}^2(x, E) - |x|^2 = \inf_{y \in E} |x - y|^2 - |x|^2 = \inf_{y \in E} |y|^2 - 2\langle x, y \rangle$$

show that $\mathrm{dist}^2(x, E)$ is semiconcave in $\mathbf{R}^n$ with semiconcavity constant less than 2.

Exercise 6. Verify that $u \in S\Omega$ if and only if $D^2 u \geq -cI$ in the sense of distributions in Ω, i.e.

$$\int_\Omega u \frac{\partial^2 \phi}{\partial \xi \partial \xi} \, dx \geq -c \int_\Omega \phi \, dx \qquad \forall \phi \in C_c^\infty(\Omega, \mathbf{R}^+), \ \xi \in \mathbf{S}^{n-1}. \qquad (59)$$

Moreover, the smallest constant $c \geq 0$ in (59) is $\mathrm{sc}(u, \Omega)$.

Exercise 7. Verify that $u(x) := \mathrm{dist}(x, E)$ is locally semiconcave in $\Omega := \mathbf{R}^n \setminus \overline{E}$; moreover, by the identity

$$D^2 u^2 = 2u D^2 u + 2\nabla u \otimes \nabla u \, \mathcal{L}^n$$

in any ball $B \subset\subset \Omega$ the semiconcavity constant can be estimated with $2 \left[\inf_B u\right]^{-1}$.

Exercise 8. Verify that $u \mapsto \mathrm{sc}(u, \Omega)$ (set to ∞ if $u \notin S\Omega$) is lower semicontinuous with respect to convergence in the sense of distributions in Ω.

In the following theorem we see that the first order sub and super jets of semiconvex functions have special properties. This leads to a simple characterization of semiconvex viscosity solutions of (44), stated in Remark 14.

Theorem 15 (first order properties of semiconvex functions). *Let u be a semiconvex function in Ω. Then*

(i) $u \in W_{\mathrm{loc}}^{1,\infty}(\Omega)$, *hence u is differentiable almost everywhere in Ω and $Du = \nabla u \mathcal{L}^n$;*

(ii) *for any $x \in \Omega$ the super jet $J^{1,+} u(x)$ is not empty if and only if u is differentiable at x;*

(iii) *for any $x \in \Omega$ the sub jet $J^{1,-}u(x)$ is not empty and can be represented as the closed convex hull of the set*

$$\nabla_* u(x) := \{p \in \mathbf{R}^n \,|\, p = \lim_{h \to \infty} \nabla u(x_h), \ x_h \to x\}$$

of reachable gradients. Moreover u is differentiable at x if and only if $J^{1,-}u(x)$ is a singleton;

(iv) *the graph of the (set valued) map $x \mapsto J^{1,-}u(x)$ is closed in $\Omega \times \mathbf{R}^n$. In particular, any differentiability point is a continuity point for ∇u.*

Proof. Being all the statements local, we will assume that Ω is convex; moreover, possibily adding to u a quadratic perturbation, we can assume that u is convex too. Under these assumptions, it is easy to see $J^{1,-}u$ reduces to the *subdifferential* of u, i.e.

$$J^{1,-}u(x) = \partial^- u(x) = \{p \in \mathbf{R}^n : u(y) \geq u(x) + \langle p, y - x \rangle \ \forall y \in \Omega\} \qquad (60)$$

Statement (i) follows by Exercise 9, the fact that $\partial^- u$ is nowhere empty is a simple application of Hahn–Banach theorem to the epigraph of u and statement (iv) easily follows by (60), because u is continuous.

(ii) If $p \in J^{1,+}u(x)$, we can choose $q \in \partial^- u(x)$ to get

$$\langle q, y - x \rangle \leq \langle p, y - x \rangle + o(|y - x|).$$

Hence $p = q$ and u is differentiable at x. The opposite implication is trivial.

(iii) By the closure of the graph of ∂^-, $\nabla_* u(x)$ is contained in $\partial^- u(x)$ as well as its closed convex hull. If some vector $p \in \partial^- u(x)$ does not belong to the closed convex hull, possibly subtracting from u a linear function, we can assume $p = 0$. By Hahn–Banach theorem we can find a unit vector q and $\alpha > 0$ such that

$$\langle z, q \rangle \geq \alpha \qquad\qquad \forall z \in \nabla_* u(x).$$

The definition of $\nabla_* u$ yields the existence of a ball $B_r(x) \subset \Omega$ such that $\langle \nabla u(y), q \rangle \geq \alpha/2$ for any differentiability point $y \in B_r(x)$. Since, by (i), $\nabla(u * \rho_\epsilon) = \nabla u * \rho_\epsilon$, a smoothing argument shows that

$$t \mapsto u(x + tq)$$

is strictly increasing in $(-r, r)$. On the other hand, since $0 \in \partial^- u(x)$, the same function attains its minimum at 0, a contradiction.

Finally, we prove that u is differentiable at x if $\partial^- u(x) = \{p\}$ is a singleton. Indeed, by the nonsmooth mean value theorem (Exercise 10) we have

$$u(y) - u(x) = \langle p_y, y - x \rangle = \langle p, y - x \rangle + \langle p_y - p, y - x \rangle$$

for suitable vectors $p_y \in \partial^- u(x + t_y(y - x))$ with $t_y \in (0, 1)$. Since $\partial^- u(x)$ contains p only, p_y converges to p as $y \to x$, so that $\langle p_y - p, y - x \rangle = o(|y - x|)$. $\square$

Remark 14. Let us consider a first order equation (44) with a continuous function $H(x, s, p)$, concave in p. Then, Theorem 15 implies that a semiconvex function u is a viscosity solution in Ω if and only if $H(x, u(x), \nabla u(x)) = 0$ almost everywhere in Ω. Indeed, if u is a viscosity solution then the equation is satisfied at any differentiability point, hence almost everywhere. Conversely, if the equation is satisfied almost everywhere, (by statement (iv)) it is also satisfied at any differentiability point; by statement (ii) we infer that u is a subsolution and

$$H(x, u(x), p) = 0 \qquad \forall p \in \nabla_* u(x), \ x \in \Omega.$$

By statement (iii) we get

$$H(x, u(x), p) \geq 0 \qquad \forall p \in J^{1,+} u(x), \ x \in \Omega$$

and this proves that u is a supersolution too.

A similar statement (with convex H) is true for semiconcave functions. By Theorem 1, this provides another proof of the fact that distance functions satisfy $|\nabla d|^2 - 1 = 0$ in $\{d > 0\}$ in the viscosity sense.

Exercise 9. Show the estimate

$$\sup_{B_r} |\nabla u| \leq \frac{\operatorname{osc}(u, B_R)}{R - r} \qquad 0 < r < R$$

for convex functions in the ball B_R.

Exercise 10. (nonsmooth mean value theorem) Let $\Omega \subset \mathbf{R}^n$ be a convex set and let $u : \Omega \to \mathbf{R}$ be convex. Then, for any pair of points $x, y \in \Omega$ there exist $t \in (0, 1)$ and $p \in \partial^- u(x + t(y - x))$ such that

$$u(y) - u(x) = \langle p, y - x \rangle.$$

Hint: first show the existence of $p \in \partial^- u(z)$ for some z on the closed segment S joining x to y, by a smoothing argument. Then, show that if $z = x$ or $z = y$ then $p \in \partial^- u(z)$ for any $z \in S$.

Using Theorem 15(ii) we can now verify that the distance function is also a supersolution of a suitable second order equation; this will be useful in §13.

Lemma 2. *The distance function from a nonempty set satisfies*

$$-\langle \nabla^2 d \nabla d, \nabla d \rangle \geq 0 \qquad \text{in } \{d > 0\}$$

in the viscosity sense.

Proof. Let $x_0 \in \{d > 0\}$ and let $\phi \in C^\infty(B_r(x_0))$ be such that $d - \phi$ has a relative minimum, equal to 0, at x_0. Since

$$d(x) \geq d(x_0) - \phi(x_0) + \phi(x) = \phi(x_0) + \langle \nabla \phi(x_0), x - x_0 \rangle + o(|x - x_0|)$$

we find that $J^{1,-}d(x_0)$ is not empty, therefore d is differentiable at x_0 and $\nabla d(x_0) = \nabla\phi(x_0)$.

Let $p = \nabla\phi(x_0)$; since $|p| = 1$ (cf. Theorem 1), the function $\Phi(t) = \phi(x_0 + tp) - \phi(x_0) - t$ satisfies $\Phi(0) = 0$ and

$$\Phi(t) \leq d(x_0 + tp) - d(x_0) - t \leq 0$$

for any t sufficiently small. Hence, $\Phi''(0) \leq 0$, which means $\langle\nabla^2\phi(x_0)p, p\rangle \leq 0$.□

The following theorem concerning second order differentials of (semi)convex functions is essentially due to Aleksandroff (see [Ale93] and also [CIL92, AA99]).

Theorem 16 (second order properties of semiconvex functions). *Let $u \in S\Omega$.*

(i) *$\nabla u \in BV_{\text{loc}}(\Omega)$, i.e., the distributional derivative D^2u of ∇u is a Radon measure in Ω;*

(ii) *for almost every $x \in \Omega$, u has a second order Taylor expansion at x:*

$$u(y) = u(x) + \langle\nabla u(x), y - x\rangle + \frac{1}{2}\langle\nabla^2 u(x)(y - x), (y - x)\rangle + o(|y - x|^2). \quad (61)$$

Moreover, $\nabla^2 u$ is the gradient of ∇u and coincides with the density of the absolutely continuous part of $D^2 u$.

Proof. (i) As in Theorem 15, it is not restrictive to assume that Ω and u are convex. Since

$$D_{\xi\eta}u = \frac{1}{4}\left[D_{(\xi+\eta)(\xi+\eta)}u - D_{(\xi-\eta)(\xi-\eta)}u\right] \qquad \xi, \eta \in \mathbf{R}^n$$

we need only to prove that $D_{\xi\xi}u$ is a Radon measure for any unit vector $\xi \in \mathbf{R}^n$. A smoothing argument shows that $D_{\xi\xi}u \geq 0$ in the sense of distributions. Let $A \subset\subset \Omega$ and let $\phi \in C_c^\infty(\Omega)$ be a function equal to 1 in a neighbourhood of A; for any $\varphi \in C_c^\infty(A)$ it holds

$$D_{\xi\xi}u(\varphi) \leq \|\varphi\|_\infty D_{\xi\xi}u(\phi) = c(A)\|\varphi\|_\infty.$$

This shows that the distribution is bounded in A, hence locally bounded in Ω. (ii) In the proof of this statement we make a stronger (and not restrictive) assumption, namely $D^2u \geq I$. Under this assumption the subdifferential is strictly monotone, i.e.

$$\langle p - q, y - x\rangle \geq |y - x|^2 \qquad \forall p \in \partial^-u(y), \; q \in \partial^-u(x). \quad (62)$$

Notice that (62) implies that for any p there exists at most one x such that $p \in \partial^-u(x)$; we will denote by $\psi : L \to \mathbf{R}^n$ the inverse function of ∂^-u, so that $p \in \partial^-u(x)$ if and only if $x = \psi(p)$. From

$$|p - q||\psi(p) - \psi(q)| \geq |\langle p - q, \psi(p) - \psi(q)\rangle| \geq |\psi(p) - \psi(q)|^2 \qquad p, q \in L$$

it follows that ψ is a Lipschitz function in L, hence differentiable almost everywhere in L. Let

$$N := \{p \in L \,|\, \text{either } \not\exists \nabla\psi(p) \text{ or } \det \nabla\psi(p) = 0\}$$

and let $\Omega' = \psi(N)$. Using the area formula for Lipschitz functions (see for instance [Sim84]) it can be proved that Ω' is negligible.

We now claim the existence, for every differentiability point $x \notin \Omega'$, of a matrix $A(x) \in \text{Sym}^n$ such that

$$\lim_{y \to x, \, p \in \partial^- u(y)} \frac{|p - \nabla u(x) - A(x)(y - x)|}{|y - x|} = 0. \tag{63}$$

Indeed, let $x = \psi(p)$ with $p \notin N$ and let A be the inverse of $\nabla\psi(p)$; for $q \in \partial^- u(y)$ it holds

$$A^{-1}(q - p - A(y - x)) = A^{-1}(q - p - A(\psi(q) - \psi(p))) =$$
$$= -[\psi(q) - \psi(p) - \nabla\psi(p)(q - p)]$$
$$= o(|q - p|) = o(|\psi(q) - \psi(p)|) = o(|y - x|).$$

We have used the fact that $|\psi(q) - \psi(p)|$ has the same order of $|q - p|$ as $q \to p$, because $\det \nabla\psi(p) \neq 0$. This proves (63) (the symmetry of A follows by the last statement, proved below, and from the symmetry of $D^2 u$).

To prove (61) we choose a point differentiability $x \in \Omega$ where (63) holds and apply the nonsmooth mean value theorem (Exercise 10) to

$$v(y) := u(y) - u(x) - \langle p, y - x \rangle - \frac{1}{2} \langle A(x)(y - x), (y - x) \rangle$$

to get, for suitable z_y on the segment joining x to y and vectors $p_y \in \partial^- u(z_y)$

$$v(y) = v(y) - v(x) = \langle p_y - p - A(x)(z_y - x), y - x \rangle = o(|z_y - x|)|y - x| = o(|y - x|^2).$$

Now we prove the last statement of the theorem by a blow-up argument; we denote by $D^a \nabla u + D^s \nabla u$ the Radon–Nikodym decomposition of $D\nabla u$ in absolutely continuous and singular part with respect to Lebesgue measure $\mathcal{L}^n$ and denote by $\nabla^2 u$ the density of $D^a \nabla u$. Let $x \in \Omega$ and let

$$v_\varrho(y) := \frac{\nabla u(x + \varrho y) - \nabla u(x)}{\varrho} \qquad y \in B_1.$$

If (63) is fulfilled at x, then v_ϱ converges in $L^\infty(B_1; \mathbf{R}^n)$ to $A(x)y$, hence the distributional derivatives Dv_ϱ converge to the distribution $A(x)\mathcal{L}^n$ as $\varrho \downarrow 0$. On the other hand, since (by a change of variables)

$$Dv_\varrho(\phi) = \varrho^{-n} \int_\Omega \phi((y - x)/\varrho) \, d(D^a \nabla u + D^s \nabla u)(y)$$

$$= \int_{B_1} \phi(y)\nabla^2 u(x + \varrho y) \, dy + \varrho^{-n} \int_\Omega \phi((y - x)/\varrho) \, dD^s \nabla u(y)$$

for any $\phi \in C_c^\infty(\Omega)$, if we choose a Lebesgue point x for $\nabla^2 u$ where $D^s \nabla u$ has zero density (these conditions are fulfilled almost everywhere) we obtain that $Dv_\varrho(\phi)$ converges as $\varrho \downarrow 0$ to

$$\nabla^2 u(x) \int_{B_1} \phi(y)\,dy$$

This proves that $A(x) = \nabla^2 u(x)$ almost everywhere in Ω.   □

Theorem 17 (maximum principle for semiconvex functions). *Let $u \in S\Omega$, $x_0 \in \Omega$ and assume that x_0 is a relative maximum point of u. Then, there exist a sequence $(x_h) \subset \Omega$ converging to x_0 and an infinitesimal sequence $(\epsilon_h) \subset (0,\infty)$ such that u is twice differentiable at x_h, has the second order Taylor expansion (61) at x_h and*

$$\lim_{h\to\infty} \nabla u(x_h) = 0, \qquad \nabla^2 u(x_h) \leq \epsilon_h I.$$

Proof. Let $B \subset\subset \Omega$ be an open ball centered at x_0 such that $u \leq u(x_0)$ on $\overline{B}$ and let $w(x) = u(x) - |x - x_0|^4$. We can apply Jensen's lemma below to w in B to find a sequence $(x_h) \subset B$ and vectors $p_h \in \mathbf{R}^n$ converging to 0 such that w is twice differentiable, has a second order Taylor expansion at x_h and

$$w(y) \leq w(x_h) + \langle p_h, y - x_h \rangle \qquad \forall y \in \overline{B}.$$

Any limit point of (x_h) obviously is a maximum for w in $\overline{B}$; since x_0 is the only maximum point for w in $\overline{B}$, this implies that x_h converges to x_0. The conclusion follows by the identity $\nabla u(x_h) = p_h + 4|x_h - x_0|^2(x_h - x_0)$ and by

$$\nabla^2 u(x_h) - 4|x_h - x_0|^2 I - 8(x_h - x_0) \otimes (x_h - x_0) = \nabla^2 w(x_h) \leq 0$$

taking $\epsilon_h := 12|x_h - x_0|^2$.   □

The following lemma shows that if a semiconvex function w in a ball B has a strict maximum point, then there exists a set of positive measure, whose size can be estimated from below, of maximizers of small linear perturbations of w.

Lemma 3 (Jensen). *Let $B \subset \mathbf{R}^n$ be an open ball of radius R, $w \in C(\overline{B}) \cap SB$ and assume that*

$$\max_{\overline{B}} w > \max_{\partial B} w \tag{64}$$

Then, setting

$$G_w^\delta := \{x \in B \mid \exists p \in \overline{B}_\delta \text{ s.t. } w(y) \leq w(x) + \langle p, y - x \rangle \ \forall y \in B\}$$

we have

$$\mathcal{L}^n(G_w^\delta) \geq \frac{\omega_n \delta^n}{sc^n(w, B)} \quad for \quad 0 < \delta < [\max_{\overline{B}} w - \max_{\partial B} w]/(2R). \tag{65}$$

Proof. Assume first $w \in C^\infty(\overline{B})$ and notice that (64) implies $sc(w, B) > 0$. We claim that for any δ fulfilling the condition in (65) we have $\nabla w(G_w^\delta) = \overline{B}_\delta$. Indeed, the inclusion $\subset$ is trivial; to show the opposite one we choose $p \in \overline{B}_\delta$ and observe that

$$\max_{x \in \overline{B}} w(x) - \langle p, x \rangle \geq \max_{x \in \overline{B}} w - R\delta, \qquad \max_{x \in \partial B} w(x) - \langle p, x \rangle \leq \max_{x \in \partial B} w + R\delta.$$

Hence, by our choice of δ, any maximizer of $w(x) - \langle p, x \rangle$ lies in B and belongs to G_w^δ. Using the change of variables formula we get

$$I := \int_{G_w^\delta} |\det \nabla^2 w| \, dx \geq \mathcal{L}^n(\{\nabla w(x) \,|\, x \in G_w^\delta\}) = \mathcal{L}^n(\overline{B}_\delta) = \omega_n \delta^n.$$

On the other hand, since $-sc(w, B)I \leq \nabla^2 w \leq 0$ at any point in G_w^δ, we can estimate the Jacobian determinant with $sc^n(w, B)$, so that

$$I \leq sc^n(w, B)\mathcal{L}^n(G_w^\delta).$$

Coupling these two estimates we infer (65).

In the general case the proof can be achieved using the approximating functions w_h provided by Exercise 11. Indeed, the uniform convergence of w_h to w implies that for h large enough (64) is fulfilled by w_h; moreover, any limit point of $G_{w_h}^\delta$ belongs to $G_w^\delta \cup \partial B$. Therefore, passing to the limit as $h \to \infty$ in

$$\mathcal{L}^n(G_{w_h}^\delta) \geq \frac{\omega_n \delta^n}{sc^n(w_h, B)} \quad \text{for} \quad 0 < \delta < [\max_{\overline{B}} w_h - \max_{\partial B} w_h]/(2R).$$

we recover (65). $\qquad\qquad\qquad\qquad\qquad\qquad\qquad\qquad\qquad\qquad\qquad\qquad\qquad$ $\square$

Exercise 11. Let $B \subset \mathbf{R}^n$ be an open ball and let $w \in C(\overline{B}) \cap \mathcal{S}B$. Then, we can find a sequence $(w_h) \subset C^\infty(\overline{B})$ uniformly converging to w in $\overline{B}$ such that

$$\lim_{h \to \infty} sc(w_h, B) = sc(w, B).$$

Hint: use convolutions and homotheties.

Remark 15. Ideas related to the proof of Jensen's lemma have been used by Ambrosio, Alberti, Cannarsa and Soner in [AAC92, ACS93] to study the dimension and the geometry of the singular sets

$$\Sigma^k(u) := \{x \in \Omega \,|\, \dim J^{1,-}u(x) = k\} \qquad k = 1, \ldots, n$$

of a function $u \in \mathcal{S}\Omega$.

Now we introduce Jensen's sup convolutions (see [Jen88, JLS88]), a natural way to approximate subsolutions by semiconvex functions, still keeping the subsolution property.

Definition 11 (sup convolutions). Let $u : A \subset \mathbf{R}^n \to \mathbf{R}$ be an upper semi-continuous function. The *sup convolution* u^ϵ of u is defined by

$$u^\epsilon(x) := \sup_{y \in A} u(y) - \frac{1}{\epsilon}|x - y|^2 \qquad x \in \mathbf{R}^n. \tag{66}$$

Proposition 4 (properties of sup convolutions). *Let A, u as in Definition 11 and let us assume that $u(x) \leq K(1 + |x|)$ for some constant K. Then*

(i) $u^\epsilon \in S\mathbf{R}^n$ and $\mathrm{sc}(u^\epsilon, \mathbf{R}^n) \leq 2/\epsilon$;
(ii) $u^\epsilon \geq u$ and $u^\epsilon \downarrow u$ in A as $\epsilon \downarrow 0$, with uniform convergence on compact subsets of A if u is continuous;
(iii) if A is locally compact and u is a viscosity subsolution of $E(\nabla u, \nabla^2 u) = 0$ in A, then u^ϵ is a viscosity subsolution of the same equation in the interior of

$$A^\epsilon := \big\{ x \in \mathbf{R}^n \mid \text{the "sup" in (66) is attained} \big\}. \tag{67}$$

Proof. (i) For any $x \in \mathbf{R}^n$ it holds

$$u^\epsilon(x) + \frac{1}{\epsilon}|x|^2 = \sup_{y \in A}\left(u(y) - \frac{1}{\epsilon}|x - y|^2 + \frac{1}{\epsilon}|x|^2\right) = \sup_{y \in A}\left(u(y) - \frac{1}{\epsilon}|y|^2 + \frac{2}{\epsilon}\langle x, y\rangle\right)$$

and this proves (i), because the supremum of any family of convex functions is convex.

(ii) Let $x \in A$ and let $y_\epsilon \in A$ such that

$$u(y_\epsilon) - \frac{1}{\epsilon}|y_\epsilon - x|^2 \geq u^\epsilon(x) - \epsilon \geq u(x) - \epsilon. \tag{68}$$

In particular, the linear growth of u implies

$$|y_\epsilon - x|^2 \leq \epsilon(\epsilon + u(y_\epsilon) - u(x)) \leq \epsilon\big[\epsilon + K + K|y_\epsilon| - u(x)\big]. \tag{69}$$

This inequality implies that y_ϵ tends to x as $\epsilon \downarrow 0$. From (68) we infer $u(y_\epsilon) \geq u^\epsilon(x)$, and the upper semicontinuity of u yields

$$u(x) \geq \limsup_{\epsilon \to 0^+} u(y_\epsilon) \geq \lim_{\epsilon \to 0^+} u^\epsilon(x).$$

This shows that $u^\epsilon(x)$ decreases to $u(x)$ for any $x \in A$. Then, since $(u^\epsilon) \subset C(A)$ decreases to u, if u is continuous the uniform convergence on compact subsets of A follows by Dini's lemma.

(iii) Let x_0 in the interior of A^ϵ and let $y_0 \in A$ a corresponding maximizer in (66). Assume that $u^\epsilon - \phi$ has a local maximum at x_0 with $u^\epsilon(x_0) = \phi(x_0)$ (see Remark 8(4)) and define

$$\psi(y) := \phi(y + x_0 - y_0).$$

We now claim that $u - \psi$ has a local maximum at y_0, where it attains the value $|x_0 - y_0|^2/\epsilon$. Indeed,

$$u(y_0) - \psi(y_0) = u^\epsilon(x_0) + \frac{1}{\epsilon}|x_0 - y_0|^2 - \phi(x_0) = \frac{1}{\epsilon}|x_0 - y_0|^2.$$

On the other hand, starting from $u^\epsilon(z) \leq \phi(z)$ for any $z \in B_r(x_0)$ we infer

$$u(y) - \frac{1}{\epsilon}|y - z|^2 \leq \phi(z) \qquad \forall y \in A, \ z \in B_r(x_0).$$

Putting $z = y + x_0 - y_0$ in the inequality above, we get

$$u(y) - \psi(y) \leq \frac{1}{\epsilon}|x_0 - y_0|^2 \qquad \forall y \in A \cap B_r(y_0)$$

as claimed. From

$$E_*\big(\nabla\psi(y_0), \nabla^2\psi(y_0)\big) \leq 0$$

we infer $E_*\big(\nabla\phi(x_0), \nabla^2\phi(x_0)\big) \leq 0$, as desired. $\qquad\square$

Remark 16. (a) Clearly $A^\epsilon = \mathbf{R}^n$ if A is closed; in general, we notice that (69) gives

$$|y_\epsilon - x|^2 \leq \epsilon\big(\epsilon + K + K|x| - u(x) + K|y_\epsilon - x|\big)$$

therefore it can be proved that

$$u^\epsilon(x) = \sup_{y \in A \cap \bar{B}_r(x)} u(y) - \frac{1}{\epsilon}|x - y|^2$$

for ϵ sufficiently small (depending on r, K, $|x|$ and $u^-(x)$). Since A is locally compact, this proves that any point of A belongs to A^ϵ for ϵ small enough; if, in addition, u is locally bounded from below, then A^ϵ contains for ϵ small enough any compact subset of A.

(b) Similar results hold for the *inf convolution* u_ϵ of u, defined by

$$u_\epsilon(x) := \inf_{y \in A} u(y) + \frac{1}{\epsilon}|x - y|^2 \qquad x \in \mathbf{R}^n. \tag{70}$$

Notice that $u_\epsilon = -(-u)^\epsilon$ for any $\epsilon > 0$.

(c) for equations depending on u, statement (iii) is true provided

$$s \mapsto E_*(s, p, X)$$

is nondecreasing.

Exercise 12. Let $f : \mathbf{R}^n \to \mathbf{R} \cup \{\infty\}$ be a lower semicontinuous and convex function. Show that the inf convolutions f_ϵ are of class $C^{1,1}$ in $\mathbf{R}^n$. Hint: use the monotonicity of $\partial^- f$.

10   Parabolic equations

In this section we will consider viscosity solutions of equations of parabolic type

$$u_t + F(\nabla u, \nabla^2 u) = 0 \qquad (71)$$

in locally compact domains $A \subset (0, \infty) \times \mathbf{R}^n$. We denote by ∇u and $\nabla^2 u$ the space first and second order gradients respectively of $u(t, x)$, and by u_t its time first order derivative. Equation (71) is a particular second order equation (37) in $(n + 1)$ variables (t, x), therefore $u : A \to \mathbf{R}$ is a viscosity subsolution of (71) if

$$\phi_t + F_*(\nabla \phi, \nabla^2 \phi) \leq 0 \qquad \text{at } (t_0, x_0)$$

whenever $\phi \in C^\infty$ in a neighbourhood U of $(t_0, x_0) \in A$ and $u^* - \phi$ has a relative maximum at (x_0, t_0). Analogously, $u : A \to \mathbf{R}$ is a viscosity supersolution of (71) if

$$\phi_t + F^*(\nabla \phi, \nabla^2 \phi) \geq 0 \qquad \text{at } (t_0, x_0)$$

whenever $\phi \in C^\infty$ in a neighbourhood U of $(t_0, x_0) \in A$ and $u_* - \phi$ has a relative minimum at (t_0, x_0).

As we already observed, these definitions could also be given using C^2 test functions ϕ; in addition, since the second order time derivative does not appear in the equation, one can use the *parabolic super and sub jets*

$$\mathcal{P}_A^+ u(t_0, x_0) := \{(\tau, p, X) \,|\, u(t, x) \leq u^*(t_0, x_0) + \tau(t - t_0) + \langle p, x - x_0 \rangle$$
$$+ \frac{1}{2} \langle X(x - x_0), (x - x_0) \rangle + o(|t - t_0| + |x - x_0|^2) \}, \qquad (72)$$

$$\mathcal{P}_A^- u(t_0, x_0) := \{(\tau, p, X) \,|\, u(t, x) \geq u_*(t_0, x_0) + \tau(t - t_0) + \langle p, x - x_0 \rangle$$
$$+ \frac{1}{2} \langle X(x - x_0), (x - x_0) \rangle + o(|t - t_0| + |x - x_0|^2) \} \qquad (73)$$

to say that u is a viscosity subsolution of (71) if and only if

$$\tau + F_*(p, X) \leq 0 \qquad \forall (t_0, x_0) \in A, \ (\tau, p, X) \in \mathcal{P}_A^+ u(t_0, x_0) \qquad (74)$$

and u is a viscosity supersolution of (71) if and only if

$$\tau + F^*(p, X) \geq 0 \qquad \forall (t_0, x_0) \in A, \ (\tau, p, X) \in \mathcal{P}_A^- u(t_0, x_0). \qquad (75)$$

However, we will not use (74) and (75) in the following. We will make the following assumptions on F:
(a) $F(p, X)$ is degenerate elliptic, according to Definition 8;
(b) $F(p, X)$ is defined in $(\mathbf{R}^n \setminus \{0\}) \times \mathrm{Sym}^n$, continuous and bounded on bounded sets;
(c) $F_*(0, O) = F^*(0, O)$, i.e., F is continuous at $(0, O)$.
Conditions (a), (b), (c) ensure the validity of the fundamental *parabolic comparison theorem*: if, at time 0, a subsolution is less than a supersolution, then the same property is valid for later times.

Theorem 18 (comparison). *Assume that F fulfils (a), (b), (c) above and let $T \in (0, \infty)$. Let u be an upper semicontinuous subsolution of (71) in $(0, T] \times \mathbf{R}^n$, let v be a lower semicontinuous supersolution of (71) in $(0, T] \times \mathbf{R}^n$ and assume that*

(i) $|u(t, x)| + |v(t, x)| \le K(1 + |x|)$ for any $(t, x) \in (0, T] \times \mathbf{R}^n$ for some constant K;

(ii) $u^*(0, x) \le v_*(0, x)$ for any $x \in \mathbf{R}^n$;

(iii) either $u^*(0, \cdot)$ or $v_*(0, \cdot)$ are uniformly continuous in $\mathbf{R}^n$.

Then, $u(t, x) \le v(t, x)$ for any $(t, x) \in (0, T] \times \mathbf{R}^n$.

Proof. We will prove the theorem under an extra assumption, namely, the existence of $R > 0$ such that

$$u(t, x) \equiv u_0, \quad v(t, x) \equiv v_0 \qquad \text{whenever } |x| + t \ge R. \qquad (76)$$

Hints about the proof of the theorem without this assumption are given in Remark 17, and a complete proof is available in [GGIS91]. We also notice that condition (76) is not greatly restrictive in many geometric applications (in which the equation is invariant under relabelling of level sets, cf. Theorem 21) as long as one considers only compact initial sets and, correspondingly, initial functions which are constant at infinity (see [ES91, CGG91]).

For simplicity, we will also assume that the time interval is $(0, \infty)$. We extend u and v to $[0, \infty) \times \mathbf{R}^n$ by upper semicontinuous and lower semicontinuous relaxation, respectively.

(i) Assume by contradiction that the statement is false. Then, there exists $(t_0, x_0) \in (0, \infty) \times \mathbf{R}^n$ such that $u(t_0, x_0) > v(t_0, x_0)$. In particular, we can find strictly positive numbers γ_0, m_0 such that, setting

$$w^\sigma(t, x, s, y) := u(t, x) - v(s, y) - \frac{1}{\sigma}\left[|x - y|^4 + |t - s|^2\right] - \gamma_0 s$$

it holds

$$\max_{t, x, s, y} w^\sigma(t, x, s, y) \ge m_0 \qquad \forall \sigma > 0. \qquad (77)$$

Indeed, we need only to evaluate the function at (t_0, x_0, t_0, x_0) to find

$$\max_{t, x, s, y} w(t, x, s, y) \ge u(t_0, x_0) - v(t_0, x_0) - \gamma_0 t_0 = m_0 > 0$$

for γ_0 small enough.

(ii) Let u^ϵ, v_ϵ be respectively the sup and inf convolutions of u and v and let

$$w^{\epsilon, \sigma}(t, x, s, y) := u^\epsilon(t, x) - v_\epsilon(s, y) - \frac{1}{\sigma}\left[|x - y|^4 + |t - s|^2\right] - \gamma_0 s.$$

Since $w^{\epsilon, \sigma} \ge w$, it holds

$$\sup_{t, x, s, y} w^{\epsilon, \sigma}(t, x, s, y) \ge m_0 \qquad \forall \sigma > 0. \qquad (78)$$

Since u and v are globally bounded, using (76) it is easy to see that there exists $\epsilon_0 > 0$ such that

$$u^\epsilon(t, x) \equiv u_0, \quad v_\epsilon(t, x) \equiv v_0 \qquad \text{whenever } |x| + t \geq 2R \qquad (79)$$

for $\epsilon \in (0, \epsilon_0)$. In particular, for any $\epsilon \in (0, \epsilon_0)$ and any $\sigma > 0$ the function $w^{\epsilon, \sigma}$ attains a maximum greater than m_0.

(iii) Let $U^\epsilon \subset (0, \infty) \times \mathbf{R}^n$ be the interior of the set

$$\left\{ (t, x) \in \mathbf{R} \times \mathbf{R}^n \mid \exists (s, y) \in (0, \infty) \times \mathbf{R}^n \text{ s.t. } u^\epsilon(t, x) = u(s, y) - \frac{1}{\epsilon}\left(|t - s|^2 + |x - y|^2\right) \right\}.$$

According to Proposition 4(iii), u^ϵ is a viscosity subsolution of (71) in U^ϵ. Let V_ϵ be the analogous set for v_ϵ. We now claim that for ϵ and σ small enough the maximum in step (ii) is attained at a point $P^{\epsilon, \sigma} := (\bar{t}, \bar{x}, \bar{s}, \bar{y})$ such that $(\bar{t}, \bar{x})$ is in U^ϵ and $(\bar{s}, \bar{y})$ is in V_ϵ.

In fact, if this were not true we could find a sequence of maximizers $P^{\epsilon_h, \sigma_h} = (t_h, x_h, s_h, y_h)$ with (σ_h) and (ϵ_h) infinitesimal and either $(t_h, x_h) \notin U^{\epsilon_h}$ or $(s_h, y_h) \notin V^{\epsilon_h}$. By (79), u^{ϵ_h} and $-v_{\epsilon_h}$ are uniformly bounded from above in $[0, \infty) \times \mathbf{R}^n$, so that from

$$\frac{1}{\sigma_h}\left[|x_h - y_h|^4 + |t_h - s_h|^2\right] \leq u^{\epsilon_h}(t_h, x_h) - v_{\epsilon_h}(s_h, y_h)$$

we infer $x_h - y_h \to 0$ and $t_h - s_h \to 0$. In addition, for h large enough it holds $|x_h| + t_h \leq 3R$; if not, for infinitely many h it holds

$$|x_h| + t_h > 2R, \qquad |y_h| + s_h > 2R$$

and from (79) the maximum would be nonpositive for any such h.

Since, by Remark 16(a), U^{ϵ_h} and V_{ϵ_h} definitely contain any compact subset of $(0, \infty) \times \mathbf{R}^n$, $t_h \wedge s_h$ must converge to 0, hence t_h, s_h are infinitesimal and x_h and y_h converge (up to subsequences) to the same vector x. Since

$$u^{\epsilon_h}(t_h, x_h) - v_{\epsilon_h}(s_h, y_h) \geq m_0$$

by Remark 16(a) we can find points (t_h', x_h', s_h', y_h') such that

$$u(t_h', x_h') - v(s_h', y_h') \geq m_0$$

and

$$\lim_{h \to \infty} \left\{ |t_h - t_h'| + |x_h - x_h'| + |s_h - s_h'| + |y_h - y_h'| \right\} = 0.$$

Passing to the limit as $h \to \infty$ we find $u(0, x) - v(0, x) \geq m_0 > 0$, which contradicts our assumption on u and v at time 0.

(iv) Choosing $\epsilon \in (0, \epsilon_0)$ and σ sufficiently small as in (iii), and denoting by $P := (\bar{t}, \bar{x}, \bar{s}, \bar{y})$ the corresponding maximizer of $w := w^{\epsilon, \sigma}$ we verify in this step that $\bar{x} \neq \bar{y}$. Indeed, if this were not true, from

$$w(\bar{t}, \bar{x}, \bar{s}, \bar{x}) \geq w(t, x, \bar{s}, \bar{x}) \qquad \forall x \in \mathbf{R}^n, t \geq 0$$

we would get

$$u^\epsilon(t,x) \le u^\epsilon(\bar{t}, \bar{x}) + \frac{1}{\sigma}\left[|x - \bar{x}|^4 - |\bar{t} - \bar{s}|^2 + |t - \bar{s}|^2 \right]$$

hence

$$\frac{2}{\sigma}(\bar{t} - \bar{s}) + F_*(0, O) \le 0.$$

On the other hand, from

$$w(\bar{t}, \bar{x}, \bar{s}, \bar{x}) \ge w(\bar{t}, \bar{x}, s, y) \qquad \forall y \in \mathbf{R}^n, \, s \ge 0$$

we would get

$$v_\epsilon(s,y) \ge v_\epsilon(\bar{s}, \bar{x}) - \frac{1}{\sigma}\left[|y - \bar{x}|^4 - |\bar{t} - \bar{s}|^2 + |\bar{t} - s|^2 \right] - \gamma_0(s - \bar{s})$$

hence

$$-\gamma_0 + \frac{2}{\sigma}(\bar{t} - \bar{s}) + F^*(0, O) \ge 0.$$

Since, by assumption (c), $F_*(0, O) = F^*(0, O)$, we get $\gamma_0 \le 0$, a contradiction.
(v) With the same notations of (iv), we will apply the maximum principle for semiconvex functions Theorem 17 to w to find a sequence of points

$$P_h := (t_h, x_h, s_h, y_h) \to (\bar{t}, \bar{x}, \bar{s}, \bar{y})$$

such that w has a second order Taylor expansion at P_h, $\nabla w(P_h)$ tends to 0 and $\nabla^2 w(P_h) \le \delta_h I$ for a suitable infinitesimal sequence δ_h. Indeed, w is semiconvex in its variables because

$$w(t,x,s,y) + \frac{1}{\epsilon}\left[t^2 + |x|^2 + s^2 + |y|^2 \right] + \frac{1}{\sigma}\left[|x - y|^4 + |t - s|^2 \right] + \gamma_0 s$$

is equal to the sum of $u^\epsilon(t, x) + (t^2 + |x|^2)/\epsilon$ and $-v_\epsilon(s, y) + (s^2 + |y|^2)/\epsilon$, which are both convex by Proposition 4(i). Since

$$\begin{cases} \nabla_x w(P_h) = \nabla u^\epsilon(t_h, x_h) - \dfrac{4}{\sigma}|x_h - y_h|^2(x_h - y_h) \\ \nabla_y w(P_h) = -\nabla v_\epsilon(s_h, y_h) - \dfrac{4}{\sigma}|x_h - y_h|^2(y_h - x_h) \end{cases}$$

the convergence to 0 of $\nabla_{x,y} w(P_h)$ yields

$$\lim_{h \to \infty} \nabla u^\epsilon(t_h, x_h) = \lim_{h \to \infty} \nabla v_\epsilon(s_h, y_h) = \bar{p} \ne 0. \tag{80}$$

with $\bar{p} := 4|\bar{x} - \bar{y}|^2(\bar{x} - \bar{y})/\sigma$. Notice that $\bar{p} \ne 0$, by step (iv). Analogously, from

$$w_t(P_h) = u_t^\epsilon(t_h, x_h) - \frac{2}{\sigma}(t_h - s_h), \qquad w_s(P_h) = -v_{\epsilon,s}(s_h, y_h) - \frac{2}{\sigma}(s_h - t_h) - \gamma_0$$

we infer

$$\lim_{h\to\infty} u_t^\epsilon(t_h, x_h) = \frac{2}{\sigma}(\bar{t} - \bar{s}), \qquad \lim_{h\to\infty} v_{\epsilon,s}(s_h, y_h) = \frac{2}{\sigma}(\bar{t} - \bar{s}) - \gamma_0. \qquad (81)$$

Now we examine the spatial second order derivatives. Let ξ be any unit vector and let $\eta = (\xi, \xi)$; since

$$\langle \nabla^2 u^\epsilon(t_h, x_h)\xi, \xi \rangle - \langle \nabla^2 v_\epsilon(s_h, y_h)\xi, \xi \rangle = \langle \nabla^2 w(P_h)\eta, \eta \rangle \le 2\delta_h$$

and since u^ϵ and v_ϵ are semiconvex, by a compactness argument we can assume that

$$\lim_{h\to\infty} \nabla^2 u^\epsilon(t_h, x_h) = X, \qquad \lim_{h\to\infty} \nabla^2 v_\epsilon(s_h, y_h) = Y \qquad (82)$$

with $Y \ge X$. By step (iii), for h large enough $(t_h, x_h) \in U^\epsilon$, hence

$$u_t^\epsilon(t_h, x_h) + F\big(\nabla u^\epsilon(t_h, x_h), \nabla^2 u^\epsilon(t_h, x_h)\big) \le 0.$$

Analogously, for h large enough $(s_h, y_h) \in V_\epsilon$, hence

$$v_{\epsilon,s}(s_h, y_h) + F\big(\nabla v_\epsilon(s_h, y_h), \nabla^2 v_\epsilon(s_h, y_h)\big) \ge 0.$$

Passing to the limit as $h \to \infty$ and using (80), (81) and (82) and the continuity of F we get

$$\frac{2}{\sigma}(\bar{t} - \bar{s}) + F(\bar{p}, X) \le 0, \qquad -\gamma_0 + \frac{2}{\sigma}(\bar{t} - \bar{s}) + F(\bar{p}, Y) \ge 0.$$

Subtracting the two inequalities gives

$$F(\bar{p}, Y) - F(\bar{p}, X) \ge \gamma_0 > 0.$$

Since $Y \ge X$, this contradicts the degenerate ellipticity of the equation. $\qquad\square$

Remark 17. For bounded time intervals $(0, T]$ one can replace $-\gamma_0 s$ by $\gamma_0/(T-s)$ in w. Without assumption (76), an additional barrier term $-\delta|x|^2$ must be added. The argument in this case is similar, but uniform estimates with respect to δ on the matrices X, Y of step (v) are needed. These can be obtained by the arguments in [CI90] (see also [CIL92]).

Now we can state a "parabolic version" of Perron's existence theorem (see Theorem 11). Given $T \in (0, \infty]$ and a uniformly continuous function u_0 defined in $\mathbf{R}^n$, we look for continuous function $u : [0, T) \times \mathbf{R}^n \to \mathbf{R}$ satisfying the Cauchy problem

$$u_t + F\big(\nabla u, \nabla^2 u\big) = 0 \quad \text{in } (0, T) \times \mathbf{R}^n, \qquad u(0, \cdot) = u_0 \qquad (83)$$

in the viscosity sense.

Theorem 19 (parabolic existence). *Let $f, g : (0, T) \times \mathbf{R}^n$ be respectively a viscosity subsolution and supersolution of (71) in $(0, T) \times \mathbf{R}^n$, such that $g^*(0, \cdot) = u_0 = f_*(0, \cdot)$ and*

$$-K(1 + |x|) \leq f(t, x) \leq g(t, x) \leq K(1 + |x|) \qquad \forall (t, x) \in (0, T) \times \mathbf{R}^n$$

for some constant K. Then, problem (83) has a continuous viscosity solution u satisfying

$$f \leq u \leq g.$$

Proof. By Theorem 11, there exists a viscosity solution w of (71) in $(0, T) \times \mathbf{R}^n$ lying between f and g. Since $w^* \leq g^* = u_0 = f_* \leq w_*$ at time 0, by applying Theorem 18 with $u := w^*$, $v := w_*$ we obtain that $u \leq v$ for any $t \geq 0$, i.e., w is continuous. $\qquad\square$

Remark 18. By applying Theorem 18 to any pair of solutions, we obtain that the solution of (83) is also unique in the class of functions with linear growth.

Hence, existence of solutions of (83) is reduced to a construction of suitable subsolutions and supersolutions of the problem which satisfy the initial condition. We will consider, as a model example, a function G_k suggested by De Giorgi in [Gio94b] and related (see Theorem 24) to k-dimensional flow by mean curvature. Let $p \in \mathbf{R}^n \setminus \{0\}$ and let $P_p = I - p \otimes p/|p|^2$ be the orthogonal projection on the hyperplane $p^\perp$ orthogonal to p. For any $X \in \mathrm{Sym}^n$, we can consider the symmetric matrix $Y := P_p X P_p$; notice that p is a eigenvector of Y, corresponding to the eigenvalue 0 (heuristically, we are "projecting" X on $p^\perp$). Then, we define

$$G_k(p, X) := -\sum_{i=1}^k \lambda_i(Y) \tag{84}$$

where $\lambda_1(Y), \ldots, \lambda_{n-1}(Y)$ are the eigenvalues of Y on $p^\perp$, in increasing order (in other words, we remove the top $(n - 1 - k)$ eigenvalues). In the case $k = (n-1)$, which corresponds to the mean curvature flow for hypersurfaces, $G_k(p, X)$ reduces to

$$-\mathrm{trace}\, Y = -\mathrm{trace}\, (P_p X P_p) = -\mathrm{trace}\, (P_p X) = -\mathrm{trace}\, X + \frac{\langle Xp, p\rangle}{|p|^2}. \tag{85}$$

By (15) and Remark 4, the equation corresponding to this choice of G_k flows each codimension 1 level set with velocity equal to the sum of the smallest k principal curvatures. If $u \geq 0$ we will see that, under regularity assumptions, this forces the 0 level set to flow by mean curvature. A simple and useful example of classical solution to the parabolic equation associated to G_k is

$$u(t, x) := |x - x_0|^2 + 2kt \qquad (t, x) \in [0, \infty) \times \mathbf{R}^n \tag{86}$$

corresponding to functions $u(t, \cdot)$ whose level sets are spheres.

Let us check now that G_k fulfils (a), (b), (c). In fact, only the verification of (a) requires some work, because G_k is clearly continuous, bounded on bounded sets and $G_{k*}(0, O) = G_k^*(0, O) = 0$. The degenerate ellipticity essentially depends on the following fact: let Y, Y' be symmetric bilinear forms in $\mathbf{R}^q$ and let $\lambda_i(Y)$, $\lambda_i(Y')$ be the ordered lists of their eigenvalues. Then

$$Y \leq Y' \quad \Longrightarrow \quad \lambda_i(Y) \leq \lambda_i(Y') \quad i = 1, \ldots, q.$$

This implication, trivial if Y and Y' commute, follows in general by the following representation formula for λ_i, due to Courant.

Lemma 4 (Courant formula). *Given any subspace $L \subset \mathbf{R}^q$, define*

$$\lambda_L(Y) := \min_{\nu \in L \setminus \{0\}} \frac{\langle Y\nu, \nu \rangle}{|\nu|^2}.$$

Then, for any $i = 1, \ldots, q$ it holds

$$\lambda_i(Y) = \max\{\lambda_L(Y) \mid L \subset \mathbf{R}^q, \ \dim(L) \geq q - i + 1\}.$$

Proof. Inequality $\leq$ simply follows choosing as L the subspace spanned by the eigenvectors corresponding to $\lambda_i(Y), \ldots, \lambda_q(Y)$, because $Y \geq \lambda_i(Y)I$ on L. To prove the opposite inequality, let L be any subspace with dimension at least $q - i + 1$ and let M be the i-dimensional vector space spanned by the eigenvectors corresponding to $\lambda_1(Y), \ldots, \lambda_i(Y)$. Then, choosing a unit vector $v \in L \cap M$ we find

$$\lambda_L(Y) \leq \langle Yv, v \rangle \leq \lambda_i(Y)$$

because $Y \leq \lambda_i(Y)I$ on M. $\qquad\square$

Now we will construct subsolutions and supersolutions for

$$u_t + G_k(\nabla u, \nabla^2 u) = 0 \quad \text{in } (0, \infty) \times \mathbf{R}^n, \qquad u(0, \cdot) = u_0. \qquad (87)$$

Let $m : [0, \infty) \to [0, \infty)$ be a modulus of continuity for u_0, i.e., a continuous function such that $m(0) = 0$ and

$$|u_0(x) - u_0(y)| \leq m(|x - y|) \qquad \forall x, y \in \mathbf{R}^n. \qquad (88)$$

Possibly replacing m by

$$\tilde{m}(t) := \sup\{|u_0(x) - u_0(y)| \mid |x - y| \leq t\} \qquad t \in [0, \infty)$$

we see that we can assume m to be uniformly continuous and subadditive:

$$m(s + t) \leq m(s) + m(t) \qquad s, t \in [0, \infty).$$

By Theorem 21 below and (86), the function

$$(t, x) \mapsto m\left(\sqrt{|x - x_0|^2 + 2kt}\right)$$

is a viscosity solution of (87) in $(0, \infty) \times \mathbf{R}^n$ for any $x_0 \in \mathbf{R}^n$. Set

$$\begin{cases} b(t, x; x_0) := u_0(x_0) - m\big(\sqrt{|x - x_0|^2 + 2kt}\big) \\ b'(t, x; x_0) := u_0(x_0) + m\big(\sqrt{|x - x_0|^2 + 2kt}\big) \end{cases}$$

$$f(t, x) := \sup_{x_0 \in \mathbf{R}^n} b(t, x; x_0), \qquad g(t, x) := \inf_{x_0 \in \mathbf{R}^n} b'(t, x; x_0).$$

Then, using the subadditivity of m, it can be easily verified that the functions b and b' are equicontinuous in (x, t) with respect to the parameter x_0, hence f and g are continuous. By (88) we infer $f(0, x) = g(0, x) = u_0(x)$ for any $x \in \mathbf{R}^n$ and Proposition 2 implies that f is a viscosity subsolution and g is a viscosity supersolution. Finally, the inequality $f \leq g$ follows by $b(t, x; x_0) \leq b'(t, x; y_0)$ which in turn follows by

$$m(|x_0 - y_0|) \leq m\big(\sqrt{|x - x_0|^2 + 2kt}\big) + m\big(\sqrt{|x - y_0|^2 + 2kt}\big)$$

and by the subadditivity of m. Hence, Theorem 19 can be applied with f and g for any $T \in (0, \infty)$. Summing up, we proved the following result:

Theorem 20. *Let G_k be defined in (84) and let u_0 be a uniformly continuous function in $\mathbf{R}^n$. Then, the Cauchy problem (87) has a unique continuous solution in $[0, \infty) \times \mathbf{R}^n$ satisfying*

$$|u(t, x)| \leq K(1 + |x| + \sqrt{t}) \qquad \forall (t, x) \in [0, \infty) \times \mathbf{R}^n$$

for some constant K.

Remark 19. In the codimension 1 case the equation in (87) can be (formally) put in the divergence form $u_t = |\nabla u| \operatorname{div}(\nabla u / |\nabla u|)$. A "viscosity" approximation of the solution of (87) is given by the solutions u^ϵ of

$$u_t = \sqrt{\epsilon^2 + |\nabla u|^2} \operatorname{div}\left(\frac{\nabla u}{\sqrt{\epsilon^2 + |\nabla u|^2}}\right) = \Delta u - \frac{\langle \nabla^2 u \nabla u, \nabla u \rangle}{\epsilon^2 + |\nabla u|^2}$$

if the initial function u_0 is $C^{1,1}$ and constant at infinity (see [ES92a] for details and for a geometric interpretation of this approximation).

Remark 20. Using Theorem 18 and the translation invariance in (x, u) of (87) it is easy to check that any modulus of continuity of u_0 is a modulus of continuity of $u(t, \cdot)$ for any $t \geq 0$. More precisely, comparing

$$u(t_0, x_0) \pm m\big(\sqrt{|x - x_0|^2 + 2k(t - t_0)}\big) \qquad \text{with} \qquad u(t, x)$$

one obtains $|u(t, x) - u(t_0, x_0)| \leq m\big(\sqrt{|x - x_0|^2 + 2k(t - t_0)}\big)$ for $t \geq t_0 \geq 0$. In particular, if u_0 is a Lipschitz function then $u(t, \cdot)$ is still Lipschitz and $u(\cdot, x)$ is 1/2-Hölder continuous.

Having in mind the geometric interpretation of (87) for *classical* solutions with nonvanishing gradient, we can now define a generalized level set evolution by mean curvature based on the interpretation of the same equation in the *viscosity* sense.

Definition 12 (level set flow). Let $\Gamma_0 \subset \mathbf{R}^n$ be a closed set and let u_0 be any uniformly continuous function such that

$$\Gamma_0 = \{x \in \mathbf{R}^n \mid u_0(x) = 0\}$$

(for instance, $u_0(x) = \mathrm{dist}(x, \Gamma_0)$ if $\Gamma_0 \neq \emptyset$). Let $u(t, x)$ be the solution of the Cauchy problem (87) given by Theorem 20. We will call the sets

$$\Gamma_t := \{x \in \mathbf{R}^n \mid u(t, x) = 0\}$$

k dimensional *level set flow* starting from Γ_0.

Of course, we have to verify that the definition above is well posed, i.e., Γ_t depends on Γ_0 but it does not depend on u_0. To see this, we will check that our equation falls in the class of so-called *geometric equations*, invariant under relabelling of level sets (see Theorem 21 below).

Definition 13 (geometric evolution equations). We say that the equation (71) is *geometric* if

$$F(\lambda p, \lambda X) = \lambda F(p, X) \quad \forall \lambda > 0, \qquad\qquad F(p, X + \sigma p \otimes p) = F(p, X) \quad \forall \sigma \in \mathbf{R}$$
$$(89)$$

for any pair (p, X) with $p \neq 0$. In particular, F_* and F^* satisfy (89) for any pair (p, X).

Noticing that

$$Y := P_p X P_p = P_p(X + \sigma p \otimes p) P_p \qquad\qquad \forall \sigma \in \mathbf{R}$$

we obtain that any $F(p, X)$ which actually depends only on Y, as the function G_k in (84), fulfils the second condition in (89). It is evident that G_k satisfies the first one too. Another example of geometric evolution equation is

$$u_t - t^\alpha |\nabla u| \, \mathrm{trace} \left(\frac{P_{\nabla u} \nabla^2 u P_{\nabla u}}{|\nabla u|} \right)^\beta = 0.$$

In the two dimensional case, this equation has been studied by Alvarez, Lions and Morel (see [ALCM93, ALM92]) as a model for non isotropic smoothing of images. In this case the parameter t plays the rôle of a scale parameter (roughly speaking, $\sqrt{t}$ is the diameter of the support of the convolution kernel); if $\alpha = \beta = 1/3$, this equation flows ellipses in ellipses, leaving their eccentricity unchanged.

Now we can state the main property of geometric evolution equations, whose meaning is that only the geometry of level sets plays a rôle in the equations, not their labels.

Theorem 21 (invariance property). *Assume that F is geometric, $A \subset (0, \infty) \times \mathbf{R}^n$ is locally compact and $\theta : \mathbf{R} \to \mathbf{R}$ is a continuous nondecreasing function. Then, for any upper semicontinuous subsolution (lower semicontinuous supersolution) u of (71) in A the function $\theta(u)$ is still a subsolution (supersolution) of the same equation in A.*

Proof. Assume first that u is bounded and θ is smooth with $\theta' > 0$ on $\mathbf{R}$. Let ψ be the inverse function of θ; if $\theta(u) - \phi$ has a relative maximum at $(t_0, x_0) \in A$, it follows that $u - \psi(\phi)$ has a relative maximum at (t_0, x_0), hence

$$\psi'(\phi)\phi_t + F_*\big(\psi'(\phi)\nabla\phi, \psi'(\phi)\nabla^2\phi + \psi''(\phi)\nabla\phi \otimes \nabla\phi\big) \leq 0$$

at (t_0, x_0). Using (89) with $\lambda = \psi'(\phi(t_0, x_0))$ and $\sigma = \psi''(\phi(t_0, x_0))$ we infer

$$\phi_t + F_*\big(\nabla\phi, \nabla^2\phi\big) \leq 0 \qquad \text{at } (t_0, x_0).$$

If u is bounded, we can use Theorem 14 and the fact that any continuous nondecreasing function θ can be approximated locally uniformly by smooth functions with strictly positive derivative to show that $\theta(u)$ is a subsolution. Finally, if u is not bounded we notice that the functions

$$v_h(t, x) := \theta\big(-h \vee u(t, x)\big)$$

Γ^+ converge to $\theta(u)$ as $h \to \infty$ and we apply Theorem 14 again. $\square$

Remark 21. If we strengthen the first condition in (89), requiring equality for any real λ, then Theorem 21 holds for any continuous function θ, without monotonicity assumptions. The proof is based on a similar argument: any continuous function can be locally uniformly approximated by functions which are locally either nonincreasing or nondecreasing (see [ES91] for details). The function G_k in (84) fulfils this stronger condition only in the case $k = (n - 1)$.

Coming back to the well posedness of Definition 12, we will first consider the extreme cases $\Gamma_0 = \mathbf{R}^n$, $\Gamma_0 = \emptyset$. If $\Gamma_0 = \mathbf{R}^n$ then $u_0 \equiv 0$ and consequently $u(t, \cdot) \equiv 0$ for any $t \geq 0$, i.e., $\Gamma_t = \mathbf{R}^n$ for any $t \geq 0$. If $\Gamma_0 = \emptyset$ we can find a smooth uniformly continuous function $\omega : [0, \infty) \to (0, \infty)$ such that either

$$u_0(x) \geq \omega(|x|) \quad \forall x \in \mathbf{R}^n \quad \text{or} \quad u_0(x) \leq -\omega(|x|) \quad \forall x \in \mathbf{R}^n.$$

Assuming the first inequality to be true, since $\omega(\sqrt{|x|^2 + 2kt})$ is a smooth solution of (87), the comparison theorem yields

$$u(t, x) \geq \omega(\sqrt{|x|^2 + 2kt}) \quad \forall t \geq 0, \, x \in \mathbf{R}^n.$$

In particular, $u(t, \cdot)$ is nowhere equal to 0 and Γ_t is empty for any $t \geq 0$. If u_0 is negative the argument is similar.

In the general case we will prove the following result:

Theorem 22 (independence on u_0). *Let $\Gamma_0 \subset \mathbf{R}^n$, $\Gamma_0 \neq \mathbf{R}^n$, $\Gamma_0 \neq \emptyset$. Let u_0, $u(t,x)$, Γ_t be as in Definition 12, let $u'(x,t)$ be the solution of (87) with the initial condition*

$$u_0'(x) := \bar{d}(x, \Gamma_0)$$

and let Γ_t' be the zero level set of $u'(t, \cdot)$. Then, $\Gamma_t = \Gamma_t'$ for any $t \geq 0$.

Proof. We will first prove that the sets $\{u(t, \cdot) \leq 0\}$ coincide with $\{u'(t, \cdot) \leq 0\}$. To this end, define

$$m(t) := \sup\left\{ u_0(x) \,|\, \mathrm{dist}(x, \{u_0 \leq 0\}) \leq t \right\}.$$

Since u_0 is uniformly continuous, m is non decreasing, uniformly continuous and $m(0) = 0$. Moreover,

$$u_0(x) \leq m\big(\mathrm{dist}(x, \{u_0 \leq 0\})\big), \qquad \forall x \in \mathbf{R}^n.$$

By Theorem 21, $u(t, x)$ and $m\big(u'(t, x) \vee 0\big)$ are solutions of (87) and by Theorem 18, we conclude that

$$u(t, x) \leq m\big(u'(t, x) \vee 0\big) \qquad \forall (t, x) \in [0, \infty) \times \mathbf{R}^n.$$

Hence, $\{u'(t, \cdot) \leq 0\} \subset \{u(t, \cdot) \leq 0\}$ for any $t \geq 0$.

Now we prove the opposite inclusion. Let $\chi(t, x)$ be the characteristic function of the zero sublevel set of u and let $w = 1 - \chi$, i.e., $w(t, x) = 0$ if $u(t, x) \leq 0$ and $w(t, x) = 1$ otherwise. Then, the functions $h_\epsilon := 1 \wedge u^+/\epsilon \; \Gamma^-$ converge as $\epsilon \downarrow 0$ to w, according to Proposition 3(iii).

By Theorem 21 and Theorem 14, w is a viscosity supersolution of (87). Since $\{u_0' \leq 0\} = \{\bar{d}(0, \cdot) \leq 0\}$, we have

$$u_0'(x) \wedge 1 \leq w(0, x).$$

Hence by Theorem 18 we conclude that $u'(t, x) \wedge 1 \leq w(t, x)$. Consequently, $\{u(t, \cdot) \leq 0\}$ is included in $\{u'(t, \cdot) \leq 0\}$.

The same argument also gives $\{u(t, \cdot) \geq 0\} = \{u'(t, \cdot) \geq 0\}$; indeed, in the proof above we only used the fact that the parabolic equation fulfilled by u and u' is geometric and degenerate elliptic, so the same argument can be applied to $-u$, $-u'$, which satisfy

$$v_t + \tilde{G}_k\big(\nabla v, \nabla^2 v\big) = 0$$

where

$$\tilde{G}_k(p, X) := -G_k(-p, -X) = -\sum_{i=1}^{k} \lambda_i\big(P_p X P_p\big),$$

and the sum runs among the largest k eigenvalues.   □

Remark 22. The proof above also shows that the "interior" and "exterior" sets

$$I_t := \{x \in \mathbf{R}^n \,|\, u(t,x) < 0\}, \qquad E_t := \{x \in \mathbf{R}^n \,|\, u(t,x) > 0\}$$

do not depend on the choice of u_0 but only on $\{u_0 < 0\}$ and $\{u_0 > 0\}$.

We can now state some elementary properties of level set flows.

Theorem 23 (properties of level set flow).

(i) the level set flow has the semigroup property, i.e., Γ_{t+s} is the evolution at time t starting from Γ_s whenever $t, s \geq 0$;

(ii) the level set flow has the inclusion property, i.e.

$$\Gamma_0 \subset \Gamma_0' \qquad \Longrightarrow \qquad \Gamma_t \subset \Gamma_t' \;\; \forall t \geq 0;$$

(iii) if Γ_0 is compact and not empty, then there exists a extinction time $T_* \in [0, \infty)$ such that $\Gamma_t \neq \emptyset$ for $t \leq T_*$ and $\Gamma_t = \emptyset$ for $t > T_*$.

Proof. (i) is a straightforward consequence of Theorem 22: one needs only to choose as initial function $u(s, \cdot)$.

(ii) If $\Gamma_0 \subset \Gamma_0'$ we can choose as initial functions the distance from Γ_0 and the distance from Γ_0' to get by Theorem 18 $u(t,x) \geq u'(t,x) \geq 0$. Consequently,

$$\Gamma_t = \{x \in \mathbf{R}^n \,|\, u(t,x) = 0\} \subset \{x \in \mathbf{R}^n \,|\, u'(t,x) = 0\} = \Gamma_t'$$

for any $t \geq 0$.

(iii) Let

$$T_* := \inf\{t \in [0, \infty) \,|\, \Gamma_s = \emptyset \;\; \forall s \in (t, \infty)\}.$$

Using statement (ii) and the evolution of spheres, it is easy to see that $T_* \leq \mathrm{diam}^2 \Gamma_0 / 2k$. Assuming $T_* > 0$, we will first prove that $\Gamma_t \neq \emptyset$ for any $t \in [0, T_*)$. First of all notice that we can choose an initial function u_0 such that $u_0(x) \geq |x| - K$ for a suitable constant K; by the comparison theorem we get

$$u(t,x) \geq \sqrt{|x|^2 + 2kt} - K \qquad \forall (t,x) \in [0, \infty) \times \mathbf{R}^n$$

so that $u(t, \cdot)$ is a coercive function for any $t \geq 0$. Now, if Γ_t were empty for some $t \in [0, T_*)$ by the comparison theorem we would get

$$u(s,x) \geq \min_{y \in \mathbf{R}^n}(t,y) > 0 \qquad \forall x \in \mathbf{R}^n, \; s \geq t$$

and Γ_s would be empty for $s \in [t, \infty)$, contradicting the minimality of T_*. Finally, $\Gamma_{T_*} \neq \emptyset$ by a compactness argument, because of the continuity of u. $\square$

There are remarkable examples of instantaneous extinction, i.e., sets Γ_0 such that $T_* = 0$. For instance this happens for any proper subset Γ_0 of the boundary of a smooth, bounded open set U. Heuristically, Γ_0 has so much curvature along its (relative) boundary in ∂U that it contracts with infinite speed. The proof, based on the strong maximum principle, is given in [ES91, ES92a].

Another pathological behaviour of level set flows is the so-called *fattening phenomenon*, i.e., Γ_t can develop an interior even though Γ_0 has no interior. This phenomenon is related to the nonuniqueness of different weak notions of mean curvature flow (for instance Brakke's ones, see [Bra78]) which are all contained (cf. [Ilm94, AS96]) inside the level set flow, forcing it to be large when there is no uniqueness. The simplest example of this situation occurs when Γ_0 is the union of the two axes in $\mathbf{R}^2$, which "fattens" instantaneously. This phenomenon is discussed in [ACI, AIVa, BP94, Ilm94, Ilm93b]. The following exercise shows that nonfattening is a generic condition.

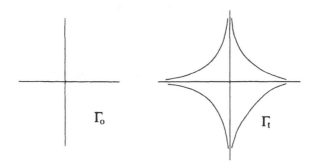

Fig. 5. Fattening phenomenon

Exercise 13. Let $u : \mathbf{R}^n \to \mathbf{R}$ be a continuous function such that $u(x) \to \infty$ as $|x| \to \infty$ and let Γ_t^τ be the level set flows starting from $\{u = \tau\}$. Show that the set

$$\mathcal{T} := \left\{ \tau \in \mathbf{R} \,|\, \mathrm{Int}\,(\Gamma_t^\tau) \neq \emptyset \ \text{for some } t \geq 0 \right\}$$

is at most countable. Hint: first prove that $\Gamma_t^\tau \cap \Gamma_t^{\tau'} = \emptyset$ whenever $t \geq 0$ and $\tau \neq \tau'$. Then, prove that

$$\mathcal{T} = \left\{ \tau \in \mathbf{R} \,|\, \mathrm{Int}\left(\bigcup_{t \geq 0} \{t\} \times \Gamma_t^\tau \right) \neq \emptyset \right\}.$$

11   Agreement with smooth flows

This section is devoted to the proof of a consistency result: the level set flow starting from a smooth manifold Γ_0 coincides with the smooth flow as long as the latter is defined. The basic idea of the proof is to compare the solution of

$$u_t + G_k(\nabla u, \nabla^2 u) = 0 \tag{90}$$

with initial condition $u_0(x) = \text{dist}(x, \Gamma_0)$ with the distance function from Γ_t, using the equation

$$\eta_t = \sum_{\lambda < 1} \frac{\lambda}{1 - \lambda} \qquad \text{(here } \lambda \text{ are the eigenvalues of } \nabla^2 \eta)$$

of Theorem 10(ii), satisfied by the squared distance function.

Lemma 5 (smooth flows subsolutions and supersolutions). *Let $(\Gamma_t)_{t \in [0,T]}$ be a smooth mean curvature flow of k dimensional manifolds, according to Definition 5, and let $d(t, \cdot)$ be the distance function from Γ_t. Then*

(i) d is a viscosity supersolution of (90) in $(0, T) \times \mathbf{R}^n$;
(ii) $[e^{-Ct}d] \wedge \sigma$ is a viscosity subsolution of (90) in $(0, T) \times \mathbf{R}^n$ for suitable constants $C, \sigma > 0$.

Proof. (i) We first notice that, by (91), η is a classical supersolution in a tubular neighbourhood $U = \{d \leq r\}$. Since the equation is geometric, d is a viscosity supersolution in U (hee we use a local version of Theorem 21).
Assume that $d - \phi$ has a local minimum, equal to 0, at (t_0, x_0); let $y_0 \in \Gamma_{t_0}$ be the least distance point from x_0 and let us set

$$\psi(t, y) := \phi(t, y + x_0 - y_0), \qquad r_0 := |x_0 - y_0|.$$

We now claim that $d - \psi$ has a local minimum at (t_0, y_0). Indeed, it holds

$$d(t_0, y_0) - \psi(t_0, y_0) = -\phi(t_0, x_0) = -r_0.$$

On the other hand, for $|y - y_0|$ and $|t - t_0|$ sufficiently small it holds

$$d(t, y) - \psi(t, y) \geq d(t, y + x_0 - y_0) - \psi(t, y) - r_0$$
$$= d(t, y + x_0 - y_0) - \phi(t, y + x_0 - y_0) - r_0 \geq -r_0$$

proving the claim. Since $(t_0, y_0) \in U$, we know that $\psi_t + G_k^*(\nabla\psi, \nabla^2\psi) \geq 0$ at (t_0, y_0), and this yields the same inequality with ϕ in place of ψ at (t_0, x_0).
(ii) By Theorem 10(ii), we know that the squared distance function $\eta := d^2/2$ satisfies

$$\eta_t - \sum_{\lambda < 1} \frac{\lambda}{1 - \lambda} = 0$$

in a tubular neighbourhood

$$U := \{(t, x) \in (0, T) \times \mathbf{R}^n \mid d(t, x) \leq 2r\}.$$

with $r > 0$. Let D such that $\lambda \leq Dd(t, x)$ for any $(t, x) \in U$ and any eigenvalue λ of $\nabla^2\eta(t, x)$ less than 1; possibily reducing r we can assume $2Dr < 1/2$, so that $1 - \lambda > 1/2$ for any $(t, x) \in U$ and any eigenvalue $\lambda < 1$ of $\nabla^2 \eta(t, x)$. Using Theorem 4 to compute $G_k(\nabla \eta, \nabla^2 \eta)$, we get

$$\eta_t + G_k(\nabla \eta, \nabla^2 \eta) = \sum_{\lambda < 1} \left(\frac{\lambda}{1 - \lambda} - \lambda \right) = \sum_{\lambda < 1} \frac{\lambda^2}{1 - \lambda} \leq 4kD^2\eta, \qquad (91)$$

so that, setting $C = 2kD^2$, $e^{-2Ct}\eta$ is a classical subsolution in U. Let

$$\sigma := e^{-CT}r, \qquad w := [e^{-2Ct}\eta] \wedge \sigma^2/2$$

and let us check that w is a global viscosity subsolution in $(0, T) \times \mathbf{R}^n$. Let (t_0, x_0), $\phi \in C^\infty$ near (t_0, x_0) such that $w - \phi$ has a local maximum at (t_0, x_0). If $e^{-Ct_0}d(t_0, x_0) \leq \sigma$, then $(x_0, t_0) \in U$ (because $\sigma e^{Ct_0} < e^{CT}\sigma = r$) and w is a viscosity subsolution near u by a local version of Theorem 21. On the other hand, if $e^{-Ct_0}d(t_0, x_0) > \sigma$ then $w \equiv \sigma^2/2$ near (t_0, x_0) and the inequality

$$\phi_t + G_{k*}(\nabla \phi, \nabla^2 \phi) \leq 0 \qquad \text{at } (t_0, x_0).$$

is trivial because $\phi_t(t_0, x_0) = \nabla \phi(t_0, x_0) = 0$, $\nabla^2 \phi(t_0, x_0) \geq 0$. Since the equation is geometric, this proves that $[e^{-Ct}d] \wedge \sigma = \sqrt{2w}$ is a global viscosity subsolution.
□

Theorem 24 (agreement with smooth flows). *Let $(\Gamma_t)_{t \in [0,T]}$ be a smooth flow and let $(\Gamma'_t)_{t \geq 0}$ be the level set flow starting from Γ_0. Then, $\Gamma_t = \Gamma'_t$ for any $t \in [0, T]$.*

Proof. We choose $u_0(x) := \text{dist}(x, \Gamma_0)$ as initial condition in (90), and we denote by u the corresponding solution. We recall that, according to Definition 12, Γ'_t are the zero level sets of u. By Theorem 18 and Lemma 5 we infer

$$[e^{-Ct}d(t, x)] \wedge \sigma \leq u(t, x) \leq d(t, x) \qquad \forall (t, x) \in [0, T) \times \mathbf{R}^n.$$

Consequently, $\Gamma_t = \{d(t, \cdot) = 0\} = \{u(t, \cdot) = 0\} = \Gamma'_t$ for any $t \in [0, T)$. Equality is extended up to $t = T$ by a continuity argument.   □

In general, no result of this type can be expected for any time: indeed, Angenent, Chopp, Ilmanen and Velasquez proved that level set flows starting from that initially smooth manifolds can develop an interior for positive times (see [AIVa, ACI]).

12 Barriers

In this section we compare the level set approach with a purely geometric approach based on the notion of barriers, introduced by De Giorgi in [Gio94b]. In addition to the general definition of barriers, [Gio94b] also contains the characterization of the smooth mean curvature flow as a system of equations for η (cf. Theorem 6) and the idea to compare the flow of hypersurfaces by the sum of the smallest k principal curvatures (cf. (84)) with k dimensional mean curvature flows.

We start with De Giorgi's general definition of barriers:

Definition 14 (De Giorgi). Let $(S, \leq)$ be a partially ordered set and let $\mathcal{F}$ be a class of functions defined on intervals $[a, b] \subset [0, \infty)$, with values in S. We say that $\phi : [0, \infty) \to S$ is a *barrier* relative to $\mathcal{F}$, and we write $\phi \in \mathcal{B}(\mathcal{F})$, provided that the following implication holds for any $f \in \mathcal{F}$,

$$f : [a, b] \to S, \quad f(a) \leq \phi(a) \qquad \Longrightarrow \qquad f(t) \leq \phi(t) \quad \forall t \in [a, b].$$

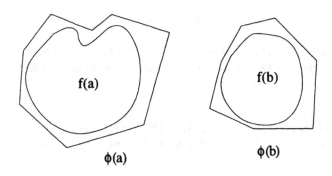

$f(a)$

$\phi(a)$

$f(b)$

$\phi(b)$

Fig. 6. Barrier property

If S is a complete lattice, then the infimum any family of barriers is still a barrier. For any $s \in S$, this suggests the following definition of the least barrier, $\mathcal{M}(\mathcal{F}, s)$, that is greater than s at time 0:

$$\mathcal{M}(\mathcal{F}, s)(t) := \inf \left\{ \phi(t) : \phi \in \mathcal{B}(\mathcal{F}), \ s \leq \phi(0) \right\}.$$

Heuristically, we think of $\mathcal{F}$ as the set of all classical solutions and $\mathcal{B}(\mathcal{F})$ as the set of all supersolutions. Then in analogy with Perron's method, $\mathcal{M}(\mathcal{F}, s)$ is a *weak* solution with initial data s.

In this section we work with the family of all subsets of $\mathbf{R}^n$, partially ordered by inclusion, and we choose as class $\mathcal{F}$ all families $\{\overline{U}_t\}_{t \in [a,b]}$ such that $\{\partial U_t\}_{t \in [a,b]}$ is a smooth mean curvature flow of hypersurfaces, defined in Definition 4 and Definition 5 up to a translation in time. Using the signed distance function $\overline{d}(t, \cdot)$ from U_t, the flow can be characterized by the equation (cf. (32))

$$\overline{d}_t(t, x) = \Delta \overline{d}(t, x) \qquad \forall t \in [a, b], \ x \in \partial U_t. \tag{92}$$

Using inner and outer approximation by tubular neighbourhoods we can also define regularized barriers as follows:

$$\mathcal{M}^*(\mathcal{F}, E)(t) := \bigcap_{\varrho > 0} \mathcal{B}\big(\mathcal{F}, N_\varrho(E)\big)(t)$$

$$\mathcal{M}_*(\mathcal{F}, E)(t) := \bigcup_{\varrho > 0} \mathcal{B}\big(\mathcal{F}, I_\varrho(E)\big)(t)$$

where

$$N_\varrho(E) := \big\{x \in \mathbf{R}^n \,|\, \mathrm{dist}(x, E) < \varrho\big\}, \qquad I_\varrho(E) := \big\{x \in \mathbf{R}^n \,|\, \mathrm{dist}(x, \mathbf{R}^n \backslash E) > \varrho\big\}. \tag{93}$$

Given a compact initial set Γ_0, in [Gio94b] De Giorgi uses the class of smooth flows $\mathcal{F}$ to give a direct definition of weak solution of the codimension 1 mean curvature flow starting from Γ_0. His definition (compare with §10 of [Ilm94] and [Ilm93b, BS98]) is:

$$\mathcal{M}^*(\mathcal{F}, \Gamma_0)(t) \qquad\qquad \forall t \geq 0. \tag{94}$$

The main difference between the level set–viscosity approach and the barrier approach is that the first one is based on a local comparison property for functions, while the second one is based on a global comparison property for sets. For this reason, existence and comparison theorems are quite easy in the theory of barriers; on the other hand, passages to limits as in Theorem 14 can be more easily handled with the viscosity approach. We will show in Theorem 25 below that in codimension 1 the two approaches produce the same generalized evolution. The connections between the level set approach and the barrier approach have been investigated by Bellettini and Paolini in [BP95b], [BP95c]; these results have been further extended by Bellettini and Novaga in [BN98, BN97a] to anisotropic and (x, t)–dependent flows.

In the following lemma we prove elementary topological properties of least barriers, which easily follow by the translation invariance in space and in time of the family $\mathcal{F}$.

Lemma 6. *Let $A \subset \mathbf{R}^n$ be an open set and let*

$$\phi(t) := \mathcal{M}\big(\mathcal{F}, A\big)(t), \qquad\qquad K(t) = \mathbf{R}^n \setminus \phi(t).$$

Then, $\phi(t)$ is open for any $t \geq 0$ and the map $t \mapsto K(t)$ is upper semicontinuous from the left, i.e., $(x_h, t_h) \to (x, t)$, $x_h \in K(t_h)$ and $t_h < t$ implies $x \in K(t)$.

Proof. The translation invariance of the class $\mathcal{F}$ easily implies that the interior of a barrier is still a barrier. Hence the minimality of ϕ forces $\phi(t)$ to coincide with its interior for any t. To check the upper semicontinuity property we define

$$\tilde{K}(t) := \bigcap_{0 < \tau < t} \overline{\bigcup_{s \in (t-\tau, t]} K(s)}, \qquad \tilde{K}(0) = K(0)$$

and verify that $\tilde{K}(t) = K(t)$ for any t. Since $\tilde{K}(t)$ contains $K(t)$ and $\phi(t)$ is the least barrier, we will prove that $\tilde{\phi}(t) := \mathbf{R}^n \setminus \tilde{K}(t)$ is a barrier. Let $\{\overline{U}_t\}_{t\in[a,b]} \in \mathcal{F}$ be such that $\overline{U}_a \subset \tilde{\phi}(a)$. For τ small enough we can assume that $y+\overline{U}_{a+\epsilon} \subset \phi(a)$ for any $\epsilon \in [0,\tau)$ and any y such that $|y| < \tau$. The barrier property implies

$$N_\tau(\overline{U}_t) \subset \phi(s) \qquad \forall s \in (t-\tau, t]$$

for any $t \geq a$. Hence,

$$\overline{U}_t \subset \text{Interior}\left(\bigcap_{s\in(t-\tau,t]} \phi(s) \right).$$

In particular, since the set on the right is contained in $\tilde{\phi}(t)$, we obtain $\overline{U}_t \subset \tilde{\phi}(t)$.
□

Theorem 25 (level set flows and barriers). *Let $\Gamma_0 \subset \mathbf{R}^n$ be a compact set and let $(\Gamma_t)_{t\geq 0}$ be the level set flow defined in §10. Then*

$$\mathcal{M}^*(\mathcal{F}, \Gamma_0)(t) = \Gamma_t \qquad \forall t \geq 0.$$

The proof of Theorem 25 we give in this notes is, in the spirit of the theory of barriers, based on global comparison arguments (the strategy of the paper [BN98] is local and provides more general results). In particular, we will need the following two facts (see [Ilm93b] and [ES92a, Hui84]) whose proof is omitted:

Lemma 7 (interposition lemma). *Let $V \subset \mathbf{R}^n$ be an open set and let $\Gamma \subset V$ be a compact set. Then, there exists a bounded open set U with $C^{1,1}$ boundary such that $\Gamma \subset U \subset\subset V$ and*

$$\text{dist}(\Gamma, \partial U) + \text{dist}(\partial U, \partial V) = \text{dist}(\Gamma, \partial V).$$

Theorem 26 (short time existence of smooth flows). *Let U be a bounded open set with smooth boundary. Then, there exists a smooth mean curvature flow $(\Gamma_t)_{t\in[0,T]}$ starting from ∂U. In addition, T can be estimated from below with the maximum of the second fundamental form of ∂U and with*

$$\rho(U) := \sup\left\{ r > 0 \,|\, \overline{d}(\cdot, U) \text{ smooth in } \{|\overline{d}|(\cdot, U) < r\} \right\}.$$

The interposition lemma plays the same rôle of $|x-y|^2$ in Theorem 12: indeed, knowing that level set flows and barriers are both well behaved with respect to smooth flows, we will use U to show that if the barrier V initially contains the level set K, then the property is true for later times.

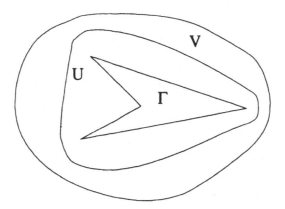

Fig. 7. Interposition lemma

Proof (Theorem 25). Let $u(t, x)$ be the unique viscosity solution of

$$u_t - \text{trace}(P_{\nabla u} \nabla^2 u) = 0 \qquad (95)$$

with the initial condition $u_0 = \text{dist}(x, \Gamma_0)$; by definition, we know that $\Gamma_t = \{u(t, \cdot) = 0\}$.

(i) To prove the inclusion $\mathcal{M}^*(\mathcal{F}, \Gamma_0)(t) \subset \Gamma_t$, it suffices to show that $\phi_\varrho(t) := \{u(t, \cdot) \leq \varrho\}$ is a barrier relative to $\mathcal{F}$, for any $\varrho > 0$. Because if ϕ_ϱ is a barrier, then

$$\mathcal{B}(\mathcal{F}, N_\varrho(\Gamma_0))(t) \subset \phi_\varrho(t) \qquad \forall \varrho > 0,$$

and the inclusion follows by letting $\varrho \downarrow 0$. The fact that ϕ_ϱ is a barrier follows from Theorem 18.

Indeed, let $\{\overline{U}_t\}_{t \in [a,b]}$ be a function in $\mathcal{F}$, such that $\overline{U}_a \subset \phi_\varrho(a)$. Set $\delta(t, x) := \text{dist}(x, \overline{U}_t)$; arguing as in Lemma 5(i), we can prove that δ is a supersolution of (95) in $(a, b) \times \mathbf{R}^n$. On the other hand, since $\overline{U}_a \subset \phi_\varrho(a)$, there exists a nondecreasing uniformly continuous function $w(t)$ such that $w(0) \leq 0$ and $w(\delta(a, \cdot)) \geq u(a, \cdot) - \varrho$ (see Theorem 22 for the construction of w). By Theorem 18, Theorem 21 and the continuity of u, δ, we get $w(\delta(t, \cdot)) \geq u(t, \cdot) - \varrho$ for $t \in [a, b]$. Hence

$$x \in \overline{U}_t \; \Rightarrow \; \delta(t, x) = 0 \; \Rightarrow \; u(t, x) \leq \rho \; \Rightarrow \; x \in \phi_\varrho(t)$$

for any $t \in [a, b]$. This shows that ϕ_ϱ is a barrier.

(ii) To prove the opposite inclusion $\subset$ we define $K(t) := \mathbf{R}^n \setminus \phi(t)$ and we show that the function $f(t) := \text{dist}(\Gamma_t, K(t))$ is nondecreasing. It is easy to see that the monotonicity of f follows by the following two properties:

$$\liminf_{s \uparrow t} f(s) \geq f(t) \qquad \forall t > 0 \qquad (96)$$

and

$$\forall t \geq 0 \text{ s.t. } f(t) > 0 \quad \exists T > t \text{ s.t. } f(s) \geq f(t) \; \forall s \in [t, T]. \tag{97}$$

Inequality (96) follows by the upper semicontinuity of $t \mapsto K(t)$ (see Lemma 6): if (x_h, y_h, s_h) is a sequence converging to (x, y, t) such that $s_h < t$, $x_h \in \Gamma_{s_h}$, $y_h \in K(s_h)$ and

$$\lim_{h \to +\infty} |x_h - y_h| = \liminf_{s \uparrow t} f(s)$$

we have $x \in \Gamma_t$, $y \in K(t)$ and the inequality follows.

(iii) In this step we begin the proof of (97). By applying Lemma 7 with $V = \phi(t)$ and $\Gamma = \Gamma_t$ we obtain a bounded open set U_t containing Γ_t whose boundary $C^{1,1}$ and satisfies

$$\text{dist}(\Gamma_t, \partial U_t) + \text{dist}(\partial U_t, K(t)) = \text{dist}(\Gamma_t, K(t)). \tag{98}$$

Let r^t be the signed distance function from U_t and let $r(s, x)$ be the unique viscosity solution of (95) with the initial condition $r(t, \cdot) = r^t$. Setting $U_s := \{r(s, \cdot) < 0\}$, Theorem 18 yields the inclusion

$$y + \Gamma_s \subset U_s \quad \forall s \geq t$$

for any $y \in \mathbf{R}^n$ such that $|y| < \text{dist}(\Gamma_t, \partial U_t)$. In particular

$$\text{dist}(\Gamma_s, \partial U_s) \geq \text{dist}(\Gamma_t, \partial U_t) \qquad \forall s \geq t. \tag{99}$$

Taking into account that

$$\text{dist}(\Gamma_s, \partial U_s) + \text{dist}(\partial U_s, K(s)) \leq \text{dist}(\Gamma_s, K(s)) \qquad \forall s \geq t$$

and (98), (99), in order to prove (97) we need only to prove in the next step the existence of $T > t$ such that

$$\text{dist}(\partial U_s, K(s)) \geq \text{dist}(\partial U_t, K(t)) \qquad \forall s \in [t, T]. \tag{100}$$

(iv) Let U^h be a sequence of open sets with smooth boundary such that the signed functions r^h from U^h uniformly converge to the signed distance function from U^t and $\rho(U^h)$ and the second fundamental forms of U^h are equibounded. Let $\overline{U}^h_{s \in [t,T]}$ be the corresponding smooth mean curvature flows starting at time t from $\overline{U}^h$ (at this point we use Theorem 26). We also denote by r^h the viscosity solutions of (95) with initial condition, at time t, given by the signed distance function from U^h.

Since the functions r_h are equicontinuous (cf. Remark 20) we can assume, extracting a subsequence if necessary, that r^h converges locally uniformly in $[t, \infty) \times \mathbf{R}^n$ to a solution of (95). By the uniqueness and convergence of the initial conditions, we infer that r^h converges to the function r of step (iii).

Since $\phi(t)$ is a barrier, we get

$$\text{dist}(U^h_s, K(s)) \geq \text{dist}(\partial U^h_t, K(t)) \tag{101}$$

for any $s \in [t, T]$, $h \in \mathbf{N}$. By the consistency Theorem 24, we know that

$$\partial U_s^h = \{x \in \mathbf{R}^n \,|\, r^h(s, x) = 0\} \qquad \forall s \in [t, T], \ h \in \mathbf{N}.$$

By the convergence of r^h to r, any point in ∂U_s can be approximated by points in U_s^h; therefore, passing to the limit as $h \to \infty$ in (101) we obtain (100).   □

Remark 23. More generally, it can be proved (see [BP95b, BP95c, BN98, BN97a]) that

$$\mathcal{M}^*\big(\mathcal{F}, \{u_0 \leq \tau\}\big)(t) = \{u(t, \cdot) \leq \tau\}, \qquad \mathcal{M}_*\big(\mathcal{F}, \{u_0 < \tau\}\big)(t) = \{u(t, \cdot) < \tau\}$$

for any $t \in [0, \infty)$ and any $\tau \in \mathbf{R}$.

13   Distance formulation of level set flows

In this section we see how the level set flow can be characterized in codimension 1 by the properties of the signed distance function from it. This was first done by Soner in [Son93b] (see also [ESS92, BSS93]). To simplify several statements, we will assume that the initial set Γ_0 is compact and nonempty. In this section, unless otherwise stated, all differential equalities and inequalities are understood in the viscosity sense.

Let us recall the definition of level set flow: given any uniformly continuous function u_0 such that $\Gamma_0 = \{u_0 = 0\}$, the level set flow $(\Gamma_t)_{t \geq 0}$ is defined by

$$\Gamma_t := \{x \in \mathbf{R}^n \,|\, u(t, x) = 0\}$$

where u is the unique solution of

$$u_t + F\big(\nabla u, \nabla^2 u\big) = 0 \tag{102}$$

with initial condition $u(0, \cdot) = u_0$ and

$$F(p, X) = -\text{trace}(P_p X) = -\text{trace}(X) + \frac{\langle Xp, p \rangle}{|p|^2} \ \forall (p, X) \in (\mathbf{R}^n \setminus \{0\}) \times \text{Sym}^n.$$

We set $A := (0, T] \times \mathbf{R}^n$, where T is the extinction time of the flow (i.e., the maximum $t \geq 0$ such that $\Gamma_t \neq \emptyset$) and we assume that T is strictly positive. Finally, for any $t \in [0, T]$ we will denote by $d(t, \cdot)$ the distance function from Γ_t.

Lemma 8 (distance supersolution). *The function $d(t, x)$ is a viscosity supersolution of (102) in A.*

Proof. We have seen in the proof of Theorem 22 that $w(t, x) := 1 - \chi_{\Gamma_t}(x)$ is a viscosity supersolution of (102) in $(0, \infty) \times \mathbf{R}^n$; since u is continuous, w is lower semicontinuous. Then, the same argument of Proposition 4(iii) shows that the functions

$$w_\lambda(t, x) := \min\left\{w(t, y) + \frac{1}{\lambda}|x - y| \,|\, y \in \mathbf{R}^n\right\}$$

are still viscosity supersolutions in $(0, \infty) \times \mathbf{R}^n$, as well as λw_λ. A simple computation shows that

$$\lambda w_\lambda(t, x) = d(t, x) \wedge \min\{\lambda + |x - y| \,|\, y \notin \Gamma_t\}.$$

By Theorem 23(iii), $d(t, \cdot) < \infty$ for any $t \in [0, T]$. Since $\lambda w_\lambda \uparrow d$ as $\lambda \uparrow \infty$, by Proposition 3 we have also

$$d = \Gamma^- \lim_{\lambda \to \infty} (\lambda w_\lambda).$$

Then, the statement follows by the stability Theorem 14.   □

The same result, with the same proof, is true for higher codimension flows, in which the function F is defined as in (84). The result stated in Lemma 8 can be formulated as

$$d_t - \Delta d + \frac{\langle \nabla^2 d \nabla d, \nabla d \rangle}{|\nabla d|^2} \geq 0 \tag{103}$$

in A. On the other hand, from Lemma 2 we know that any distance function satisfies

$$-\langle \nabla^2 d \nabla d, \nabla d \rangle \geq 0 \tag{104}$$

in $A \cap \{d > 0\}$. In particular,

$$d_t - \Delta d \geq 0 \quad \text{in } A \cap \{d > 0\}. \tag{105}$$

Let us define

$$d_1(t, x) := \text{dist}\big(x, \{u(t, \cdot) \leq 0\}\big), \qquad d_2(t, x) := -\text{dist}\big(x, \{u(t, \cdot) \geq 0\}\big). \tag{106}$$

Since, by the invariance theorem, $\{u(t, \cdot) \leq 0\}$ is the level set flow starting from $\{u_0 \leq 0\}$, from (105) and Remark 8(3) we infer the following result:

Proposition 5. *The functions d_1 and d_2 respectively satisfy*

$$d_t - \Delta d \geq 0 \quad \text{in } A \cap \{d_1 > 0\}, \qquad d_t - \Delta d \leq 0 \quad \text{in } A \cap \{d_2 < 0\}. \tag{107}$$

Notice that if the zero level set is smooth, then $\bar{d} := d_1 + d_2$ is the signed distance function from $\{u(t, \cdot) \leq 0\}$ (see also Definition 15). It is natural to investigate to what extent the conditions of Proposition 5 characterize the level set flow. The following theorem, proved in [BSS93], provides an answer to this problem, showing that d_1 (resp. d_2) is the minimal (maximal) function with this property.

Theorem 27 (heat equation sub and supersolutions). *Let* $(\Lambda_t)_{t\in[0,T]}$ *be closed subsets of* $\mathbf{R}^n$.

(i) Assume that $z(t,x) := \mathrm{dist}(x,\Lambda_t)$ *is lower semicontinuous in* $[0,T] \times \mathbf{R}^n$, $z(\cdot,x)$ *is left continuous on* $\{z(\cdot,x) = 0\}$ *and*

$$(z \wedge k)_t - \Delta(z \wedge k) \geq 0 \qquad in \ A \cap \{z > 0\} \tag{108}$$

for any $k > 0$. *Then*

$$\Lambda_0 \subset \{u_0 \leq 0\} \qquad \Longrightarrow \qquad \Lambda_t \subset \{u(t,\cdot) \leq 0\} \quad \forall t \in [0,T].$$

In particular $z \geq d_1$ *in* $[0,T] \times \mathbf{R}^n$.
(ii) Assume that $z(t,x) := -\mathrm{dist}(x,\Lambda_t)$ *is upper semicontinuous in* $[0,T] \times \mathbf{R}^n$, $z(\cdot,x)$ *is left continuous on* $\{z(\cdot,x) = 0\}$ *and*

$$(z \vee m)_t - \Delta(z \vee m) \leq 0 \qquad in \ A \cap \{z < 0\} \tag{109}$$

for any $m < 0$. *Then*

$$\Lambda_0 \subset \{u_0 \geq 0\} \qquad \Longrightarrow \qquad \Lambda_t \subset \{u(t,\cdot) \geq 0\} \quad \forall t \in [0,T].$$

In particular, $z \leq d_2$ *in* $[0,T] \times \mathbf{R}^n$.

Proof. We can assume, of course, that the set $\mathcal{T}$ of all $t \in (0,T]$ such that $\Lambda_t \neq \emptyset$ is not empty. We will prove statement (i), because the proof of the second one can be obtained by applying (i) to $-z$ and replacing u by $-u$ (cf. Remark 8(3)). We need only to prove that $v := z \wedge d_1$ is a viscosity supersolution of (102) in A. In fact, if this is true the comparison theorem yields $v \geq d_1$ in A, hence $z \geq d_1$ and $\{z(t,\cdot) = 0\} \subset \{d_1(t,\cdot) = 0\}$.

By Lemma 9 below, we need only to prove that v is a viscosity supersolution in $A \cap \{v > 0\}$ (notice that condition (i) of the lemma follows by the left continuity hypothesis and that condition (iii) is trivially satisfied). Let $(t_0, x_0) \in A \cap \{v > 0\}$ and assume that $v - \phi$ has a strict global minimum point at (t_0, x_0) for some smooth function ϕ. If $z(t_0, x_0) \geq d_1(t_0, x_0)$ we know that $d_1 - \phi$ has a relative minimum at (t_0, x_0) and we can apply the supersolution property of d_1. Therefore, in the following we can assume that

$$0 < z(t_0, x_0) < d_1(t_0, x_0) < \infty$$

and that $z - \phi$ has a strict global minimum at (t_0, x_0).

Recall that $z(t_0, \cdot)$ is locally semiconcave in $\mathbf{R}^n$ (cf. Exercise 7), therefore Thm. 15(ii) yields that $z(t_0, \cdot)$ is differentiable at x_0 and $\nabla z(t_0, x_0) = \nabla \phi(t_0, x_0)$. By Theorem 1, $\nabla z(t_0, x_0)$ is a unit vector. In order to gain semiconcavity in (t, x), we approximate z by the functions

$$z_\epsilon(t,x) := \inf_{s\in\mathcal{T}}\left\{z(s,x) + \frac{1}{\epsilon}|s - t|^2\right\} \qquad (t,x) \in \mathbf{R} \times \mathbf{R}^n.$$

Let K be a closed ball centered at (t_0, x_0) such that $m := \min_{K \cap A} z > 0$; using Exercise 7 and the identity

$$z_\epsilon(t, x) = \inf_{s \in T} \inf_{y \in A_s} \left\{ |x - y| + \frac{1}{\epsilon} |s - t|^2 \right\}$$

it is not hard to see that

$$(t, x) \mapsto z_\epsilon(t, x) - \frac{1}{\epsilon} |t|^2 - \frac{1}{m} |x|^2$$

is concave in K, hence z_ϵ is semiconcave in the interior of K.

By using Proposition 4(iii) and Remark 16(a), and noticing that

$$z_\epsilon(t, x) = \inf_{s \in [0,T]} \left\{ z(s, x) \wedge k + \frac{1}{\epsilon} |s - t|^2 \right\} \qquad (t, x) \in K$$

for k large enough, it can be proved that definitely the functions z_ϵ are viscosity supersolutions of the heat equation in K; in additions, the minimizers (t_ϵ, x_ϵ) of $z_\epsilon - \phi$ in K converge to (t_0, x_0) as $\epsilon \downarrow 0$. The semiconcavity of z_ϵ also yields

$$z_{\epsilon t}(t_\epsilon, x_\epsilon) = \phi_t(t_\epsilon, x_\epsilon), \qquad \nabla z_\epsilon(t_\epsilon, x_\epsilon) = \nabla \phi(t_\epsilon, x_\epsilon). \qquad (110)$$

The functions $z_\epsilon(t, \cdot)$ have the same Lipschitz constant of $z(t, \cdot)$, hence $|\nabla z_\epsilon|^2 - 1 \le 0$ almost everywhere; on the other hand, for the same reasons mentioned above, definitely

$$|\nabla z_\epsilon|^2 - 1 \ge 0 \qquad \text{in } K$$

because $z(t, \cdot)$ satisfies $|\nabla z|^2 - 1 = 0$ in $\{z(t, \cdot) > 0\}$ for any t such that $\Lambda_t \ne \emptyset$. Since z_ϵ are locally Lipschitz functions, it turns out that $|\nabla z_\epsilon|^2 = 1$ almost everywhere in K.

Let us temporarily fix ϵ sufficiently small to ensure all the conditions mentioned above; by applying Theorem 17 to $w := \phi - z_\epsilon$ we obtain a sequence (t_h, x_h) converging to (t_ϵ, x_ϵ) such that w (hence z_ϵ) is twice differentiable, has a second order Taylor expansion at (t_h, x_h) and

$$z_{\epsilon t}(t_h, x_h) \to z_{\epsilon t}(t_\epsilon, x_\epsilon),$$
$$\nabla z_\epsilon(t_h, x_h) \to \nabla z(t_\epsilon, x_\epsilon),$$
$$\nabla^2 \phi(t_h, x_h) \le \nabla^2 z_\epsilon(t_h, x_h) + \eta_h I$$

with (η_h) infinitesimal. Since $|\nabla z_\epsilon|^2 = 1$ almost everywhere in K, it turns out that

$$\langle \nabla^2 z_\epsilon(t_h, x_h) \nabla z_\epsilon(t_h, x_h), \nabla z_\epsilon(t_h, x_h) \rangle = 0,$$

so that the supersolution property of z_ϵ at (x_h, t_h) and the degenerate ellipticity of F^* yield

$$0 \le z_{\epsilon t}(t_h, x_h) - \Delta z_\epsilon(t_h, x_h) = z_{\epsilon t}(t_h, x_h) + F^* \left(\nabla z_\epsilon(t_h, x_h), \nabla^2 z_\epsilon(t_h, x_h) \right)$$
$$\le z_{\epsilon t}(t_h, x_h) + F^* \left(\nabla z_\epsilon(t_h, x_h), \nabla^2 \phi(x_h, t_h) - \eta_h I \right).$$

Passing to the limit as $h \to \infty$ we get

$$z_{\epsilon t}(t_\epsilon, z_\epsilon) + F^* \left(\nabla z_\epsilon(t_\epsilon, z_\epsilon), \nabla^2 \phi(t_\epsilon, z_\epsilon) \right) \ge 0.$$

The proof is concluded by using (110) and letting $\epsilon \downarrow 0$. $\qquad \square$

Remark 24. The reason for truncation in (108) is that we allow Λ_t to be empty, and therefore z to be infinite. It is easy to check that (108) is fulfilled for any $k > 0$ if and only if

$$\phi_t(t_0, x_0) - \Delta\phi(t_0, x_0) \geq 0$$

whenever $(t_0, x_0) \in A$, $z(t_0, x_0) \in (0, \infty)$, ϕ is smooth and $z - \phi$ has a relative minimum at (t_0, x_0). However, strictly speaking this does not mean that z is a viscosity supersolution of the heat equation in $A \cap \{0 < z < \infty\}$ because $\{0 < z < \infty\}$ need not be open and hence $A \cap \{0 < z < \infty\}$ need not be locally compact.

Lemma 9. *Let $v : A \to [0, \infty)$ be a lower semicontinuous function satisfying:*

(i) *for every $(t, x) \in A$ with $v(t, x) = 0$, there exists a sequence $(t_n, x_n) \to (x, t)$ such that $v(t_n, x_n) \to 0$ and $t_n < t$;*
(ii) *v is a viscosity supersolution of (102) on $A \cap \{v > 0\}$;*
(iii) *$x \mapsto v(t, x)$ is continuous in $\mathbf{R}^n$ for any $t \in (0, T]$.*

Then v is a viscosity supersolution of (102) in A.

Proof. For $\epsilon > 0$, let $m_\epsilon(r) = (r - \epsilon)^+$. We claim that $m_\epsilon(v)$ is a viscosity supersolution of (102) in A. Suppose that for some test function function ϕ, $m_\epsilon(v) - \phi$ has a strict minimum at $(t_0, x_0) \in A$. If $v(t_0, x_0) > 0$ we can use a local version of Theorem 21 and (ii) to obtain that $m_\epsilon(v)$ is a supersolution near (t_0, x_0).

If $v(t_0, x_0) = 0$, by (i) there exists a sequence $(t_h, x_h) \to (t_0, x_0)$ with $v(t_h, x_h) \to 0$ and $t_h < t_0$. Recall that v is continuous in the x variable. Therefore for sufficiently large h, $m_\epsilon(v(t_h, x_0)) = 0$. Since $t_h \uparrow t$, we conclude that $\phi_t(t_0, x_0) \geq 0$. Also $m_\epsilon(v(t_0, x)) \equiv 0$ for all x near x_0. Hence $\nabla\phi(t_0, x_0) = 0$, $\nabla^2\phi(t_0, x_0) \leq 0$ and at (t_0, x_0) we have

$$\phi_t + F^*(\nabla\phi, \nabla^2\phi) \geq 0$$

be the degenerate ellipticity of F.

Since $(v - \epsilon)^+$ is a supersolution, we can let $\epsilon \downarrow 0$ and use Theorem 14 to conclude that v is a viscosity supersolution in A. $\square$

14   Approximation by the reaction diffusion equation

In this section we will analyze the asymptotic behaviour of solutions u^ϵ of the so-called *reaction diffusion equation*

$$u_t = \Delta u + \frac{1}{\epsilon^2} u(1 - u^2) \qquad \text{on } (0, \infty) \times \mathbf{R}^n \tag{111}$$

with suitable initial conditions which ensure that $|u^\epsilon| < 1$. The equations above are the gradient flows, up to a time scaling, of the functionals

$$F_\epsilon(u) := \frac{1}{2} \int_{\mathbf{R}^n} \left[\epsilon|\nabla u|^2 + \frac{1}{\epsilon} W(u) \right] dx \qquad u \in H^1_{loc}(\mathbf{R}^n) \tag{112}$$

where $W(t) := (1 - t^2)^2/2$ is a "double well" potential with two global minima at ± 1. Precisely, if $v^\epsilon(t, x)$ is the gradient flow of F_ϵ, then $u^\epsilon(t, x) = v^\epsilon(t/\epsilon, x)$. Since $F_\epsilon(u^\epsilon(t, \cdot))$ decreases along the flow (111), if the initial energy is uniformly bounded in ϵ this forces u^ϵ to converge to ± 1 as $\epsilon \downarrow 0$, so that an "interface" between two regions is created in the limit (see for instance [BK91]).

It was conjectured by Allen & Cahn in [AC79] that the interface should flow by mean curvature; this expectation was confirmed by Keller, Rubinstein & Sternberg in [KRS89] by a formal asymptotic expansion. Also De Giorgi used in [Gio90a] this approximation to construct weak solutions of the codimension 1 mean curvature flow problem. Before stating a rigorous result in Theorem 29 below, we notice that, by (13), the mean curvature flow is the "gradient flow" of the area functional. Hence, an heuristic justification of the convergence of the reaction–diffusion equation to a mean curvature flow is also provided by the following theorem (see [Mod87, MM77]). For convenience, we state the result for W potentials vanishing in $\{0, 1\}$.

Theorem 28 (Modica–Mortola). *Let us extend the functional F_ϵ in (112) to $L^2(\mathbf{R}^n)$ setting $F_\epsilon(u) = \infty$ if $u \notin H^1(\mathbf{R}^n)$ and let us assume that $W \geq 0$ is continuous in $\mathbf{R}$ and such that $W(t) = 0$ if and only if $t \in \{0, 1\}$. Then, for any $u \in L^2(\mathbf{R}^n)$ it holds*

$$\Gamma^- \lim_{\epsilon \downarrow 0} F_\epsilon(u) = \begin{cases} \lambda |D\chi_E|(\mathbf{R}^n) & \text{if } u = \chi_E; \\ \infty & \text{otherwise,} \end{cases}$$

where

$$\lambda := \int_0^1 \sqrt{W(s)}\, ds$$

$$= \frac{1}{2} \min \left\{ \int_{-\infty}^\infty |\gamma'(t)|^2 + W(\gamma(t))\, dt \mid \gamma \in C^1(\mathbf{R}), \ \gamma(-\infty) = 1, \ \gamma(\infty) = 0 \right\}.$$

Proof. Recall that (see for instance [Giu84]), given $u \in L^1(\mathbf{R}^n)$, $|Du|(\mathbf{R}^n)$ is defined by

$$\sup \left\{ \int_{\mathbf{R}^n} u \operatorname{div} g\, dx \mid g \in C_c^1(\mathbf{R}^n, \mathbf{R}^n) \right\}$$

and coincides with the L^1 norm of ∇u if $u \in W^{1,1}(\mathbf{R}^n)$ and with the perimeter of E if $u = \chi_E$ and E has a smooth boundary. Moreover, $u \mapsto |Du|(\mathbf{R}^n)$ is lower semicontinuous with respect to $L_{\text{loc}}^1(\mathbf{R}^n)$ convergence because, given $g \in C_c^\infty(\mathbf{R}^n, \mathbf{R}^n)$,

$$u \mapsto \int_{\mathbf{R}^n} u \operatorname{div} g\, dx$$

is continuous with respect to $L_{\text{loc}}^1(\mathbf{R}^n)$ convergence.

Let us prove the liminf inequality. Possibly replacing W by

$$\tilde{W}(t) := W(t) \wedge M \quad \text{with} \quad M = \sup_{t \in [0,1]} W(t)$$

we can assume that W is bounded in $\mathbf{R}$. Let (u_ϵ) be converging in $L^2(\mathbf{R}^n)$ to u; it is not restrictive to assume that the liminf of $F_\epsilon(u_\epsilon)$ is finite, hence we can find $u_h = u_{\epsilon_h}$ such that

$$\lim_{h \to \infty} F_{\epsilon_h}(u_h) = \liminf_{\epsilon \downarrow 0} F_\epsilon(u_\epsilon) \in \mathbf{R}.$$

Since

$$\int_{\mathbf{R}^n} W(u_h) \, dx \leq 2\epsilon_h F_{\epsilon_h}(u_h)$$

the Fatou Lemma implies that the integral of $W(u)$ is zero. By our assumption on W, this means that $u \in \{0,1\}$ almost everywhere, hence we can represent u by χ_E for some Borel set $E \subset \mathbf{R}^n$. Using the Cauchy inequality we can estimate

$$F_{\epsilon_h}(u_h) \geq \int_{\mathbf{R}^n} \sqrt{W(u_h)}|\nabla u_h| \, dx = \int_{\mathbf{R}^n} |\nabla \phi(u_h)| \, dx = |D\phi(u_h)|(\mathbf{R}^n)$$

where $\phi(t) := \int_0^t \sqrt{W(s)} \, ds$. Since ϕ is a Lipschitz function, $\phi(u_\epsilon)$ converges in $L^2(\mathbf{R}^n)$ to $\phi(u)$; using the lower semicontinuity of $v \mapsto |Dv|(\mathbf{R}^n)$ we get

$$\lim_{h \to \infty} F_{\epsilon_h}(u_h) \geq \liminf_{h \to \infty} |D\phi(u_h)|(\mathbf{R}^n) \geq |D\phi(u)|(\mathbf{R}^n).$$

The liminf inequality is achieved noticing that $\phi(u) = \lambda u = \lambda \chi_E$.

To prove the limsup inequality we first assume that $u = \chi_E$ for some bounded open set E with smooth boundary; given any C^1 function γ such that $\gamma(t) = 0$ and $\gamma(-t) = 1$ for t sufficiently large we set $u_\epsilon = \gamma(\bar{d}/\epsilon)$, where $\bar{d}$ is the signed distance function from E and using the co-area formula (see for instance [EG92]) we estimate

$$\begin{aligned} F_\epsilon(u_\epsilon) &= \frac{1}{2\epsilon} \int_{\mathbf{R}^n} [|\gamma'|^2(\bar{d}/\epsilon) + W(\bar{d}/\epsilon)] \, dx \\ &= \frac{1}{2\epsilon} \int_{-\infty}^{\infty} g(s)[|\gamma'(s/\epsilon)|^2 + W(s/\epsilon)] \, ds \\ &= \frac{1}{2} \int_{-\infty}^{\infty} g(\epsilon t)[|\gamma'(t)|^2 + W(t)] \, dt \end{aligned}$$

where $g(t) = |D\chi_{\{\bar{d} \leq t\}}|(\mathbf{R}^n)$. By the smoothness of ∂E, $g(t)$ converges to $|D\chi_E|(\mathbf{R}^n)$ as $\epsilon \downarrow 0$; moreover, by our choice of γ, u_ϵ converges in $L^2(\mathbf{R}^n)$ to χ_E, therefore

$$\Gamma^- \limsup_{\epsilon \downarrow 0} F_\epsilon(u) \leq \limsup_{\epsilon \downarrow 0} F_\epsilon(u_\epsilon) \leq \frac{1}{2}|D\chi_E|(\mathbf{R}^n) \int_{-\infty}^{\infty} [|\gamma'(t)|^2 + W(t)] \, dt.$$

Taking the infimum over all admissible γ we get

$$\Gamma^- \limsup_{\epsilon \downarrow 0} F_\epsilon(\chi_E) \leq \lambda |D\chi_E|(\mathbf{R}^n).$$

In the general case we notice (see for instance [Giu84]) that any set $E \subset \mathbf{R}^n$ such that $\mathcal{L}^n(E) + |D\chi_E|(\mathbf{R}^n) < \infty$ can be approximated in area and perimeter by bounded open sets E_h with smooth boundary, hence the lower semicontinuity of Γ^- limits yields

$$\Gamma^- \limsup_{\epsilon \downarrow 0} F_\epsilon(\chi_E) \leq \liminf_{h \to \infty} \left[\Gamma^- \limsup_{\epsilon \downarrow 0} F_\epsilon(\chi_{E_h}) \right]$$

$$\leq \limsup_{h \to \infty} \lambda |D\chi_{E_h}|(\mathbf{R}^n)$$

$$= \lambda |D\chi_E|(\mathbf{R}^n).$$

$\square$

Coming back to the solutions of (111), we begin with the analysis of one dimensional, stationary solutions of the equation. The stationary, one dimensional problem corresponds to the ODE

$$v''(t) = \frac{1}{2\epsilon^2} W'(v(t)) = \frac{1}{\epsilon^2} v(t)(v^2(t) - 1) \qquad v(-\infty) = -1, \ v(\infty) = 1 \tag{113}$$

whose unique solution (up to translations in time) is the inverse function of

$$s \mapsto \int_0^s \frac{\epsilon}{\sqrt{W(\tau)}} \, d\tau.$$

A simple computation shows that

$$v(t) = q(t/\epsilon) \qquad \text{where} \qquad q(t) := \frac{\exp(\sqrt{2}t) - 1}{\exp(\sqrt{2}t) + 1}. \tag{114}$$

The function q is called *standing wave* of the equation; a variational property of q is stated in Exercise 14 below.

Exercise 14. Show that q is the unique solution up to translations in time of the variational problem

$$\min \left\{ \int_{-\infty}^{+\infty} |\gamma'(t)|^2 + W(\gamma(t)) \, dt \mid \gamma \in C^1(\mathbf{R}), \ \gamma(-\infty) = -1, \ \gamma(\infty) = 1 \right\}.$$

Let $E \subset \mathbf{R}^n$ be a set (not necessarily open, or bounded); assuming that neither $E = \emptyset$ nor $E = \mathbf{R}^n$, let us choose as initial condition in (111) the function

$$u_{0,\epsilon}(x) := q(\bar{d}(x, E)/\epsilon).$$

Then, the theory of semilinear parabolic equations, outlined in §15 (see in particular Corollary 2 and Remark 29), provides a unique bounded solution u_E^ϵ of (111) which can still be represented by $q(z^\epsilon/\epsilon)$ for suitable functions

$$z^\epsilon \in C^\infty\big((0, \infty) \times \mathbf{R}^n\big) \cap C^0\big([0, \infty) \times \mathbf{R}^n\big) \tag{115}$$

satisfying $z^\epsilon(0, \cdot) = \overline{d}(\cdot, E)$ and

$$z_t^\epsilon - \Delta z^\epsilon = \frac{\sqrt{2}}{\epsilon} q(z^\epsilon/\epsilon)(1 - |\nabla z^\epsilon|^2) \qquad \text{in } (0, \infty) \times \mathbf{R}^n \qquad (116)$$

$$\text{Lip}\,(z^\epsilon(t, \cdot), \mathbf{R}^n) \le 1 \qquad \forall t \ge 0. \qquad (117)$$

Setting $u_0 \equiv 1$ and $u_{\mathbf{R}^n} \equiv -1$, by the properties (16) of the signed distance function and by Theorem 32 we infer

$$u_{\mathbf{R}^n \setminus E}^\epsilon = -u_E^\epsilon, \qquad E \subset F \implies u_E^\epsilon \ge u_F^\epsilon \qquad (118)$$

for any set $E \subset \mathbf{R}^n$ and any $\epsilon > 0$. Since q is strictly increasing, similar inequalities are satisfied by z^ϵ.

Now we can state the first result about the convergence of the functions u^ϵ, proved by De Mottoni & Schatzman in [MS89a, MS95] using a rigorous asymptotic expansion technique. Independently, Chen obtained a similar result in [Che92], but using suitable subsolutions and supersolutions. Moreover, using the supersolution properties of distance functions from the flow, Evans, Soner & Souganidis proved in [ESS92] a more general statement for the level set flow. We will prove this first convergence result using Chen's original argument; only later we will see how viscosity theory can be used to get more general results.

Theorem 29 (short time behaviour of u^ϵ). *Let us assume E bounded, open, with smooth boundary, and let $T > 0$ be the maximal existence time of a smooth mean curvature flow $(\partial E_t)_{t \in [0,T)}$ starting from ∂E. Then*

$$u^\epsilon \to -1 \quad \text{in} \quad \bigcup_{t \in [0,T)} \{t\} \times E_t, \qquad u^\epsilon \to 1 \quad \text{in} \quad \bigcup_{t \in [0,T)} \{t\} \times (\mathbf{R}^n \setminus \overline{E}_t)$$

with local uniform convergence.

Proof. Let $\overline{d}(t, x)$ be the signed distance function from E_t; since, by (32), $\overline{d}_t = \Delta \overline{d}$ on $\{\overline{d} = 0\}$, for any $T' \in [0, T)$ we can find $r, D > 0$ depending only on T' such that

$$|\overline{d}_t(t, x) - \Delta \overline{d}(t, x)| \le D|\overline{d}(t, x)| \qquad (119)$$

for any $(t, x) \in [0, T'] \times \mathbf{R}^n$ such that $|\overline{d}(t, x)| \le 6r$ (D can be estimated using the equation (34)). We will prove local uniform convergence of u^ϵ for $t \in [0, T']$.

Chen's strategy for building subsolutions of (111) is to look for functions of the form

$$\overline{u}^\epsilon(t, x) := q_\lambda\left(\frac{\tilde{z}(t, x)}{\epsilon}\right)$$

where $\tilde{z}$ is a suitable regularization of $\overline{d}(x, E_t)$ and q_λ is the unique solution of the problem

$$q'' - C_\lambda q' + q - q^3 - \lambda = 0, \qquad q(-\infty) = z_\lambda^-, \; q(0) = z_\lambda^0, \; q(\infty) = z_\lambda^+ \qquad (120)$$

for $\lambda > 0$ sufficiently small, where $z_\lambda^- < z_\lambda^0 < z_\lambda^+$ are the zeros of $(t - t^3 - \lambda)$. Existence of solutions of (120) for a unique constant C_λ has been proved in [AW75]; moreover, by Lemma 2 in the appendix of [FH88], there exist strictly positive constants λ_*, α, A such that

$$z_\lambda^+ + \alpha\lambda \leq 1, \qquad\qquad z_\lambda^- + \alpha\lambda \leq -1 \qquad\qquad (121)$$

$$|C_\lambda| + \left\|\frac{q_\lambda''}{q_\lambda'}\right\|_\infty \leq A \qquad\qquad (122)$$

$$0 < q_\lambda'(t) < Ae^{-\alpha|t|} \qquad\qquad \forall t \in \mathbf{R} \qquad\qquad (123)$$

$$z_\lambda^- < q_\lambda(t) < z_\lambda^- + Ae^{\alpha t} \;\; \forall t \leq 0, \qquad z_\lambda^+ - Ae^{-\alpha t} < q_\lambda(t) < z_\lambda^+ \;\; \forall t \geq 0 \;\; (124)$$

for any $\lambda \in [0, \lambda_*]$. In the following the constants r, D (depending on the flow) and λ_*, α, A (depending on the equation) will be fixed; we also fix $M > 4D$ and a smooth function $h : \mathbf{R} \to \mathbf{R}$ satisfying

$$h(t) = t \;\; \text{in } (-1, 1), \qquad h(t) = 2 \;\; \text{in } (3, \infty), \qquad h(t) = -2 \;\; \text{in } (-\infty, 3) \tag{125}$$

and such that $0 \leq h' \leq 1$, $0 \leq h'' \leq 1$.

(i) Let $\varrho \in (0, r)$ such that $\varrho e^{MT'} < r$; we will check in this step that the function

$$\tilde{z}(t, x) := 2rh\left(\frac{\bar{d}(t, x)}{2r}\right) - \varrho e^{Mt}$$

satisfies

$$\tilde{z}_t - \Delta\tilde{z} - \frac{C_\lambda}{\epsilon} + \frac{A}{\epsilon}(1 - |\nabla\tilde{z}|^2) \leq \frac{5A}{\epsilon}\chi_{\{|\tilde{z}|\geq\varrho\}} \qquad \text{in } [0, T'] \times \mathbf{R}^n \qquad (126)$$

for $\epsilon > 0$ satisfying

$$\max\left\{\frac{1}{2r}, \frac{Mr}{2}, 6rD\right\} < \frac{A}{\epsilon} \qquad\qquad (127)$$

and $\lambda \in [0, \lambda_*]$ such that $|C_\lambda| \leq M\epsilon\varrho/2$. Notice that, by our choice of ϱ, $|\tilde{z}(t, x)| \geq r$ whenever $|\bar{d}(t, x)| \geq 2r$ and $(t, x) \in [0, T'] \times \mathbf{R}^n$. Using (119) we estimate

$$\tilde{z}_t - \Delta\tilde{z} - \frac{C_\lambda}{\epsilon} + \frac{A}{\epsilon}(1 - |\nabla\tilde{z}|^2)$$

$$= h'(\bar{d}_t - \Delta\bar{d}) - M\varrho e^{Mt} - \frac{C_\lambda}{\epsilon} + \frac{1}{2r}h''|\nabla\bar{d}|^2 + \frac{A}{\epsilon}(1 - [h'\nabla\bar{d}]^2)$$

$$\leq (D|\bar{d}| - M\varrho e^{Mt} + \frac{M\varrho}{2}e^{Mt})\chi_{\{|\bar{d}|\leq 2r\}} + h'D|\bar{d}|\chi_{\{2r\leq|\bar{d}|\leq 6r\}}$$

$$+ \left(\frac{M\varrho}{2} + \frac{1}{2r} + \frac{A}{\epsilon}\right)\chi_{\{|\bar{d}|\geq 2r\}}$$

$$\leq (D|\tilde{z}| - (\frac{M}{2} - D)e^{Mt}\varrho)\chi_{\{|\bar{d}|\leq 2r\}} + \frac{4A}{\epsilon}\chi_{\{|\bar{d}|\geq 2r\}}$$

$$\leq 3rD\chi_{\{|\tilde{z}|\geq(M/(2D)-1)\varrho\}} + \frac{4A}{\epsilon}\chi_{\{|\tilde{z}|\geq r\}}$$

where h and all its derivatives are evaluated at $\bar{d}/2r$. Since $M > 4D$ and $\varrho < r$, (126) follows.

(ii) Now we will build a subsolution $\tilde{u}^\epsilon$ of (111) of the form $q_\lambda(\tilde{z}/\epsilon)$, where $\tilde{z}$ has been previously defined and λ is suitably chosen, as a function of ϵ. Using (120) and (122) we estimate

$$\tilde{u}_t^\epsilon - \Delta\tilde{u}^\epsilon - \frac{1}{\epsilon^2}\tilde{u}^\epsilon(1 - (\tilde{u}^\epsilon)^2) = \frac{1}{\epsilon}q_\lambda'\left(\tilde{z}_t - \Delta\tilde{z} - \frac{C_\lambda}{\epsilon}\right) + \frac{1}{\epsilon^2}q_\lambda''(1 - |\nabla\tilde{z}|^2) - \frac{\lambda}{\epsilon^2}$$

$$\leq \frac{1}{\epsilon}q_\lambda'\left(\tilde{z}_t - \Delta\tilde{z} - \frac{C_\lambda}{\epsilon} + \frac{A}{\epsilon}(1 - |\nabla\tilde{z}|^2)\right) - \frac{\lambda}{\epsilon^2}$$

where q and all its derivatives are evaluated at $\tilde{z}/\epsilon$. Assuming that $\tilde{z}$ satisfies the differential inequality (126) in $[0, T'] \times \mathbf{R}^n$ (by (127), this is true for ϵ small enough and $|C_\lambda| \leq M\epsilon\varrho/2$) we infer

$$\tilde{u}_t^\epsilon - \Delta\tilde{u}^\epsilon - \frac{1}{\epsilon^2}\tilde{u}^\epsilon(1 - (\tilde{u}^\epsilon)^2) \leq \frac{1}{\epsilon^2}\left[5Aq_\lambda'(\tilde{z}/\epsilon)\chi_{\{|\tilde{z}|\geq\varrho\}} - \lambda\right] \leq 0.$$

The last inequality is satisfied by choosing any

$$\lambda \geq 5A^2e^{-\alpha\varrho/\epsilon} \tag{128}$$

because, by (123), $q_\lambda'(\tilde{z}/\epsilon) \leq Ae^{-\alpha\varrho/\epsilon}$ as soon as $|\tilde{z}| \geq \varrho$. We will choose $\lambda = \epsilon^2$; this choice of λ ensures, by (122), the validity of the additional condition $|C_\lambda| \leq M\epsilon\varrho/2$ for ϵ small enough.

(iii) We have proved in the previous two steps that for any $\varrho \in (0, r)$ the function $\tilde{u}^\epsilon = q_\lambda(\tilde{z}/\epsilon)$ is a subsolution of (111) provided $\lambda = \epsilon^2$ and ϵ is small enough. Now we claim that for ϵ small enough $\tilde{u}^\epsilon(0, \cdot) \leq u^\epsilon(0, \cdot)$. To check this inequality we will consider two cases: if $\tilde{z}(0, x) \leq -\varrho/2$, then (124) and (121) give

$$\tilde{u}^\epsilon(0, x) \leq z_\lambda^- + Ae^{-\alpha\varrho/(2\epsilon)} \leq z_\lambda^- + \alpha\lambda \leq -1 \leq u^\epsilon(0, x)$$

for ϵ small enough. On the other hand, if $\tilde{z}(0, x) \geq -\varrho/2$, then $2rh(\bar{d}(0, x)/2r) \geq \varrho/2$, hence $\bar{d}(0, x) \geq \varrho/2$ and (124) and (121) give

$$u^\epsilon(0, x) \geq 1 - Ae^{-\alpha\varrho/(2\epsilon)} \geq 1 - \alpha\lambda \geq z_\lambda^+ \geq \tilde{u}^\epsilon(0, x)$$

for ϵ small enough. By a parabolic comparison theorem (cf. Remark 28(1)) we obtain that $\tilde{u}^\epsilon \leq u^\epsilon$ in $[0, T'] \times \mathbf{R}^n$ for ϵ small enough (depending on ϱ, T').

(iv) In this step we show that u^ϵ converges to $+1$ at any $(t, x) \in [0, T'] \times \mathbf{R}^n$ such that $\bar{d}(t, x) > 0$. We need only to choose $\varrho \in (0, r)$ small enough to have

$$\tilde{z}(t, x) = 2rh\left(\frac{\bar{d}(t, x)}{2r}\right) - \varrho e^{Mt} > 0.$$

Then,

$$\liminf_{\epsilon\downarrow 0} u^\epsilon(t, x) \geq \liminf_{\epsilon\downarrow 0} q_{\epsilon^2}\left(\frac{\tilde{z}(t, x)}{\epsilon}\right) \geq \liminf_{\epsilon\downarrow 0} z_\epsilon^+ - Ae^{-\alpha\tilde{z}(t,x)/\epsilon} = 1.$$

The proof of convergence to -1 is similar: in this case one builds supersolutions of (111) by an analogous method, with $\lambda \in [-\lambda_*, 0]$. $\square$

Remark 25. Choosing $\lambda = \epsilon^k$ and $\varrho = m\epsilon|\ln \epsilon|$ in the above proof, with $am > 2k$ and $k \geq 2$, one obtains polynomial convergence of u^ϵ to ± 1, and the rate of convergence as $\epsilon \to 0$ of the Hausdorff distance between the fronts $\{u^\epsilon(t, \cdot) = 0\}$, ∂E_t can be estimated by $O(\epsilon|\ln \epsilon|)$ (see [Che92] for details). This estimate has been improved to $O(\epsilon^2|\ln \epsilon|)$ in [BP95a].

Using the theory of viscosity solutions we can improve Theorem 29 in two directions: first, we can say something about the behaviour of u^ϵ after the onset of singularities, removing at the same time the regularity assumption on ∂E (see [ESS92, BSS93]); then, representing u^ϵ by $q(z^\epsilon/\epsilon)$, we can prove that under the assumptions of Theorem 29 the functions z^ϵ are locally uniformly converging to the signed distance function from E_t. In particular, this gives *exponential* convergence of u^ϵ to ± 1 as $\epsilon \to 0$.

We will prove the convergence of z^ϵ first in the radial case and at time 0, and later in full generality.

Proposition 6 (convergence in the radial case). *Let z^ϵ be satisfying (116) with initial condition $z^\epsilon(0, x) = |x - x_0| - R$. Then,*

$$\lim_{\epsilon, t \downarrow 0} z^\epsilon(t, x) = |x - x_0| - R \qquad \forall x \in \mathbf{R}^n.$$

Proof. It is not restrictive to assume $x_0 = 0$. Inequality

$$\liminf_{\epsilon, t \downarrow 0} z^\epsilon(t, x) \geq |x| - R.$$

simply follows by the fact that $z_\xi(t, x) := \langle x, \xi \rangle - R$ are solutions of (116) for any unit vector ξ. Setting

$$z^*(x) = \limsup_{\epsilon, t \downarrow 0} z^\epsilon(t, x), \qquad u^*(x) := \limsup_{\epsilon, t \downarrow 0} u^\epsilon(t, x)$$

we will prove that

$$z^*(x) \leq |x| - R \qquad \forall x \in \mathbf{R}^n. \tag{129}$$

From (117) and Theorem 29 we get

$$z^*(x) \leq z^*(y) + |x - y| \qquad \forall x, y \in \mathbf{R}^n \tag{130}$$

$$x \in B_R \quad \Longrightarrow \quad u^*(x) = -1. \tag{131}$$

Notice that (131) and (130) imply that $z^* \leq 0$ in $\overline{B}_R$; moreover, since $z^* \geq z_* \geq 0$ on ∂B_R, z^* vanishes on ∂B_R. In particular

$$z^*(x) \leq z^*\left(\frac{Rx}{|x|}\right) + |x| - R = |x| - R \qquad \forall x \in \mathbf{R}^n \setminus B_R$$

so that (129) is satisfied in $\mathbf{R}^n \setminus B_R$.

Equation (116) and the theory of viscosity solutions show that $w := -z^*$ satisfies

$$|\nabla w|^2 - 1 = 0 \quad \text{in } \{u^* < 0\}.$$

Indeed, inequality $\leq$ follows by (130) and Lemma 1, while inequality $\geq$ will be proved in step (iii) of the proof of Theorem 30. Since, by (131), B_R is a subset of $\{u^* < 0\}$, the conditions

$$|\nabla w|^2 - 1 = 0 \quad \text{in } B_R, \qquad\qquad w = 0 \quad \text{in } \partial B_R$$

and the characterization of distance functions through the eikonal equation (cf. Theorem 12) give $w(x) = R - |x|$ in B_R, i.e., $u^*(x) = |x| - R$ in B_R. $\qquad\square$

Now we want to relate the behaviour of u^ϵ with the level set flow, thus getting convergence of the reaction diffusion equation for "generic" initial data (see Remark 26). Following essentially [ESS92, BSS93], we will state two results: the first one (cf. Theorem 30) is only concerned with the behaviour of u^ϵ on $\{u \neq 0\}$ and it does not require any regularity assumption on the flow. Under a mild regularity assumption on the flow, the second one (cf. Corollary 1) provides the convergence of z^ϵ.

Let $E \subset \mathbf{R}^n$, let $\Gamma_0 = \partial E$ and let us assume for simplicity Γ_0 compact and not empty. Choosing as initial condition in (90) any uniformly continuous function whose zero level set is Γ_0, we will denote by $u : [0,\infty) \times \mathbf{R}^n \to \mathbf{R}$ the corresponding level set solution and by $T_* \in [0,\infty)$ the maximal existence time of the flow. We also set $B = (0,\infty) \times \mathbf{R}^n$.

Theorem 30 (convergence of u^ϵ). *Let $(u^\epsilon)_{\epsilon>0}$ be the solutions of (111) with the initial condition*

$$u_{0,\epsilon}(x) := q\big(\overline{d}(x, E)/\epsilon\big).$$

Then,

$$\begin{cases} u^\epsilon \to -1 \;\; in \; \big\{(t,x) \in [0,T_*] \times \mathbf{R}^n \,|\, u(t,x) < 0\big\}, \\[2mm] u^\epsilon \to \;\; 1 \;\; in \; \big\{(t,x) \in [0,T_*] \times \mathbf{R}^n \,|\, u(t,x) > 0\big\} \end{cases}$$

with local uniform convergence.

Proof. Let us represent u^ϵ by $q(z^\epsilon/\epsilon)$, where the functions z^ϵ satisfy (115), (116), (117). The proof is heavily based on viscosity theory, on the characterization of distance functions through the eikonal equation (cf. Theorem 12) and on Theorem 27. We will also use Proposition 6, in order to work out some comparison arguments.

(i) Recall that $|\nabla z^\epsilon| \leq 1$ in $(0,\infty) \times \mathbf{R}^n$. In particular, (116) gives

$$z_t^\epsilon - \Delta z^\epsilon \geq 0 \quad \text{in } B \cap \{z^\epsilon > 0\}, \qquad\qquad z_t^\epsilon - \Delta z^\epsilon \leq 0 \quad \text{in } B \cap \{z^\epsilon < 0\}. \quad (132)$$

(ii) Following Barles & Perthame [BP87], for any $(t,x) \in [0,\infty) \times \mathbf{R}^n$ we define

$$z_*(t,x) := \liminf_{s \to t,\, \epsilon \to 0+} z^\epsilon(s,x), \qquad\qquad z^*(t,x) = \limsup_{s \to t,\, \epsilon \to 0+} z^\epsilon(s,x).$$

By the gradient bound on z^ϵ, it follows that

$$
\begin{cases}
z_*(t,x) = \displaystyle\liminf_{(s,y)\to(t,x),\,\epsilon\downarrow 0} z^\epsilon(s,y) = \Gamma^- \liminf_{\epsilon\downarrow 0} z^\epsilon(t,x) \\[2mm]
z^*(t,x) = \displaystyle\limsup_{(s,y)\to(t,x),\,\epsilon\downarrow 0} z^\epsilon(s,y) = \Gamma^+ \limsup_{\epsilon\downarrow 0} z^\epsilon(t,x).
\end{cases}
\tag{133}
$$

Passing to the limit as $\epsilon \downarrow 0$ in (132) and using Theorem 14 we get

$$
\begin{aligned}
(z_* \wedge k)_t - \Delta(z_* \wedge k) \geq 0 \quad \text{in } B \cap \{z_* > 0\}, \\
(z^* \vee m)_t - \Delta(z^* \vee m) \leq 0 \quad \text{in } B \cap \{z^* < 0\}
\end{aligned}
\tag{134}
$$

for any $k > 0$, $m < 0$, in the viscosity sense. Again by the gradient bound, z_* and z^* satisfy

$$
z(t,x) \leq z(t,y) + |x-y| \qquad \forall t \in [0,\infty),\ x,y \in \mathbf{R}^n.
\tag{135}
$$

(iii) We define also $z_1 := z_* \vee 0$ and $z_2 := z^* \wedge 0$; we will see that z_1 and z_2 satisfy the assumptions of Theorem 27. To this end, we first notice that (134) gives

$$
\begin{aligned}
(z_1 \wedge k)_t - \Delta(z_1 \wedge k) \geq 0 \quad \text{in } B \cap \{z_1 > 0\}, \\
(z_2 \vee m) - \Delta(z_2 \vee m) \leq 0 \quad \text{in } B \cap \{z_2 < 0\}
\end{aligned}
\tag{136}
$$

for any $k > 0$, $m < 0$. Moreover, since z_1 and z_2 satisfy (135), it holds

$$
|\nabla z_1(t,\cdot)|^2 - 1 \leq 0, \qquad\qquad 1 - |\nabla z_2(t,\cdot)|^2 \geq 0
\tag{137}
$$

in the viscosity sense, for any $t \in [0,\infty)$ such that $z_1(t,\cdot) < \infty$ (respectively, $z_2(t,\cdot) > -\infty$). We have to prove that $z_1(t,\cdot)$ and $-z_2(t,\cdot)$ are actually distance functions from (possibly empty) closed sets Λ_t, Λ'_t for $t \in [0,\infty)$. We will prove this only for z_1, the proof for z_2 being analogous. By Theorem 12, we need only to show that z_1 satisfies $|\nabla z_1|^2 - 1 = 0$ in $\{z_1(t_0,\cdot) > 0\}$ in the viscosity sense for any $t_0 \in [0,\infty)$ such that $z_1(t_0,\cdot) < \infty$. By (137), only the supersolution property has to be proved.

Let

$$
u_*(t,x) := \liminf_{s\to t,\,y\to x} u^\epsilon(s,y).
$$

We will first prove that $|\nabla\phi(t_0,x_0)|^2 - 1 \geq 0$ whenever $u_*(t_0,x_0) > 0$, $\phi(t,x)$ is smooth and $z_1 - \phi$ has a relative minimum at (t_0,x_0). Assuming, as usual, the relative minimum to be strict, by Remark 12 we can find (t_ϵ,x_ϵ) converging to (t_0,x_0) such that $z^\epsilon - \phi$ has a local minimum at (t_ϵ,x_ϵ). If $t_\epsilon > 0$ for sufficiently small ϵ, from (116) we infer

$$
\frac{\sqrt{2}}{\epsilon} u^\epsilon(1 - |\nabla z^\epsilon|^2) = z^\epsilon_t - \Delta z^\epsilon \leq \phi_t - \Delta\phi \qquad \text{at } (t_\epsilon,x_\epsilon)
$$

hence

$$
\sqrt{2}u^\epsilon(1 - |\nabla\phi|^2) \leq C\epsilon \qquad \text{at } (t_\epsilon,x_\epsilon).
\tag{138}
$$

Since $u_*(t_0, x_0) > 0$ we can pass to the limit as $\epsilon \downarrow 0$ in (138) to get

$$|\nabla\phi(t_0, x_0)|^2 - 1 \geq 0. \tag{139}$$

On the other hand, if $t_0 = 0$ and $t_{\epsilon_i} = 0$ for some infinitesimal sequence (ϵ_i), we can use the fact that $z^{\epsilon_i}(0, \cdot)$ are distance functions to recover (139) again.

Now we will use a blow-up argument to recover the supersolution property of $z_1(t_0, \cdot)$. Let $w_h(s, x) = z_1(t_0 + h(s - t_0), x)$ with $h > 0$; by a simple scaling (in time) argument it can be proved that $|\nabla\phi(t_0, x_0)|^2 - 1 \geq 0$ whenever $u_*(t_0, x_0) > 0$, ϕ is smooth and $w_h - \phi$ has a relative minimum at (t_0, x_0). Since the functions w_h are Γ^- converging as $h \downarrow 0$ to $w(s, x) := z_1(t_0, x)$, by using Remark 12 again we obtain that $|\nabla\phi(t_0, x_0)|^2 - 1 \geq 0$ whenever $u_*(t_0, x_0) > 0$, ϕ is smooth and $w - \phi$ has a strict relative minimum at (t_0, x_0). By approximation, the same property holds if (t_0, x_0) is only a local minimum; since w is independent of s we can take test functions ϕ independent of s to obtain that $z_1(t_0, \cdot)$ is a supersolution of $|\nabla z_1|^2 - 1 = 0$ in $\{z_1(t_0, \cdot) > 0\} \subset \{u_*(t_0, \cdot) > 0\}$.

(iv) We know from the previous step that $z_1(t, x) = \mathrm{dist}(x, \Lambda_t)$ and $z_2(t, x) = -\mathrm{dist}(x, \Lambda'_t)$ for suitable closed sets Λ_t, Λ'_t for any $t \in [0, \infty)$. Now we will prove that

$$\Lambda_0 \subset E \cup \partial E, \qquad \Lambda'_0 \subset (\mathbf{R}^n \setminus E) \cup \partial E. \tag{140}$$

We will only prove the first inclusion in (140); the second one can be easily deduced from the first and from (118). Let $x_0 \notin \overline{E}$ and let us prove that $x_0 \notin \Lambda_0$, i.e., $z_*(0, x_0) > 0$. Let $\tilde{u}^\epsilon = q(\tilde{z}^\epsilon/\epsilon)$ be the solutions of (111) with initial condition

$$\tilde{u}^\epsilon(0, x) = q(\overline{d}(x, B_r(x_0))/\epsilon) \qquad 0 < r < d(x_0, E).$$

Since $B_r(x_0) \subset \mathbf{R}^n \setminus E$, from (118) and the monotonicity of q we infer $\tilde{z}^\epsilon \geq -z^\epsilon$. In particular,

$$z_*(0, x_0) \geq (-\tilde{z})_*(0, x_0) = -\tilde{z}^*(0, x_0).$$

Since, by Proposition 6, $\tilde{z}_*(0, x_0) = \tilde{z}^*(0, x_0) = -r < 0$, we obtain that $z_*(0, x_0) > 0$.

(v) We have checked in the previous steps that z_1 and z_2 satisfy all the assumptions of Theorem 27, with the exception of the left continuity hypothesis. We will give a detailed proof for z_1, the one for z_2 being analogous.

Assume that $z_1(t_0, x_0) = 0$ for some $(t_0, x_0) \in B$, and assume by contradiction that there exist a sequence $t_i \uparrow t_0$ and $r > 0$ such that $z_1(t_i, x_0) = z_*(t_i, x_0) \geq 3r$ for any $i \in \mathbf{N}$. Let $\tilde{z}^\epsilon$ be defined as in step (iv), and let $\epsilon_j \downarrow 0$, (s_j) converging to t_0 be such that

$$\lim_{j \to \infty} z^{\epsilon_j}(s_j, x_0) \leq 0.$$

Since, by Proposition 6

$$\lim_{\epsilon, s \downarrow 0} \tilde{z}^\epsilon(s, x_0) = \overline{d}(x_0, B_r(x_0)) = -r$$

we can find $\tau > 0$ such that

$$-\tilde{z}^\epsilon(s, x_0) \geq r/2 \qquad \forall s, \epsilon \in (0, \tau). \tag{141}$$

We first choose i such that $t_0 - t_i < \tau$; then, we choose j large enough, such that

$$\epsilon_j \in (0, \tau), \quad t_i < s_j < t_i + \tau, \quad z^{\epsilon_j}(t_i, x_0) \geq 2r, \quad z^{\epsilon_j}(s_j, x_0) < r/2. \qquad (142)$$

By (117), $z^{\epsilon_j}(t_i, \cdot) \geq r$ in $B_r(x_0)$, and this easily leads to the inequality

$$z^{\epsilon_j}(t_i, x) \geq -\overline{d}(x, B_r(x_0)) \qquad \qquad \forall x \in \mathbf{R}^n.$$

By Theorem 32 (taking the left composition with q and exploiting (111)) we infer

$$z^{\epsilon_j}(t, x_0) \geq -\tilde{z}^{\epsilon_j}(t - t_i, x_0) \qquad \qquad \forall t \geq t_i.$$

By (142) we can take $t = s_j$ in the inequality above and use (141) with $\epsilon = \epsilon_j$ and $s = s_j - t_i$ to get $z^{\epsilon_j}(s_j, x_0) \geq r/2$, thus contradicting the last inequality in (142).

(vi) From (140), (136) and Theorem 27 we infer

$$\Lambda_t \subset \{u(t, \cdot) \leq 0\}, \qquad \Lambda_t' \subset \{u(t, \cdot) \geq 0\} \qquad \forall t \in [0, T_*].$$

Equivalently

$$z_*(t, x) \vee 0 \geq \text{dist}(x, \{u(t, \cdot) \leq 0\}), \qquad z^*(t, x) \wedge 0 \leq -\text{dist}(x, \{u(t, \cdot) \geq 0\}) \qquad (143)$$

for $(t, x) \in [0, T_*] \times \mathbf{R}^n$.

(vii) Now we can prove the stated convergence of u^ϵ. Assume that $u(t_0, x_0) > 0$; then $x_0 \notin \Lambda_{t_0}$ and $z_*(t_0, x_0) = z_1(t_0, x_0) > 0$. By the definition of z_*, this implies that

$$\min_K z^\epsilon \geq \frac{1}{2} z_*(t_0, x_0) > 0$$

in a compact neighbourhood K of (t_0, x_0), for ϵ sufficiently small. Hence, $u^\epsilon = q(z^\epsilon/\epsilon)$ tends to $q(\infty) = 1$. Analogously, assume that $u(t_0, x_0) < 0$; then $x_0 \notin \Lambda_{t_0}'$ and $z^*(t_0, x_0) = z_2(t_0, x_0) < 0$. By the definition of z^*, this implies that

$$\max_K z^\epsilon \leq \frac{1}{2} z^*(t_0, x_0) < 0$$

in a compact neighbourhood K of (t_0, x_0), for ϵ sufficiently small. Hence, $u^\epsilon = q(z^\epsilon/\epsilon)$ tends to $q(-\infty) = -1$. $\qquad \square$

Remark 26. (1) Let $u_0 : \mathbf{R}^n \to \mathbf{R}$ be a uniformly continuous function such that $u(x) \to \infty$ as $|x| \to \infty$. By the translation invariance of the problem, Theorem 30 says that the solutions of (111) with initial condition

$$u_{0,\epsilon}(x) = q(\overline{d}(x, \{u_0 \leq \gamma\})/\epsilon)$$

converge locally uniformly to 1 in $\{u > \gamma\}$ and to -1 in $\{u < \gamma\}$. This assumptions on $u_{0,\epsilon}$ has been greatly weakened in [Son93a].

(2) Since the set of real numbers γ such that

$$\bigcup_{t \geq 0} \{t\} \times \{u(t, \cdot) = \gamma\} \subset [0, \infty) \times \mathbf{R}^n$$

is not $\mathcal{L}^{n+1}$ negligible is at most countable, (1) shows that for generic initial conditions the behaviour of u^ϵ is completely determined, at least in a L^p sense (see also [Ilm93a, Son93a] for convergence results out of a "sharp" interface).

Under a stronger regularity assumption on the level set flow we can also say more about the behaviour of z^ϵ.

Definition 15 (regular flows). Let $(\Gamma_t)_{t \geq 0}$ be a level set flow and let u be any solution of (102) whose zero level set is $\bigcup_{t \geq 0} \{t\} \times \Gamma_t$. We say that the flow is regular in $J \subset \mathbf{R}$ if

$$\bar{d}(t, \{u(t, \cdot) \leq 0\}) = -\bar{d}(t, \{u(t, \cdot) \geq 0\}) \qquad \forall t \in J.$$

By Remark 22 the definition above is well posed and does not depend on the choice of u. Notice that the flow of spheres fails to be regular at the extinction time. Exercise 15 below shows that regularity is actually a topological condition on the sets $\{u(t, \cdot) > 0\}$ and $\{u(t, \cdot) < 0\}$ for $t \in J$, certainly not true when the fattening phenomenon occurs or at the extinction time.

Exercise 15. Let $w : \mathbf{R}^n \to \mathbf{R}$ be a continuous function. Show that the following three conditions are equivalent:
(a) $\bar{d}(x, \{w \leq 0\}) = -\bar{d}(x, \{w \geq 0\})$ for any $x \in \mathbf{R}^n$;
(b) $\overline{\{w < 0\}} = \{w \leq 0\}$ and $\overline{\{w > 0\}} = \{w \geq 0\}$;
(c) $\partial\{w > 0\} = \partial\{w < 0\} = \{w = 0\}$.

We can now prove the following refinement of Theorem 30 and Theorem 29:

Corollary 1 (convergence of z^ϵ). *Under the same assumptions of Theorem 30, let us assume that the level set flow is regular in some interval $J \subset [0, T_*)$. Then, representing u^ϵ by $q(z^\epsilon/\epsilon)$, the functions z^ϵ converge locally uniformly in $J \times \mathbf{R}^n$ to $\bar{d}(t, \{u \leq 0\})$. In particular, if E and $T \leq T_*$ are as in the statement of Theorem 29, we can take $J = [0, T)$.*

Proof. We claim that

$$z_*(t, x) \geq \bar{d}(x, \{u(t, \cdot) \leq 0\}), \qquad z^*(t, x) \leq -\bar{d}(x, \{u(t, \cdot) \geq 0\})$$

for any $t \in J$. Let us prove the first inequality (the proof of the second one is analogous). By (143), the inequality is clearly true at (t, x) if $u(t, x) > 0$; by a density argument (here we use regularity) it is still true if $u(t, x) = 0$. If $u(t, x) < 0$ we can choose a least distance point y of x from $\{u(t, \cdot) > 0\}$; since $u(t, y) = 0$ it holds $z_*(t, y) \geq 0$, so that

$$z_*(t, x) \geq z_*(t, y) - |x - y| = z_*(t, y) - \text{dist}(x, \{u(t, \cdot) > 0\})$$
$$\geq -\text{dist}(x, \{u(t, \cdot) > 0\}) = \bar{d}(x, \{u(t, \cdot) \leq 0\}).$$

By the regularity assumption we get

$$z^*(t, x) \leq -\bar{d}(x, \{u(t, \cdot) \geq 0\}) = \bar{d}(x, \{u(t, \cdot) \leq 0\}) \leq z_*(t, x).$$

Therefore $z_* = z^*$ in $J \times \mathbf{R}^n$ and the local uniform convergence follows by (133) and Exercise 5. $\square$

A simple example shows that the regularity condition, unlike fattening, is not generic, not even before the extinction time: it is enough to consider a uniformly continuous function $u : \mathbf{R}^2 \to \mathbf{R}$ having

$$\{|x + 1|^2 + y^2 < r^2\} \cup \{|x - 1|^2 + y^2 < 4r^2\} \qquad r \in [0, 1/2)$$

as sublevel sets, corresponding to suitable real numbers τ_r. Then, for any $r \in [0, 1/2)$ it happens that at time $T = r^2/2 < 2r^2$ the point $(-1, 0)$ belongs to $\{u(T, \cdot) \leq \tau_r\}$ but not to $\overline{\{u(T, \cdot) < \tau_r\}}$.

15   Semilinear evolution equations

In this section we will recall several results concerning semilinear parabolic equations modeled on (111). While we will give a short and self contained proof of existence and uniqueness of solutions, we will state for simplicity without proof more advanced results, as the regularity of weak solutions and the strong maximum principle.

We will be concerned with the equation

$$u_t = \Delta u - F'(u) \qquad \text{in } (0, \infty) \times \mathbf{R}^n \tag{144}$$

in the distribution sense. In (144), $F : \mathbf{R} \to [0, \infty)$ is a smooth function satisfying

$$F'' \in L^\infty(\mathbf{R}^n), \qquad F(t) \geq \alpha t^2 + \beta \quad \forall t \in \mathbf{R} \tag{145}$$

for some constants $\alpha > 0$, $\beta \in \mathbf{R}$. Following [Ilm94], we will first consider periodic initial conditions. To this aim, we will denote by $L^2_\tau(\mathbf{R}^n)$ the class of τ-periodic (in all variables) functions belonging to $L^2((0, \tau)^n)$ and by $H^1_\tau(\mathbf{R}^n)$ the Sobolev space of all functions $u \in L^2_\tau(\mathbf{R}^n)$ whose distributional derivative belongs to $L^2_\tau(\mathbf{R}^n)$.

Given a separable Hilbert space H and $T \in (0, \infty]$, we will denote by $H^1((0, T); H)$ the vector space of all Borel functions $u \in L^2((0, T); H)$ such that

$$\int_a^b v(t)\, dt = u(b) - u(a) \qquad \forall a, b \in (0, T)$$

for some function $v \in L^2((0, T); H)$. The function v is uniquely determined and denoted, as usual, by u'. Notice that (with the natural choice of norms)

$$H^1((0, T); H) \hookrightarrow C^{0, 1/2}([0, T]; H)$$

and that $L^2((0, T); L^2(\Omega)) \cong L^2((0, T) \times \Omega)$. Additional properties of these spaces are listed in Exercise 16 below.

Exercise 16. Let $\Omega \subset \mathbf{R}^n$ be open, $u \in H^1\big((0,T); L^2(\Omega)\big)$. Show that

(i) the function $u_t(t,x) := u'(t)(x)$ is the time distributional derivative of u, viewed as a function in $L^2\big((0,T) \times \Omega\big)$;

(ii) for any bounded Borel function $\phi : \Omega \to \mathbf{R}$ the function

$$t \mapsto \int_\Omega u^2(t,x)\phi(x)\,dx$$

is absolutely continuous in $(0,T)$ and its derivative equals

$$2\int_\Omega u(t,x)u_t(t,x)\phi(x)\,dx$$

for almost every $t \in (0,T)$;

(iii) for almost every $t \in (0,T)$ it holds

$$\lim_{\sigma \to 0}\left\|\frac{u(t+\sigma) - u(t)}{\sigma} - u'(t)\right\| = 0.$$

Hint: (i), (ii) follow by a smoothing argument; (iii) is satisfied at any Lebesgue point of u'.

Following a time discretization and minimization approach (the so–called *implicit Euler discretization*), based on the fact that (144) is the gradient flow of

$$\mathcal{E}(u) := \int_{(0,\tau)^n} \frac{1}{2}|\nabla u|^2 + F(u)\,dx \qquad u \in H^1_\tau(\mathbf{R}^n).$$

we are now able to state our first existence result.

Theorem 31 (existence and uniqueness). *For any $u_0 \in H^1_\tau(\mathbf{R}^n)$ there exists a unique*

$$u \in L^\infty\big([0,\infty); H^1_\tau(\mathbf{R}^n)\big) \cap H^1_{\mathrm{loc}}\big([0,\infty); L^2_\tau(\mathbf{R}^n)\big)$$

satisfying (144) and such that $u(0) = u_0$. Moreover, $u(t)$ satisfies the energy identity

$$\int_s^t |u'(r)|^2\,dr = \mathcal{E}(u(s)) - \mathcal{E}(u(t)) \qquad \forall s,t \in (0,\infty),\ s \le t. \qquad (146)$$

Proof. (Existence) We will denote by $Y = (0,\tau)^n$ the periodicity cube of the functions we will deal with. The differential operator $-\Delta v + F'(v)$ will be denoted by $\mathcal{E}'(v)$; notice that $\mathcal{E}'(v)$ is the Gateaux derivative of $\mathcal{E}$ with respect to $L^2_\tau(\mathbf{R}^n)$ norm, that $\mathcal{E}'(v)$ is defined if and only if $\Delta v \in L^2_\tau(\mathbf{R}^n)$ in the sense of distributions, and that (144) is formally equivalent to $u'(t) = -\mathcal{E}'(u(t,\cdot))$.

The monotonicity of $-\Delta v$ and the first condition in (145) give

$$\langle \mathcal{E}'(v) - \mathcal{E}'(w), v - w\rangle = \langle -\Delta v + \Delta w, v - w\rangle + \langle F'(v) - F'(w), v - w\rangle \ge -M\|v - w\|_{L^2_\tau}^2 \qquad (147)$$

with $M := \|F''\|_\infty$. This weaker monotonicity property will be used in the end of the proof, to obtain the energy identity.

(i) For any integer $h \geq 1$ we define functions $u_h^j \in L_r^2(\mathbf{R}^n)$ by a recursive minimization process: we set $u_h^0 = u_0$ and, given u_h^j, u_h^{j+1} is a minimizer of

$$v \ni H_r^1(\mathbf{R}^n) \mapsto \mathcal{E}(v) + \frac{h}{2}\int_Y |v - u_h^j|^2 \, dx.$$

Notice that a minimizer exists because, by the second condition in (145), F has quadratic growth, hence minimizing sequences are bounded in $H_r^1(\mathbf{R}^n)$ and $u \mapsto \mathcal{E}(u)$ is lower semicontinuous with respect to the weak topology of $H_r^1(\mathbf{R}^n)$. Any minimizer satisfies the Euler equation

$$\mathcal{E}'(u_h^{j+1}) + h(u_h^{j+1} - u_h^j) = 0. \tag{148}$$

Using (147) it is not hard to see that u_h^{j+1} is uniquely determined by (148) as soon as $h > M$.

The minimality of u_h^{j+1} enables to estimate the distance between u_h^{j+1} and u_h^j with the energy loss:

$$\frac{h}{2}\|u_h^{j+1} - u_h^j\|_{L_r^2}^2 \leq \mathcal{E}(u_h^j) - \mathcal{E}(u_h^{j+1}) \qquad \forall j \in \mathbf{N}. \tag{149}$$

For later use in the proof of (146), we note that

$$\sum_{j=1}^\infty \|\mathcal{E}'(u_h^j)\|_{L_r^2}^2 = h^2 \sum_{j=1}^\infty \|u_h^j - u_h^{j-1}\|_{L_r^2}^2 \leq 2h\sum_{j=1}^\infty \mathcal{E}(u_h^{j-1}) - \mathcal{E}(u_h^j) \leq 2h\mathcal{E}(u_0). \tag{150}$$

Given u_h^j, we can now define $v_h(t) \in L_{\text{loc}}^2([0,\infty); H_r^1(\mathbf{R}^n))$ as the piecewise linear function such that $v_h(j/h) = u_h^j$, i.e.

$$v_h(t) = u_h^{[ht]} + (ht - [ht])(u_h^{[ht+1]} - u_h^{[ht]}) \qquad \forall t \geq 0. \tag{151}$$

(ii) In this step we find uniform bounds on $v_h(t)$, before passing to the limit in the next step. By (149), the map $j \mapsto \mathcal{E}(u_h^j)$ is nonincreasing in j; taking into account the definition of $v_h(t)$ and the second condition in (145) we find

$$\int_Y \frac{1}{2}|\nabla v_h(t)|^2 \, dx + \alpha\|v_h(t)\|_{L_r^2}^2 \leq \sup_{j \in \mathbf{N}}\left\{\int_Y \frac{1}{2}|\nabla u_h^j|^2 \, dx + \alpha\|u_h^j\|_{L_r^2}^2\right\} \leq \mathcal{E}(u_0) - \beta\tau^n$$

for any $t \geq 0$. This provides a uniform bound in $L^\infty([0,\infty); H_r^1(\mathbf{R}^n))$ for v_h. The uniform bound in $H_{\text{loc}}^1([0,\infty); L_r^2(\mathbf{R}^n))$ can be obtained from (149) as follows:

$$\int_0^\infty \|v_h'(t)\|_{L_r^2}^2 \, dt = \frac{1}{h}\sum_{j=0}^\infty h^2\|u_h^{j+1} - u_h^j\|_{L_r^2}^2 \leq 2\sum_{j=0}^\infty \mathcal{E}(u_h^j) - \mathcal{E}(u_h^{j+1}) \leq 2\mathcal{E}(u_0).$$

These bounds imply, by Exercise 18, the existence of a subsequence $(h(k))$ and of a function $v(t)$ such that $v_{h(k)}(t)$ weakly converges in $H_r^1(\mathbf{R}^n)$ to $v(t)$ for any

$t \in [0, \infty)$. Clearly $v(0) = u_0$, and we will prove in the next step that v is a solution of (144).

(iii) From (148) and the definition of $v_h(t)$ we obtain

$$v'_h(t) = -\mathcal{E}'(v_h([ht + 1]/h)) \qquad \forall t \in (0, \infty) \setminus \mathbf{N}/h \qquad (152)$$

By Exercise 16(i) we can rewrite the identity above as

$$\int_0^\infty \left(\int_{\mathbf{R}^n} F'(v_h([ht + 1]/h, x))\phi(t, x) - v_h(t, x)\phi_t(t, x)\, dx \right) dt$$

$$+ \int_0^\infty \left(\int_{\mathbf{R}^n} \langle \nabla v_h([ht + 1]/h, x), \nabla \phi(t, x) \rangle\, dx \right) dt = 0$$

for any $\phi \in C_c^\infty((0, \infty) \times \mathbf{R}^n)$. The equicontinuity of $t \mapsto v_h(t)$ (with respect to L_τ^2 norm) implies that $v_{h(k)}([h(k)t + 1]/h(k))$ weakly converges in $H_\tau^1(\mathbf{R}^n)$ to $v(t)$ for any $t \in [0, \infty)$.

Therefore, setting $h = h(k)$ in the identity above and passing to the limit as $k \to \infty$ we find

$$\int_0^\infty \left(\int_{\mathbf{R}^n} F'(v(t, x))\phi(t, x) - v(t, x)\phi_t(t, x) + \langle \nabla v(t, x), \nabla \phi(t, x) \rangle\, dx \right) dt = 0.$$

As ϕ is arbitrary, this proves that v satisfies (144).

(Uniqueness) It is a straightforward consequence of Theorem 32 and Remark 28(1) below. The uniqueness of u implies that that $v_h(t)$ weakly converges in $H_\tau^1(\mathbf{R}^n)$ to $u(t)$ for any $t \geq 0$.

(Energy identity) We will first prove that

$$j \mapsto \|\mathcal{E}'(u_h^j)\|_{L_\tau^2}^2 + 4M\mathcal{E}(u_h^j) \qquad \text{is nonincreasing.} \qquad (153)$$

To prove (153) we use (147), (148) and (149):

$$\|\mathcal{E}'(u_h^{j+1})\|_{L_\tau^2}^2 = \langle \mathcal{E}'(u_h^{j+1}) - \mathcal{E}'(u_h^j), \mathcal{E}'(u_h^{j+1}) + \mathcal{E}'(u_h^j) \rangle + \|\mathcal{E}'(u_h^j)\|_{L_\tau^2}^2$$

$$= \langle \mathcal{E}'(u_h^{j+1}) - \mathcal{E}'(u_h^j), -h(u_h^{j+1} - u_h^j) + \mathcal{E}'(u_h^j) \rangle + \|\mathcal{E}'(u_h^j)\|_{L_\tau^2}^2$$

$$\leq \langle \mathcal{E}'(u_h^{j+1}) - \mathcal{E}'(u_h^j), \mathcal{E}'(u_h^j) \rangle + \|\mathcal{E}'(u_h^j)\|_{L_\tau^2}^2 + Mh\|u_h^{j+1} - u_h^j\|_{L_\tau^2}^2$$

$$\leq \|\mathcal{E}'(u_h^j)\|_{L_\tau^2}\|\mathcal{E}'(u_h^{j+1})\|_{L_\tau^2} + Mh\|u_h^{j+1} - u_h^j\|_{L_\tau^2}^2$$

$$\leq \frac{1}{2}\|\mathcal{E}'(u_h^{j+1})\|_{L_\tau^2}^2 + \frac{1}{2}\|\mathcal{E}'(u_h^j)\|_{L_\tau^2}^2 + 2M\left[\mathcal{E}(u_h^j) - \mathcal{E}(u_h^{j+1})\right].$$

Rearranging, we get

$$\|\mathcal{E}'(u_h^{j+1})\|_{L_\tau^2}^2 + 4M\mathcal{E}(u_h^{j+1}) \leq \|\mathcal{E}'(u_h^j)\|_{L_\tau^2}^2 + 4M\mathcal{E}(u_h^j)$$

which yields (153). Let $\epsilon > 0$ be given and assume that $[h\epsilon] > 1$; from (150) we infer the existence of an integer $j_\epsilon \in [1, [h\epsilon]]$ such that

$$\|\mathcal{E}'(u_h^{j_\epsilon})\|_{L_\tau^2}^2 \leq \frac{2h\mathcal{E}(u_0)}{[h\epsilon]}$$

and (153) and the monotonicity of $j \mapsto \mathcal{E}(u_h^j)$ give

$$\|\mathcal{E}'(u_h^j)\|_{L_\tau^2}^2 \leq \frac{2h\mathcal{E}(u_0)}{[h\epsilon]} + 4M\mathcal{E}(u_0) \leq 4(\frac{1}{\epsilon} + M)\mathcal{E}(u_0) \qquad \forall j \geq [h\epsilon].$$

Since $v_h(t)$ is a convex combination of $u_h^{[ht]}$ and $u_h^{[ht]+1}$, and since the L_τ^2 norms of u_h^j are equibounded in h and j, the inequality above implies

$$\sup_{[h\epsilon]>1} \sup_{t \geq \epsilon} \|\mathcal{E}'(v_h(t))\|_{L_\tau^2} < \infty \qquad \forall \epsilon > 0. \tag{154}$$

By (152), a similar estimate holds true for $\|v_h'(t)\|_{L_\tau^2}$, so that in the limit we obtain that $u' \in L_{loc}^\infty((0,\infty); L_\tau^2(\mathbf{R}^n))$.

We are now ready to prove (146). Since

$$\mathcal{E}(u(s)) - \mathcal{E}(u(t)) \tag{155}$$
$$= \frac{1}{2} \int_Y \langle \nabla u(s,x) - \nabla u(t,x), \nabla u(s,x) + \nabla u(t,x) \rangle \, dx +$$
$$\int_Y F(u(s,x)) - F(u(t,x)) \, dx$$
$$\leq -\frac{1}{2} \int_Y \langle u(s,x) - u(t,x), \Delta u(s,x) + \Delta u(t,x) \rangle \, dx$$
$$+ K \int_Y (1 + |u(s,x)| + |u(t,x)|)|u(s,x) - u(t,x)| \, dx$$

(for a suitable constant K depending only on F) the local boundedness of $\|u'(t)\|_{L_\tau^2}$, of $\|u(t)\|_{L_\tau^2}$ and of $\|\Delta u\|_{L_\tau^2}$ in $(0,\infty)$ yield that $t \mapsto \mathcal{E}(u(t))$ is locally a Lipschitz function in $(0,\infty)$. Taking $s = t + \delta$, dividing both sides by $\delta \neq 0$ and letting $\delta \to 0$ gives

$$\frac{d}{dt}\mathcal{E}(u(t)) = -\int_Y u_t(t,x)\Delta u(t,x) \, dx + \int_Y u_t(t,x)F'(u(t,x)) \, dx = -\|u'(t)\|_{L_\tau^2}^2$$

at any differentiability point of u in $(0,\infty)$. Hence, (146) follows by integration.
□

Remark 27. (1) If F is convex, then $\mathcal{E}'(u) = -\Delta u + F'(u)$ is a monotone operator and it is easy to check that

$$u_h^j = \left(I + \frac{\mathcal{E}'}{h}\right)^{-j} u_0 \qquad \forall j \in \mathbf{N}.$$

The convergence of $u_h^{[ht]}$ to the solution of the evolution problem $u' = -\mathcal{E}'(u)$ is classical, and known as *exponential formula* (see for instance [Bre73], [Cra86]). More generally, this approximation process can be built in any metric space and, under mild regularity assumptions on the energy functional $\mathcal{E}$, it can be shown

its convergence (up to subsequences) to a "steepest descent flow" (see [GMT82, GMT80, GMT85, Amb95a]). An abstract approach to evolution problems based on time discretization and minimization has been proposed by De Giorgi in [Gio93a]. As in [ATW93] this approach allows, in the recursive minimization problem, perturbations of $\mathcal{E}$ which are not the square of a distance function.
(2) If $\Delta u_0 \in L^2_r(\mathbf{R}^n)$, then (153) shows that $\|u'(t)\|_{L^2_r}$ and $\|\Delta u(t)\|_{L^2_r}$ are both globally bounded in $[0, \infty)$. In particular, the identity

$$\mathcal{E}(v) = \int_Y -\frac{1}{2}v\Delta v + F(v)\, dx$$

shows that $t \mapsto \mathcal{E}(u(t))$ is continuous up to $t = 0$.

Theorem 32 (comparison theorem). *Let u, v be bounded solutions of (144) in the class*

$$L^2_{\mathrm{loc}}\big([0, \infty); H^1(B_R)\big) \cap H^1_{\mathrm{loc}}\big([0, \infty); L^2(B_R)\big) \qquad \forall R > 0.$$

Then, $u_0 \le v_0$ implies $u \le v$ in $(0, \infty) \times \mathbf{R}^n$.

Proof. Let $\psi(t) = (t \vee 0)$, and let us choose a family $(\phi_h)_{h\ge 1}$ of functions satisfying

$$\phi_h \in C^2_c\big(B_{h+1}, [0,1]\big), \qquad \phi_h \equiv 1 \text{ in } B_h, \qquad C := \sup_{x,h} |\nabla^2 \phi_h(x)| < \infty.$$

We define

$$I_h(t) := \frac{1}{2}\int_{\mathbf{R}^n} \psi^2(u-v)\phi_h\, dx$$

where the integration is done holding t fixed. Using (144) (here we also use Exercise 16(i) and a simple variant of Exercise 16(ii)) we compute

$$
\begin{aligned}
I'_h(t) &= \int_{\mathbf{R}^n} \psi(u-v)(u_t - v_t)\phi_h\, dx \\
&= \int_{\mathbf{R}^n} \psi(u-v)\big[\Delta u - \Delta v\big]\phi_h\, dx - \int_{\mathbf{R}^n} \psi(u-v)\big[F'(u) - F'(v)\big]\phi_h\, dx \\
&\le -\int_{E_t} \langle \nabla\phi_h, \nabla u - \nabla v\rangle \psi(u-v)\, dx - \int_{E_t} |\nabla u - \nabla v|^2 \phi_h\, dx \\
&\quad + \|F''\|_\infty \int_{\mathbf{R}^n} \psi^2(u-v)\phi_h\, dx
\end{aligned}
$$

for almost every t, where $E_t := \{u(t, \cdot) > v(t, \cdot)\}$. The integral containing $\nabla\phi_h$ can be estimated multiplying and dividing by $\sqrt{\phi_h}$ and using the inequality $|ab| \le a^2 + b^2$. In this way we get

$$I'_h(t) \le \int_{\{\phi_h > 0\}} \frac{|\nabla\phi_h|^2}{\phi_h} \psi^2(u-v)\, dx + 2\|F''\|_\infty I_h(t).$$

Taking into account that $I_h \leq I_{h+1}$ and Exercise 17 we eventually get

$$I_h'(t) \leq M I_{h+1}(t) \qquad \text{with} \qquad M := C + 2\|F''\|_\infty.$$

Since u and v are bounded, the function

$$I(t) := \sum_{h=1}^{\infty} 2^{-h} h^{-n} I_h(t)$$

is absolutely continuous and satisfies $I'(t) \leq 2^n M I(t)$ for almost every t. Since $I(0) = 0$, it follows that $I(t) = 0$ (i.e., $u(t,x) \leq v(t,x)$ for almost every $x \in \mathbf{R}^n$) for any $t \geq 0$. $\qquad\qquad\square$

Remark 28. (1) A similar, simpler, proof works for periodic solutions. In this case it is enough to take $\phi \equiv 1$ and to integrate over a periodicity cube. The same proof also works assuming only that u is a subsolution and v is a supersolution. (2) A careful inspection of the proof above shows that it still works under the following weaker assumption on the growth of u and v: $|u(t,x)| + |v(t,x)| \leq g(x)$ for any (t,x), where g is independent of t and satisfies

$$\limsup_{h\to\infty} \left(\int_{B_h} g(x)\,dx \right)^{-1} \left(\int_{B_{h+1}} g(x)\,dx \right) < \infty.$$

This allows, for instance, any polynomial or exponential growth. Without *any* growth assumption the comparison theorem is false, even for the heat equation (see for instance [Fri64] for sharp growth assumptions and for a counterexample).

Exercise 17. Show that

$$\frac{|\nabla\phi|^2}{2\phi} \leq \max|\nabla^2\phi| \qquad \text{in } \mathbf{R}^n$$

for any $\phi \in C_c^2(\mathbf{R}^n, [0, \infty))$. Hint: consider the maximizers of $|\nabla\phi|^2/(\phi + \epsilon)$.

Theorem 33 (strong maximum principle). *Let $v \in C^\infty((0,\infty) \times \mathbf{R}^n)$ be a nonpositive bounded function satisfying*

$$v_t \leq \Delta v - cv \qquad \text{in } (0,\infty) \times \mathbf{R}^n$$

for some constant $c \geq 0$. Then, $v(t_0, x_0) = 0$ for some point (t_0, x_0) in $(0,\infty) \times \mathbf{R}^n$ implies that v is identically equal to 0 in $(0, t_0) \times \mathbf{R}^n$.

The strong maximum principle (see for instance [PW67], Theorem 4, page 172) can be used to show that

$$u < 1 \quad \text{in } (0,\infty) \times \mathbf{R}^n$$

for smooth solutions $u : (0, \infty) \times \mathbf{R}^n \to [-1, 1]$ of the reaction–diffusion equation

$$u_t = \Delta u + u(1 - u^2) \qquad \text{in } (0, \infty) \times \mathbf{R}^n \tag{156}$$

such that $|u(0, \cdot)| < 1$. Indeed, writing $v = u - 1$, we obtain

$$v_t = \Delta v - 2v - v^2(v + 3) \le \Delta v - 2v \qquad \text{in } (0, \infty) \times \mathbf{R}^n$$

because $v + 3 = u + 2 \ge 0$. Since (156) is odd in u, we have also

$$u > -1 \quad \text{on } (0, \infty) \times \mathbf{R}^n.$$

Lemma 10 (smoothness and bounds). *Let $u_0 \in L^2_\tau(\mathbf{R}^n)$ be a Lipschitz function and let u as in Theorem 31. Then u is smooth in $(0, \infty) \times \mathbf{R}^n$ and continuous in $[0, \infty) \times \mathbf{R}^n$. In addition, if*

$$F(t) = \frac{(1 - t^2)^2}{4} \qquad \forall t \in [-1, 1],$$

and if $q : \mathbf{R} \to (-1, 1)$ is the function in (114), the following two statements hold:

(i) if $|u_0| \le 1$, then $|u| < 1$ in $(0, \infty) \times \mathbf{R}^n$;
(ii) if $u_0 = q(z_0)$ and $\mathrm{Lip}\,(z_0, \mathbf{R}^n) \le 1$, then u can be represented by $u(t, \cdot) = q(z(t, \cdot))$, with z satisfying $\mathrm{Lip}\,(z(t, \cdot), \mathbf{R}^n) \le 1$ for any $t \ge 0$.

Proof. The smoothness of u and its continuity up to $t = 0$ follow by the regularity theory for semilinear equations (see for instance [Lun94], Chapter 7).
(i) The inequality $|u| \le 1$ can be obtained either from the maximum principle or directly from the construction of u in Theorem 31. Indeed, by a truncation argument, it is not hard to see that $|u^j_h| \le 1$ implies $|u^{j+1}_h| \le 1$. The strong inequality follows by the regularity of u and from strong maximum principle, as explained above.
(ii) A straightforward computation shows that $z(t, x)$ satisfies

$$z_t - \Delta z = \sqrt{2}q(z)(1 - |\nabla z|^2) \qquad \text{in } (0, \infty) \times \mathbf{R}^n. \tag{157}$$

Let us assume, by contradiction, that $z(t_0, x_0) - z(t_0, y_0) > |x_0 - y_0|$ for some t_0, x_0, y_0. Then, for λ small enough the maximum of

$$w(t, x, y) := z(t, x) - z(t, y) - \sqrt{1 + \lambda}\,|x - y|$$

is strictly positive. Let $(\bar{t}, \bar{x}, \bar{y})$ be a maximum point; since the maximum is strictly positive it holds $\bar{x} \ne \bar{y}$. In addition, our assumption on u_0 implies

$$z(0, \bar{x}) - z(0, \bar{y}) \le |\bar{x} - \bar{y}|$$

so that $\bar{t} > 0$. From the identities

$$z_t(\bar{t}, \bar{x}) = z_t(\bar{t}, \bar{y}), \qquad \nabla z(\bar{t}, \bar{x}) = -\nabla z(\bar{t}, \bar{y}) = \sqrt{1 + \lambda}\,\frac{\bar{x} - \bar{y}}{|\bar{x} - \bar{y}|}$$

and from (157) we infer

$$\Delta z(\bar{t}, \bar{x}) - \Delta z(\bar{t}, \bar{y}) = \sqrt{2}\lambda\big[q(z(\bar{t}, \bar{x})) - q(z(\bar{t}, \bar{y}))\big].$$

Since $z(\bar{t}, \bar{x}) > z(\bar{t}, \bar{y})$, it follows that $\Delta z(\bar{t}, \bar{x}) > \Delta z(\bar{t}, \bar{y})$. On the other hand, since

$$t \mapsto w(\bar{t}, \bar{x} + t\xi, \bar{y} + t\xi) = z(\bar{t}, \bar{x} + t\xi) - z(\bar{t}, \bar{y} + t\xi) - \sqrt{1 + \lambda}|\bar{x} - \bar{y}|$$

achieves its maximum at 0, it follows that

$$\langle \nabla^2 z(\bar{t}, \bar{x})\xi, \xi \rangle \leq \langle \nabla^2 z(\bar{t}, \bar{y})\xi, \xi \rangle$$

for any vector $\xi \in \mathbf{R}^n$, hence $\nabla^2 z(\bar{t}, \bar{x}) \leq \nabla^2 z(\bar{t}, \bar{y})$ and this is a contradiction.$\square$

Exploiting Theorem 31 and Lemma 10, we will now prove an existence and uniqueness theorem for the evolution problem (144) without periodicity assumptions on u_0, assuming that $F(t) = (1 - t^2)^2/4$ in $[-1, 1]$.

Corollary 2 (solutions of the reaction–diffusion equation). *Let* $E \subset \mathbf{R}^n$, $E \neq \emptyset$, $E \neq \mathbf{R}^n$. *The equation (156) has a unique bounded solution*

$$u \in L^\infty\big([0, \infty); H^1(B_R)\big) \cap H^1_{\mathrm{loc}}\big([0, \infty); L^2(B_R)\big) \qquad \forall R > 0 \qquad (158)$$

with the initial condition $u_0(x) = q\big(\bar{d}(x, E)\big)$. *Moreover,*

$$u \in C\big([0, \infty) \times \mathbf{R}^n\big) \cap C^\infty\big((0, \infty) \times \mathbf{R}^n\big)$$

and u *can be represented by* $q(z)$ *with* z *satisfying* $|\nabla z| \leq 1$ *in* $(0, \infty) \times \mathbf{R}^n$.

Proof. Let $E_h = [E \cap (0, h)^n] + h\mathbf{Z}^n$ and, for h sufficiently large to ensure that $E_h \neq \emptyset$ and $E_h \neq \mathbf{R}^n$, define $u_0^h(x) = q\big(\bar{d}(x, E_h)\big)$. Denoting by u_h the solutions of (2) given by Theorem 31 and representing, thanks to Lemma 10(ii), u_h by $q(z_h)$, we know that $|\nabla z_h| \leq 1$, hence $|\nabla u_h|$ are equibounded in $[0, \infty) \times \mathbf{R}^n$.

We now claim that

$$\int_0^T \int_{\mathbf{R}^n} u_{ht}^2 \phi \, dx \, dt \leq C(\phi, T) \qquad (159)$$

for any $\phi \in C_c^2(\mathbf{R}^n, [0, \infty))$ and any $T > 0$. The proof of (159) is based on a local version of the energy identity (146). Let

$$I_h(t) := \frac{1}{2} \int_{\mathbf{R}^n} \Big(|\nabla u_h|^2 + \frac{(1 - u_h^2)^2}{2}\Big)\phi \, dx$$

where the integration is done holding t fixed. Using the smoothness of u, (156) and Exercise 17 we estimate

$$I_h'(t) = \int_{\mathbf{R}^n} \left(\nabla u_h \nabla u_{ht} + u_h(u_h^2 - 1)u_{ht} \right) \phi \, dx \tag{160}$$

$$= -\int_{\mathbf{R}^n} u_{ht}\left(\Delta u_h + u_h(1 - u_h^2) \right) \phi \, dx - \int_{\mathbf{R}^n} \langle \nabla u_h, \nabla \phi \rangle u_t \, dx$$

$$\leq -\int_{\mathbf{R}^n} u_{ht}^2 \phi \, dx + \frac{1}{2} \int_{\mathbf{R}^n} u_{ht}^2 \phi \, dx + \frac{1}{2} \int_{\{\phi > 0\}} \frac{|\nabla \phi|^2}{\phi} |\nabla u_h|^2 \, dx$$

$$\leq -\frac{1}{2} \int_{\mathbf{R}^n} u_{ht}^2 \phi \, dx + \max |\nabla^2 \phi| \int_{\{\phi > 0\}} |\nabla u_h|^2 \, dx.$$

Notice that, by the gradient bound on ∇u_h, $I_h(t)$ can be uniformly estimated in t and in h. Hence, (159) easily follows by (160) by integration in time.

By Exercise 18, possibly extracting a subsequence, we can assume that $u_h(t, \cdot)$ weakly converges in $H^1_{\text{loc}}(\mathbf{R}^n)$ to some function $u(t, \cdot)$ for any $t \geq 0$. Passing to the limit as $h \to \infty$ in (144) and using (159) and the gradient estimate on u_h, we obtain that u fulfils (158) and satisfies (156) in the distribution sense. Moreover, the initial condition at $t = 0$ is satisfied in the $L^2_{\text{loc}}(\mathbf{R}^n)$ sense.

The regularity theory for semilinear parabolic equations (see for instance [Lun94], Chapter 7) implies that u is smooth in $(0, \infty) \times \mathbf{R}^n$ and continuous up to $t = 0$. By the strong maximum principle, $|u| < 1$ in $[0, \infty) \times \mathbf{R}^n$, hence we can still represent u by $q(z)$. Since u_h converges to u, z_h converges to z, hence $|\nabla z| \leq 1$.

The solution u is unique because of the comparison theorem. $\qquad\square$

Remark 29. Similar results hold true for the equation

$$u_t = \Delta u + \frac{1}{\epsilon^2} u(1 - u^2) \qquad \text{in } (0, \infty) \times \mathbf{R}^n$$

which can be transformed into (156) scaling by a factor ϵ in space and by a factor ϵ^2 in time.

Exercise 18. Let (u_h) be a sequence bounded in the spaces

$$H^1_{\text{loc}}\left([0, \infty); L^2(B_R)\right) \qquad \text{and} \qquad L^\infty\left([0, \infty); H^1(B_R)\right)$$

for any $R > 0$. Show that there exist functions $u(t)$ and a subsequence $h(k)$ such that $u_{h(k)}(t)$ weakly converges in $H^1_{\text{loc}}(\mathbf{R}^n)$ to $u(t)$ for any $t \in [0, \infty)$.

Variational models for phase transitions, an approach via Γ-convergence

G. Alberti

Introduction

This paper is an extended version of the lecture delivered at the Summer School on *Differential Equations and Calculus of Variations* (Pisa, September 16-28, 1996). That lecture was conceived as an introduction to the theory of Γ-convergence and in particular to the Modica-Mortola theorem; I have tried to reply the style and the structure of the lecture also in the written version. Thus first come few words on the definition and the meaning of Γ-convergence, and then we pass to the theorem of Modica and Mortola. The original idea was to describe both the mechanical motivations which underlay this result and the main ideas of its proof. In particular I have tried to describe a guideline for the proof which would adapt also to other theorems on the same line. I hope that this attempt has been successful. Notice that I never intended to give a detailed and exhaustive description of the many results proved in this field through the recent years, not even of the main ones. In particular the list of references is not meant to be complete, neither one should assume that the contributions listed here are the most relevant or significant.

The rest of this paper is organized as follows:

1. A brief introduction to Γ-convergence
2. The Cahn-Hilliard model for phase transitions and the Modica-Mortola theorem
3. The optimal profile problem and the proof of the Modica-Mortola theorem
4. Final remarks

Acknowledgements: I thank Luigi Ambrosio for many useful remarks on a preliminary version. This note was written during a one-year visit at the Max Planck Institute for Mathematics in the Science in Leipzig, whose hospitality and support I gratefully acknowledge.

1 A brief introduction to Γ-convergence

The notion of Γ-convergence was introduced by E. De Giorgi and T. Franzoni in [GF75]; even though it is mainly intended as a notion of convergence for variational functionals on function spaces, it is more convenient to give its definition and main properties in a slightly more general setting, namely as a notion of convergence for functions on a metric space. Therefore in what follows X is a metric space, u an element of X, F a function from X into $[0, +\infty]$, and ϵ is a parameter which converges to 0. In the applications X will be a space of functions

u on some open domain Ω of $\mathbf{R}^n$, and F a functional on X; typical examples
are given by integral functionals on Sobolev or L^p spaces (cf. paragraph 1.3).
What we present here is a rather simplified version of the original definition.
A detailed and sistematic treatment of the general theory Γ-convergence, and
many applications as well, can be found in G. Dal Maso's book [Mas93] (see also
[Amb99], section 8).

Warning: Throughout this paper, instead of sequences of functions (and functionals) labelled by some integer parameter which tends to infinity, we consider
families of functions labelled by a continuous parameter ϵ which tends to 0.
Nevertheless we use the term "sequence" also to denote such ordered families
(and, for instance, we write (u_ϵ) instead of $\{u_\epsilon\}$). On this line, a subsequence
of (u_ϵ) is any sequence (u_{ϵ_n}) such that $\epsilon_n \to 0$ as $n \to \infty$, and we say that (u_ϵ)
is pre-compact in the corresponding (metric) space X if every subsequence ad-
mits a sub-subsequence which converge in X. In proofs we often omit to relabel
subsequences.

Definition 1. *Let X be a metric space, and for $\epsilon > 0$ let be given $F_\epsilon : X \to$
$[0, +\infty]$. We say that F_ϵ Γ-converge to F on X as $\epsilon \to 0$, and we write $F_\epsilon \overset{\Gamma}{\to} F$,
if the following two conditions hold:*

(LB) Lower bound inequality: *for every $u \in X$ and every sequence (u_ϵ) s.t.
$u_\epsilon \to u$ in X there holds*

$$\liminf_{\epsilon \to 0} F_\epsilon(u_\epsilon) \geq F(u) ; \tag{1.1}$$

(UB) Upper bound inequality: *for every $u \in X$ there exists (u_ϵ) s.t. $u_\epsilon \to u$ in
X and*

$$\lim_{\epsilon \to 0} F_\epsilon(u_\epsilon) = F(u) . \tag{1.2}$$

Condition (LB) means that whatever sequence we choose to approximate u,
the value of $F_\epsilon(u_\epsilon)$ is, in the limit, larger than $F(u)$; on the other hand condition
(UB) implies that this bound is sharp, that is, there always exists a sequence
(u_ϵ) which approximates u so that $F_\epsilon(u_\epsilon) \to F(u)$.

Remark 1. When proving a Γ-convergence result, it is often convenient to reduce
the amount of verifications and constructions. To this aim we notice that if (LB)
holds, then equality (1.2) can be replaced by

$$\limsup_{\epsilon \to 0} F_\epsilon(u_\epsilon) \leq F(u) . \tag{1.3}$$

Assume furthermore that we can find a set $\mathcal{D} \subset X$ which satisfies the following condition: for every $u \in X$ there exists an approximating sequence $(u_n) \subset \mathcal{D}$
such that $u_n \to u$ and $F(u_n) \to F(u)$; then a simple diagonal argument shows it
is enough to verify condition (UB) for all $u \in \mathcal{D}$ only, and not for every $u \in X$.
In fact one can push this argument a bit further, and just verify that for every
$u \in \mathcal{D}$ and every $\eta > 0$ there exists a sequence $(u_\epsilon) \subset X$ (but a subsequence is
not enough!) such that $\limsup d(u_\epsilon, u) \leq \eta$ and $\limsup F_\epsilon(u_\epsilon) \leq F(u) + \eta$.

The main properties of Γ-convergence are listed in the following statement.

Proposition 1. *We have the following:*

(i) *the Γ-limit F is always lower semicontinuous on X;*

(ii) Stability under continuous perturbations: *if $F_\epsilon \xrightarrow{\Gamma} F$ and G is continuous, then $F_\epsilon + G \xrightarrow{\Gamma} F + G$;*

(iii) Stability of minimizing sequences: *if $F_\epsilon \xrightarrow{\Gamma} F$ and v_ϵ minimizes F_ϵ over X, then every cluster point of (v_ϵ) minimizes F over X.*

The proof of this proposition is left to the reader, and we pass to describe how this notion of variational convergence will be used.

1.1 Asymptotic behaviour of minimizers and compactness

Assume that for every $\epsilon > 0$ we are given a function v_ϵ which minimizes the functional F_ϵ on X, and that we want to know what happens of v_ϵ as $\epsilon \to 0$. Sometimes the minimizers v_ϵ can be written via some explicit formula from which we can deduce all information about the asymptotic behaviour of v_ϵ when ϵ tends to 0. If no such formula is available, and indeed this is often the case, then we can exploit the fact that each v_ϵ solves the Euler-Lagrange equation associated with F_ϵ and try to understand which kind of limit equation is verified by a limit point v of (v_ϵ). Another possibility is to take the Γ-limit F of the functionals F_ϵ (if any exists), and then use statement (iii) of Proposition 1 to show that any limit point v of v_ϵ is in fact a minimizer of F, and in particular solves the Euler-Lagrange equation associated with F. Notice that such a strategy makes sense only if we know *a priori* that the minimizing sequence (v_ϵ) is pre-compact in X (even the fact that F has some minimizer v does not imply that v is a limit point of v_ϵ). According to this viewpoint a Γ-convergence result for the functionals F_ϵ should always be paired with a *compactness result* for the corresponding minimizing sequences (v_ϵ). In fact one usually tries to prove the following asymptotical equi-coercivity of F_ϵ:

(C) *Compactness* : let be given sequences (ϵ_n) and (u_n) such that $\epsilon_n \to 0$ and $F_{\epsilon_n}(u_n)$ is bounded; then (u_n) is pre-compact in X.

1.2 Interesting rescalings

If v_ϵ minimizes F_ϵ, then it minimizes also $\lambda_\epsilon F_\epsilon$ for every $\lambda_\epsilon > 0$. This means that information about the limit points of (v_ϵ) can be recovered also by the Γ-limit of $\lambda_\epsilon F_\epsilon$; different choices of the scaling factors λ_ϵ generate different Γ-limits, which give different information. Notice that it may well happen that the functionals F_ϵ converge to a constant functional F, so that the fact that every limit point v minimizes F actually gives no information about v, while the Γ-limit of the functionals $\lambda_\epsilon F_\epsilon$ may be less trivial for suitable choice of λ_ϵ (see for instance Remark 8). Therefore the problem arises of finding $\lambda_\epsilon > 0$ so that the Γ-limit of the rescaled functionals $\lambda_\epsilon F_\epsilon$ gives the largest amount of information; sometimes this optimal rescaling is evident but sometimes it is not (compare for instance the situations described in paragraph 1.3 and Theorem 1).

We conclude this section with a simple but instructive example.

1.3   An example from homogenization

Let X be the class of all u in the Sobolev space $W^{1,2}(0,1)$ such that $u(0) = 0$ and $u(1) = 1$, endowed with the strong topology of $L^2(0,1)$. Let a be the 1-periodic function on $\mathbf{R}$ which is equal to α_1 on $[0,1/2)$ and to α_2 on $[1/2,1)$, with $0 < \alpha_1 < \alpha_2 < +\infty$, and set

$$F_\epsilon(u) := \int_0^1 a(x/\epsilon) \left| \dot{u}(x) \right|^2 dx . \tag{1.4}$$

Then the functional F_ϵ Γ-converge on X to

$$F(u) := \alpha \int_0^1 |\dot{u}|^2 \quad \text{where} \quad \alpha := \frac{2\alpha_1 \alpha_2}{\alpha_1 + \alpha_2} . \tag{1.5}$$

This is a simple example of *homogenization* (cf. [Mas93], chapters 24 and 25, or [BD98b]). The proof is quite instructive, but we just give a sketch, leaving the details to the interested reader.

1. Start with the constructive part of the proof, that is, with the upper bound inequality. Take u affine on $(0,1)$ and show that (1.2) can be fulfilled by suitable approximating functions u_ϵ which are affine on every interval of the type $\big[n\epsilon, (n+1/2)\epsilon \big)$ with $n = 0, 1, \ldots$ (these are the intervals where $a(x/\epsilon)$ is constant).
2. Extend the previous construction to every u which is piecewise affine on $(0,1)$.
3. Use a proper density argument to conclude the proof of the upper bound inequality (cf. Remark 1).
4. Try to understand why the approximation proposed in step 2 is optimal, and then prove the lower bound inequality.

Remark 2. The choice of the L^2-topology on X may look unnatural, and indeed an explanation is required. Since $F_\epsilon(u) \geq \alpha_1 \int |\dot{u}|^2$, when $F_\epsilon(u_\epsilon)$ is bounded in ϵ the functions u_ϵ are weakly pre-compact in $W^{1,2}(0,1)$, but not strongly. Hence the compactness condition (C) in paragraph 1.1 is verified if we endow X with the L^2-topology (recall that the weak topology of $W^{1,2}$ is not metrizable, and anyhow conditions (LB) and (UB) in Definition 1 do not change if we replace the L^2-topology with the weak $W^{1,2}$-topology).

Remark 3. The pointwise limit of $F_\epsilon(u)$ as $\epsilon \to 0$ is $\overline{F}(u) := \bar{\alpha} \int |\dot{u}|^2$ where $\bar{\alpha}$ is the average of α_1 and α_2, while the value of α in (2.5) is such that $1/\alpha$ is the average of $1/\alpha_1$ and $1/\alpha_2$; in particular $\alpha < \bar{\alpha}$.

Notice that if we endow X with the strong $W^{1,2}$-topology, the Γ-limit of F_ϵ is $\overline{F}$. This shows that the choice of the topology on X does affect the Γ-limit. In view of paragraph 1.1 the right choice is the L^2-topology, because this way the compactness property (C) is verified.

Remark 4. The minimizers v_ϵ of F_ϵ over X can be easily computed (at least for $\epsilon = 1/n$), and then also the limit of v_ϵ as $\epsilon \to 0$ can be directly computed. It is interesting to perform such a calculation and then compare with the result obtained via Γ-convergence.

2   The Cahn-Hilliard model for phase transitions and the Modica-Mortola theorem

Consider a container which is filled with two immiscible and incompressible fluids (oil and water, or if you prefer two different phases of the same fluid). In the classical theory of phase transition it is assumed that, at equilibrium, the two fluids arrange themselves in order to minimize the area of the interface which separates the two phases (we neglect the interaction of the fluids with the wall of the container and the effect of gravity).

This situation is modelled as follows: the container is represented by a bounded regular domain Ω in $\mathbf{R}^3$, and every configuration of the system is described by a function u on Ω which takes value 0 on the set which is occupied by the first fluid, and value 1 on the set occupied by the second fluid; the singular set of u (i.e., the set of discontinuity points of u) is the interface between the two fluids, and we denote it by Su. The space of all admissible configurations is given by all $u : \Omega \to \{0,1\}$ which satisfy $\int u = V$ where V is the total volume of the second fluid (we assume $0 < V < \mathrm{vol}(\Omega)$). Finally we postulate an energy of the form

$$F(u) := \sigma \, \mathcal{H}^2(Su) \tag{2.1}$$

where the parameter σ is called the *surface tension* between the two fluids and $\mathcal{H}^2$ is the two-dimensional Hausdorff measure (when A is a regular surface then $\mathcal{H}^2(A)$ is simply the total area of A). Therefore $F(u)$ is a surface energy distributed on the interface Su, and the equilibrium configuration is obtained by minimizing F over the space all admissible configurations.

An alternative way to study systems of two immiscible fluids is to assume that the transition is not given by a separating interface, but is rather a continuous phenomenon occurring in a thin layer which, on a macroscopic level, we identify with the interface. This means that we allow for a fine mixture of the two fluids. In this case a configuration of the system is represented by a function $u : \Omega \to [0,1]$ where $u(x)$ denotes the average volume density of the second fluid at the point $x \in \Omega$ (thus $u(x) = 0$ means that the first fluid only is present at x, $u(x) = 1/2$ means that both fluids are present with the same rate, and so on). The space of all admissible configuration is the class X of all $u : \Omega \to [0,1]$ such that $\int u = V$ (recall that V is the total volume of the second fluid) and to every configuration u is associated the energy

$$E_\epsilon(u) := \epsilon^2 \int_\Omega |\nabla u|^2 + \int_\Omega W(u) , \qquad (2.2)$$

where ϵ is small positive parameter and W is a continuous positive function which vanishes only at 0 and 1 (in short, a double-well potential). When we come to minimize E_ϵ, the term $\int W(u)$ favourites those configurations which take values close to 0 and 1 (phase separation), while the term $\epsilon^2 \int |\nabla u|^2$ penalizes the spatial inhomogeneity of u. When ϵ is small the first term prevails, and the minimum of E_ϵ is attained by a function u_ϵ which takes mainly values close to 0 and 1 (and takes both, because of the volume constraint $\int u = V$) and the transition from 0 to 1 occurs in a thin layer (in fact with thickness of order ϵ). This model was proposed by J.W. Cahn and J.E. Hilliard in [CH58]. Notice that the energy E_ϵ was there obtained as a first order approximation of a more general one.

A connection between the classical model and the Cahn-Hilliard model was established by L. Modica [Mod87], who proved that the minimizers of E_ϵ converge to minimizers of F. This was obtained by showing that suitable rescalings of the functionals E_ϵ Γ-converge to F. In order to state the precise Γ-convergence result we need to fix some notation. In what follows Ω is a bounded open set in $\mathbf{R}^N$ (and $N = 3$ is a particular case); we take V so that $0 < V < \text{vol}(\Omega)$ and then we denote by X the class of all measurable functions $u : \Omega \to [0, 1]$ such that $\int_\Omega u = V$, endowed with the L^1 norm. We also denote by $BV(\Omega, \{0, 1\})$ the set of all functions $u : \Omega \to \{0, 1\}$ with bounded variation, and Su is now the set of all *essential* singularities of u (for more details and precise definitions see [EG92], chapter 5).

Theorem 1 (L. Modica and S. Mortola [MM77], see also [Mod87]).
Set $\sigma := 2 \int_0^1 \sqrt{W(u)}\, du$, and for every $\epsilon > 0$ let

$$F_\epsilon(u) := \tfrac{1}{\epsilon}E_\epsilon(u) = \begin{cases} \epsilon \int_\Omega |\nabla u|^2 + \frac{1}{\epsilon} \int_\Omega W(u) & \text{if } u \in W^{1,2}(\Omega) \cap X, \\ +\infty & \text{elsewhere in } X, \end{cases} \qquad (2.3)$$

and

$$F(u) := \begin{cases} \sigma \, \mathcal{H}^{N-1}(Su) & \text{if } u \in BV(\Omega, \{0, 1\}) \cap X, \\ +\infty & \text{elsewhere in } X. \end{cases} \qquad (2.4)$$

Then the functionals F_ϵ Γ-converge to F in X, and the compactness condition (C) in paragraph 1.1 is satisfied.

Corollary 1. *If v_ϵ minimizes F_ϵ (or equivalently E_ϵ) on X, then the sequence (v_ϵ) is pre-compact in X, and every limit point v minimizes F.*

Remark 5. Each functional F_ϵ is lower semicontinuous and coercive with respect to the strong topology of X, and then it has at least one minimizer. Minimizing F

over X means finding a set $A \subset \Omega$ among those with prescribed (N-dimensional) volume V which minimizes the (($N-1$)-dimensional) area of $\partial A \cap \Omega$. In particular this implies that $\partial A \cap \Omega$ is an oriented surface with boundary in $\partial \Omega$ and constant mean curvature (by a well-known result of H. Federer [Fed70] this surface is always analytic if $N \leq 7$; if $N > 7$ singularities may appear, and the notions of surface and mean curvature must be intended in a particular weak sense which we do not specify here).

Remark 6. When u is a function in $BV(\Omega, \{0, 1\})$, Su is not the set of all points where u is discontinuous (that is, the topological boundary in Ω of the phase $\{u = 1\}$), but the set of all points where u is *essentially discontinuous*, that is, it has no approximate limit. This set agrees with the so-called *measure theoretic boundary* of $\{u = 1\}$ in Ω, and then $F(u)$ is finite if and only if the phase $\{u = 1\}$ (and the phase $\{u = 0\}$) is set of finite perimeter in Ω in the sense of Geometric Measure Theory (see [EG92], section 5.8). Notice that for such functions u, $F(u)$ is equal to the total variation $\int |Du|$ of the measure derivative Du multiplied by σ.

Remark 7. The existence of a minimizer of E_ϵ over X may be proved via standard lower semicontinuity and compactness results. This minimizer may be not unique. Indeed if Ω is a ball centered in 0 and the minimizer of E_ϵ is unique, then it must be invariant under rotation, i.e., is radially symmetric. On the other hand a simple computation shows that F has no radially symmetric minimizer, and then we deduce that the minimizers of E_ϵ cannot be radially symmetric for values of ϵ arbitrary close to 0 (recall Corollary 1), and then the minimizer of E_ϵ is not unique for every ϵ sufficently small.

Remark 8. To a certain extent the rescaling $F_\epsilon := \frac{1}{\epsilon} E_\epsilon$ given in Theorem 1 is optimal. Indeed one can easily check out the following table:

(a) $E_\epsilon \xrightarrow{\Gamma} E$ with $E(u) = \int_\Omega W(u)$;
(b) if $\lambda_\epsilon \to \infty$ and $\epsilon \lambda_\epsilon \to 0$ then $\lambda_\epsilon E_\epsilon \xrightarrow{\Gamma} E$ with $E(u) = 0$ when u takes the
    values 0 and 1 a.e. in Ω, and $E(u) = +\infty$ otherwise;
(c) if $\epsilon \lambda_\epsilon \to \infty$ then $\lambda_\epsilon E_\epsilon \xrightarrow{\Gamma} E$ with $E(u) = +\infty$ everywhere in X.

In all these cases the set of minimizers of the Γ-limit strictly includes the minimizers of F.

3   The optimal profile problem and the proof of the Modica-Mortola theorem

In this section we give the main ideas of the proof of Theorem 1, and we also try to point out the underlying technical tools. The usual proof is indeed simple and elegant, but also quite specific (see [Mod87], and paragraph 4.5); in particular it hardly adapts even to close generalizations of this statement. For these reasons I prefer to describe here a different approach, which should give a deeper insight in the structure of this theorem, and is probably more flexible. This is also an attempt to gather some important ideas which are scattered in the literature, and therefore not immediately available to non-experts. We remark that another very general approach to the proof of lower bound inequalities (in relaxation but also in Γ-convergence) based on an exthensive use of blow-up techniques has been developed in [FM92]; see also [AB98a] and [BBH94] for applications to theorems of Modica-Mortola type.

First of all we notice that Theorem 1 reduces to the following three statements:

(i) *Compactness*: let be given sequences (ϵ_n) and (u_n) such that $\epsilon_n \to 0$ and $F_{\epsilon_n}(u_n)$ is bounded; then (u_n) is pre-compact in $L^1(\Omega)$ and every limit point belongs to $BV(\Omega, \{0, 1\})$;

(ii) *Lower bound inequality*: if $u \in BV(\Omega, \{0, 1\})$, $(u_\epsilon) \subset W^{1,2}(\Omega)$ and $u_\epsilon \to u$ in $L^1(\Omega)$ then

$$\liminf_{\epsilon \to 0} F_\epsilon(u_\epsilon) \geq \sigma \mathcal{H}^{N-1}(Su) \ ; \tag{3.1}$$

(iii) *Upper bound inequality*: for every $u \in BV(\Omega, \{0, 1\})$ exists $(u_\epsilon) \subset W^{1,2}(\Omega)$ such that $u_\epsilon \to u$ in $L^1(\Omega)$, $\int u_\epsilon = \int u$ for every ϵ and

$$\limsup_{\epsilon \to 0} F_\epsilon(u_\epsilon) \leq \sigma \mathcal{H}^{N-1}(Su) \ . \tag{3.2}$$

Warning: in the following we consider only functions which take values in the interval $[0, 1]$. For sequences of such functions convergence in measure is equivalent to convergence in L^p for any $p \in [1, \infty)$, and then we simply call it *strong convergence*. Similarly, all weak L^p-topologies with $p \in [1, \infty)$ induce the same convergence, which is therefore referred to as *weak convergence*.

We first prove the three statements above in the one-dimensional case, and then we briefly show how to pass to the two-dimensional case, which requires the same amount of work as the passage to arbitrary dimension.

3a The one-dimensional case

We can assume that Ω is an open bounded interval. In this case $BV(\Omega, \{0, 1\})$ turns out to be the class of all $u : \Omega \to \{0, 1\}$ with finitely many discontinuities (that is, piecewise constant on Ω), and $\mathcal{H}^0$ is simply the measure that counts points. We often consider functions u such that $F_\epsilon(u)$ is finite; then u belongs to $W^{1,2}(\Omega)$, and in particular it admits a continuous representant on $\overline{\Omega}$; unless differently stated we always refer to this representant. The basic ingredients of the proof are given in the following three paragraphs.

3.1   Localization of F_ϵ

It is useful to consider F_ϵ also as a function of the integration domain; hence we set

$$F_\epsilon(u, A) := \epsilon \int_A |\dot{u}|^2 + \frac{1}{\epsilon} \int_A W(u) \tag{3.3}$$

(for every measurable set A and every function u whose derivative belongs to $L^2(U)$ for some open set U which contains A). In particular $F_\epsilon(u) = F_\epsilon(u, \Omega)$. The functional in (3.3) is sometimes called the *localization* of the functional F_ϵ given in (2.3). Notice that $F_\epsilon(u, A)$ is a positive measure with respect to the variable A; in the following we often use this property without further mention.

3.2   Scaling property of F_ϵ

For every A and every u we set $u^\epsilon(x) := u(\epsilon x)$ and $\frac{1}{\epsilon}A := \{x : \epsilon x \in A\}$. Then we immediately obtain the following *scaling identity*

$$F_\epsilon(u, A) = F_1\left(u^\epsilon, \tfrac{1}{\epsilon}A\right) . \tag{3.4}$$

In some sense we may say that the choice of scaling $F_\epsilon := \frac{1}{\epsilon}E_\epsilon$ is the optimal one exactly because of identity (3.4).

3.3   The optimal profile problem

We consider now the minimum problem

$$\bar{\sigma} := \inf\left\{F_1(u, \mathbf{R}) : u : \mathbf{R} \to [0,1], \lim_{x \to -\infty} u(x) = 0, \lim_{x \to +\infty} u(x) = 1\right\} . \tag{3.5}$$

The number $\bar{\sigma}$ represents the minimal cost in term of the energy F_1 for a transition from the value 0 to the value 1 on the entire real line. The minimum problem (3.5) is therefore called the *optimal profile problem*, and a solution γ is an *optimal profile for transition* (with respect to the non-scaled energy F_1). We will show that the optimal profile problem is the key to the proof of Theorem 1, and in particular $\bar{\sigma}$ turns out to be equal to the constant σ in the statement of Theorem 1 (see Proposition 2). The connection between $\bar{\sigma}$ and the cost of transition relative to the energy F_ϵ is made by the following lemma:

Lemma 1. *Let be given an interval I and a function $u : I \to [0,1]$. Assume that there exists $a, b \in I$ and $\delta > 0$ so that $u(a) \leq \delta$ and $u(b) \geq 1 - \delta$. Then for every $\epsilon > 0$ there holds*

$$F_\epsilon(u, I) \geq \bar{\sigma} - O(\delta) , \tag{3.6}$$

where the error estimate $O(\delta)$ depends on δ, but neither on ϵ nor on u.

Proof. We can assume that $a < b$ and $I = (a, b)$, otherwise we just replace I with (a, b). By identity (3.4) it suffices to prove (3.6) when $\epsilon = 1$. In order to compare $F_\epsilon(u, I)$ with $\bar{\sigma}$ we extend u to the whole of $\mathbf{R}$ as shown in the picture.

Now an immediate estimate shows that $F_1(u, \mathbf{R} \setminus I) \leq \delta + C\delta$ where C is the maximum value of W in $[0, 1]$. Hence we set $O(\delta) := (1 + C)\delta$ and we conclude

$$F_1(u, I) = F_1(u, \mathbf{R}) - F_1(u, \mathbf{R} \setminus I) \geq \bar{\sigma} - O(\delta) . \qquad \square$$

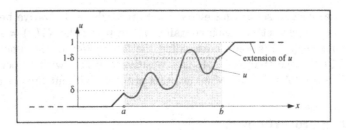

Fig. 1. Extension of the function u out of $I = (a, b)$

Proposition 2. *The minimum in (3.5) is attained and agree with the constant* σ *given in Theorem 1, that is,* $\bar{\sigma} = 2 \int_0^1 \sqrt{W(u)}\, du$.

Proof. We first prove that $F_1(u, \mathbf{R}) \geq \sigma$ for every $u : \mathbf{R} \to [0, 1]$ which tends to 1 (resp. to 0) at $+\infty$ (resp. at $-\infty$), and then we show that the equality is attained by some particular choice of u. We apply the inequality $a^2 + b^2 \geq 2ab$, with $a := \dot{u}(x)$ and $b := \sqrt{W(u(x))}$, to the definition of $F_1(u, \mathbf{R})$, and then we get

$$
\begin{aligned}
F_1(u, \mathbf{R}) &= \int_{-\infty}^{+\infty} \left[\dot{u}^2(x) + W(u(x)) \right] dx \\
&\geq 2 \int_{-\infty}^{+\infty} \sqrt{W(u(x))}\, \dot{u}(x)\, dx = 2 \int_0^1 \sqrt{W(u)}\, du =: \sigma\, .
\end{aligned}
\tag{3.7}
$$

Now we recall that equality holds in $a^2 + b^2 \geq 2ab$ when $a = b$, and then equality holds in (3.7) when u satisfies the differential equation $\dot{u} = \sqrt{W(u)}$. Thus we consider the Chauchy problem

$$
\begin{cases}
\dot{u} = \sqrt{W(u)}\, , \\
u(0) = 1/2\, .
\end{cases}
\tag{3.8}
$$

The constant functions 1 and 0 are solution of $\dot{u} = \sqrt{W(u)}$ (because $W(0) = W(1) = 0$), and since $\sqrt{W}$ is continuous the problem (3.8) admits a global (increasing) solution on $\mathbf{R}$ (we have uniqueness when W is of class C^1). Since $W(u) > 0$ for $0 < u < 1$, this global solution must converge to 1 (resp. to 0) at $+\infty$ (resp. at $-\infty$). □

Remark 9. The possibility of an explicit computation of $\bar{\sigma}$ is quite specific of the form of the functional F_1. In many generalizations such a computation is not possible (cf. paragraphs 4.2 and 4.4). For this reason it is preferable to define σ and $\bar{\sigma}$ separately, and then show that $\sigma = \bar{\sigma}$. For the rest of the proof we will refer mainly to $\bar{\sigma}$, this being the "natural" constant to consider.

3.4   Proof of statements (i) and (ii)

We first sketch the idea of the proof. If we take a sequence (u_ϵ) such that $F_\epsilon(u_\epsilon) \leq C < +\infty$, then in particular $\int_\Omega W(u_\epsilon) \leq C\epsilon$, and this implies that the functions u_ϵ take values close to 0 or 1 outside an exceptional set with measure of order ϵ (recall that W is continuous and strictly positive between 0 and 1). If the sequence (u_ϵ) converges weakly but not strongly, then it must "oscillate" between values close to 0 and 1; on the other hand Lemma 1 shows that the cost of each oscillation (in term of the localized energy F_ϵ) is roughly of order $\bar{\sigma}$, and then the bound on $F_\epsilon(u_\epsilon)$ allows only for finitely many oscillations. Hence (u_ϵ) converges strongly to a limit function u which takes values 0 or 1 almost everywhere. Let us compute now the number of transitions of u from 0 to 1 or viceversa. Passing to a subsequence, and modifying u in a set of measure 0, we may assume that u_ϵ converge to u everywhere. Then, if we take x_0 and x_1 so that $u(x_0) = 0$ and $u(x_1) = 1$, by (3.6) we get $F_\epsilon\big(u_\epsilon,(x_0,x_1)\big) \geq \bar{\sigma} - o(1)$. Hence the bound $F_\epsilon(u_\epsilon) \leq C$ implies that u has at most $C/\bar{\sigma}$ transitions from 0 to 1, that is, $\mathcal{H}^0(Su) \leq C/\bar{\sigma}$. Eventually we notice that passing to subsequences we can take the bound $C := \sup F_\epsilon(u_\epsilon)$ arbitrarily close to the lower limit of $F_\epsilon(u_\epsilon)$, and then (3.1) is proved.

The previous heuristic argument can be made rigorous in a very elegant and simple way by use of the notion of Young measure. To this aim we refer the interested reader to the coincise paper [Bal89]; a more detailed and exhaustive treatment of the subject can be found in [Val94], while a how-to-use guide will be provided in the first chapters of [Mül]. For our purposes it should be enough to recall that given an arbitrary sequence of functions $u_\epsilon : \Omega \to [0,1]$, we may extract a subsequence (not relabelled) whose asymptotic behaviour is captured by a certain family of probability measures $\{\nu_x : x \in \Omega\}$ called the *Young measure generated by* (u_ϵ). Every ν_x is a probability measure on the interval $[0,1]$ related with the asymptotic distribution of the values of u_ϵ, close to x. In particular the functions u_ϵ converge weakly to the function u which takes every $x \in \Omega$ into the center of mass of ν_x (that is, $u(x) := \int u\,d\nu_x(u)$), and we have strong convergence if and only if ν_x is a Dirac mass for a.e. $x \in \Omega$. Moreover for every test function $f : \Omega \times [0,1] \to \mathbf{R}$ which is continuous with respect to the second variable and bounded there holds

$$\int_\Omega f\big(x, u_\epsilon(x)\big)\,dx \longrightarrow \int_\Omega \Big[\int_0^1 f(x,u)\,d\nu_x(u) \Big]\,dx \ . \tag{3.9}$$

Take now a sequence (u_ϵ) such that $F_\epsilon(u_\epsilon) \leq C < +\infty$, and assume that it generates a Young measure $\{\nu_x : x \in \Omega\}$. Then $\int_\Omega W(u_\epsilon) \to 0$, and if we apply (3.9) with $f(x,u) := W(u)$ we get

$$\int_\Omega \Big[\int_0^1 W(u)\,d\nu_x(u) \Big]\,dx = 0 \ .$$

We infer that for a.e. $x \in \Omega$ the measure ν_x is supported on $\{0,1\}$, and then it can be written as

$$\nu_x := \lambda(x) \cdot \delta_0 + \big(1 - \lambda(x)\big) \cdot \delta_1 \ , \tag{3.10}$$

for suitable $\lambda(x) \in [0,1]$ (as usual δ_t denotes the Dirac mass at t). Take now any interval $I \subset \Omega$ where λ is neither a.e. equal to 0 nor a.e. equal to 1. Then the functions u_ϵ must take in I values close to 0 and values close to 1, and then Lemma 1 yields $F_\epsilon(u_\epsilon, I) \geq \bar{\sigma} - o(1)$. Therefore the bound on $F_\epsilon(u_\epsilon)$ implies that we can find at most $C/\bar{\sigma}$ disjoint intervals of this type. Hence λ agrees a.e. with a function u which takes values 0 or 1 everywhere and has finitely many transition from 0 to 1; in short, $u \in BV(\Omega, \{0,1\})$. Hence ν_x is a Dirac mass at 0 for a.e. x such that $u(x) = 0$, and a Dirac mass at 1 for a.e. x such that $u(x) = 1$. This implies that the limit of the functions u_ϵ is u, and the convergence is strong; statement (i) is proved. Since the number of transition of u from 0 to 1 is by definition $\mathcal{H}^0(Su)$, the previous argument provides the following bound:

$$C \geq \bar{\sigma} \mathcal{H}^0(Su) \ . \tag{3.11}$$

Inequality (3.11) implies (3.1) because, passing to subsequences, we can take the upper bound C arbitrarily close to the lower limit of $F_\epsilon(u_\epsilon)$.

3.5   Proof of statement (iii)

Once the proof in paragraph 3.4 and the meaning of the optimal profile problem (3.5) are well understood, the rest of the proof of Theorem 1, namely the upper bound inequality, is almost immediate. Assume for simplicity that Ω is an interval which contains the point 0 and that $u(x) = 1$ for $x \geq 0$ and $u(x) = 0$ for $x < 0$. Thus Su consists of the sole point 0, and we construct the approximating sequence (u_ϵ) which satisfies (3.2) by taking suitable scaling of the optimal profile γ; more precisely we set $u_\epsilon(x) := \gamma(x/\epsilon)$. Then $u_\epsilon(x) \to u(x)$ for every $x \neq 0$ because the limit of γ at $+\infty$ is 1 and the limit at $-\infty$ is 0, and identity (3.4) yields

$$F_\epsilon(u_\epsilon) = F_1(\gamma, \tfrac{1}{\epsilon}\Omega) \leq F_1(\gamma, \mathbf{R}) = \bar{\sigma} \ .$$

So far we did not care about the constraint $\int u_\epsilon = \int u$; in order to fit it, one has to slighlty modify the previous definition, for instance by taking carefully chosen translations of u_ϵ. The construction of (u_ϵ) in the general case is sketched in the picture below; the details are left to the reader.

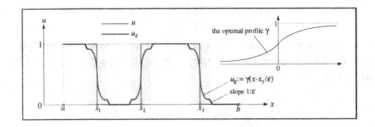

Fig. 2. Construction of the function u_ϵ, here $\Omega = (a,b)$ and $Su = \{x_1, x_2, x_3\}$

3b The general case

The proof of Theorem 1 in dimension larger than one is usually achieved by suitable adaptations of the proof for the one-dimensional case. Yet it is important to notice that statements (i) and (ii) can be deduced directly from the corresponding one-dimensional statements via a general slicing argument (recall that statement (i), (ii), (iii) are given at the beginning of Section 3 and proved in the one-dimensional case in subsection 3a). This fact shows that to some extent the nature of Theorem 1 is one-dimensional; for instance it can hardly adapt to the so-called Ginzburg-Landau functionals (see paragraph 4.3). We remark that the use of slicing arguments to reduce to lower-dimensional statements has already been applied in many different situations; we just mention here the proof of the compactness theorem for integral currents due to B. White [Whi89], the original proof of the compactness theorem for SBV functions in [Amb89], and the rank-one property of derivatives of BV functions in [Alb93]. For simplicity we give the proof only in the two-dimensional case.

3.6 Some notation

We assume for the time being that Ω is a rectangle of the form $I \times J$, with I, J open intervals with length smaller than 1, and we write every $x \in \Omega$ as $x = (y, z)$ with $y \in I$, $z \in J$. For every function u defined on Ω and every $y \in I$ we denote by u^y the function on J defined by $u^y(z) := u(y, z)$, and for every $z \in J$ we denote by u^z the function on I defined by $u^z(y) := u(y, z)$. The functions u^y and u^z are called *one-dimensional slices* of u. We denote by $\overline{F}_\epsilon(u, A)$ the one-dimensional functional given in (3.3) for every open interval A and every $u : A \to [0, 1]$. We recall now that if $u \in W^{1,2}(\Omega)$ then $u^y \in W^{1,2}(J)$ for a.e. $y \in I$ and $u^z \in W^{1,2}(I)$ for a.e. $z \in J$, and

$$\frac{\partial u}{\partial z}(x) = \dot{u}^y(z) , \quad \frac{\partial u}{\partial y}(x) = \dot{u}^z(y) \quad \text{for a.e. } x \in \Omega$$

(see [EG92], section 4.9.2). Since $|\nabla u|^2 \geq \left|\frac{\partial u}{\partial z}\right|^2$, (resp. $\left|\frac{\partial u}{\partial y}\right|^2$), we immediately obtain the following *slicing inequalities* :

$$F_\epsilon(u) \geq \int_I \overline{F}_\epsilon(u^y, J) \, dy \quad \left(\text{resp. } \int_J \overline{F}_\epsilon(u^z, I) \, dz \right) . \tag{3.12}$$

3.7 A compactness criterion via slicing

In order to deduce statement (i) from the corresponding one-dimensional compactness statement, we have to make a connection between the pre-compactness of a sequence (u_ϵ) of functions from Ω into $[0, 1]$ and the pre-compactness of the slices (u_ϵ^y) and (u_ϵ^z). The simplest result on this line is the following criterion:

(C1) *assume that $(u_\epsilon^y) \subset\subset L^1(J)$ for a.e. $y \in I$ and $(u_\epsilon^z) \subset\subset L^1(I)$ for a.e. $z \in J$; then $(u_\epsilon) \subset\subset L^1(\Omega)$*

(here $* \subset \subset **$ reads as "$*$ is pre-compact in $**$"). Unfortunately this result does not fit our purposes, but a sufficently general statement is obtained by allowing for some "perturbations": we say that a sequence $(\bar{u}_\epsilon)$ is δ-*close to* (u_ϵ) if $\|u_\epsilon - \bar{u}_\epsilon\|_1 < \delta$ for every ϵ, and then we have the following modification of (C1):

(C2) *assume that for every $\delta > 0$ there exist sequences $(u_{\delta,\epsilon})$ and $(\hat{u}_{\delta,\epsilon})$ δ-close to (u_ϵ) so that $(u_{\delta,\epsilon}^y) \subset \subset L^1(J)$ for a.e. $y \in I$ and $(\hat{u}_{\delta,\epsilon}^z) \subset \subset L^1(I)$ for a.e. $z \in J$; then $(u_\epsilon) \subset \subset L^1(\Omega)$.*

The proof of these compactness criteria essentially relies on the characterization of pre-compact sets in the strong L^1-topology given by the Fréchet-Kolmogorov theorem; for a general statement and a detailed proof we refer to [ABS98], Section 6.3.

3.8 Proof of statements (i) and (ii)

Let be given (u_ϵ) so that $F_\epsilon(u_\epsilon) \le C < +\infty$. Then (3.12) yields

$$C \ge \int_I \overline{F}_\epsilon(u_\epsilon^y, J)\, dy \quad \text{for every } \epsilon > 0 \tag{3.13}$$

(and a similar inequality holds for $\overline{F}_\epsilon(u_\epsilon^z, I)$). We fix now $\delta > 0$ and for every $\epsilon > 0$ we take $u_{\delta,\epsilon} : \Omega \to [0,1]$ so that

$$u_{\delta,\epsilon}^y := \begin{cases} u_\epsilon^y & \text{if } \overline{F}_\epsilon(u_\epsilon^y, J) \le C/\delta, \\ 0 & \text{otherwise.} \end{cases}$$

By (3.13) we have that $u_{\delta,\epsilon}^y = u_\epsilon^y$ for all $y \in I$ apart a set of measure smaller than δ, and then $\|u_\epsilon - u_{\delta,\epsilon}\|_1 \le \delta|I| \le \delta$. Hence the sequence $(u_{\delta,\epsilon})$ is δ-close to (u_ϵ). Moreover, for every $y \in I$ there holds $\overline{F}_\epsilon(u_{\delta,\epsilon}^y, J) \le C/\delta$ (recall that $\overline{F}_\epsilon(0, I) = 0$) and then the sequence $(u_{\delta,\epsilon}^y)$ is pre-compact in $L^1(J)$ by the one-dimensional version of statement (i). In a similar way we can construct a sequence $(\hat{u}_{\delta,\epsilon})$ δ-close to (u_ϵ) so that $(\hat{u}_{\delta,\epsilon}^z) \subset \subset L^1(I)$ for a.e. $z \in J$, and therefore the compactness criterion (C2) shows that the sequence (u_ϵ) is pre-compact in $L^1(\Omega)$.

Assume now that (u_ϵ) converge to u in $L^1(\Omega)$. Then $u_\epsilon^y \to u^y$ in $L^1(J)$ for a.e. $y \in I$, and if we apply Fatou's lemma to inequality (3.12) we get

$$\liminf_{\epsilon \to 0} F_\epsilon(u_\epsilon) \ge \int_I \left[\liminf_{\epsilon \to 0} \overline{F}_\epsilon(u_\epsilon^y, J)\right] dy \ .$$

Hence $\liminf \overline{F}_\epsilon(u_\epsilon^y, J)$ is finite for a.e. $y \in I$, and then u^y belongs to $BV(J, \{0,1\})$ by the one-dimensional version of (i). Moreover the the one-dimensional version of (ii) yields

$$\liminf_{\epsilon \to 0} F_\epsilon(u_\epsilon) \ge \sigma \int_I \mathcal{H}^0(Su^y)\, dy \ , \tag{3.14}$$

and recalling that $\mathcal{H}^0(Su^y)$ is the total variation of u^y in $BV(J)$, (3.14) yields $\int_I \|\dot{u}^y\| \, dy < +\infty$, and in a similar way one gets $\int_J \|\dot{u}^z\| \, dz < +\infty$. We use now the following important fact: a function $u \in L^1(\Omega)$ belongs to $BV(\Omega)$ if (and only if) $u^y \in BV(J)$ for a.e. $y \in I$, $u^z \in BV(I)$ for a.e. $z \in J$, and $\int_I \|\dot{u}^y\| \, dy$ and $\int_J \|\dot{u}^z\| \, dz$ are finite (see [EG92], section 5.10.2).

Finally we recover the lower bound inequality (3.1) from (3.14). Assume for the moment that Su is a regular curve in Ω. Then the integral at the right hand side of (3.14) is just the measure of the projection of Su on I (keeping the multiplicity into account), which in general can be smaller than the length of Su. In fact the length of a curve C in $I \times J$ is close to the measure of its projection on I if and only if the normal to C is close to the projection axis (in this case the z-axis). Keeping this in mind we cover Su (up to a subset with small measure) by pairwise disjoint squares Q_i so that within each Q_i the normal to Su is close to one of the axes of Q_i.

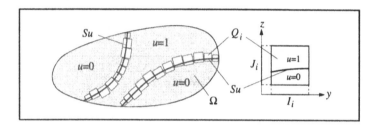

Fig. 3. Covering the set Su with squares

Then, for every Q_i, inequality (3.14) becomes

$$\liminf_{\epsilon \to 0} F_\epsilon(u_\epsilon, Q_i) \geq \sigma \int_{I_i} \mathcal{H}^0(Su^y) \, dy \simeq \sigma \mathcal{H}^1(Su \cap Q_i) \,,$$

and taking the sum over all i,

$$\liminf_{\epsilon \to 0} F_\epsilon(u_\epsilon) \geq \sum_i \liminf_{\epsilon \to 0} F_\epsilon(u_\epsilon, Q_i) \geq \sigma \mathcal{H}^1(Su) - o(1) \,.$$

This argument can be made rigorous for every singular set Su by a careful use of the Besicovitch covering theorem (see [EG92], section 1.5) and a detailed description of the pointwise property of rectifiable sets and of the measure theoretic boundary of sets with finite perimeter (cf. [EG92], chapter 5).

3.9 Proof of statement (iii)

Unlike statements (i) and (ii), the proof of statement (iii) cannot be achieved by reduction to the one-dimensional case. On the other hand by Remark 1 it is enough to prove the upper bound inequality only for a suitable dense subset $\mathcal{D}$ in

X. In this case we choose as $\mathcal{D}$ the class of all $u \in BV(\Omega, \{0,1\})$ whose singular set Su is a piecewise affine curve in $\mathbf{R}^2$ (a polyhedral surface of dimension $N - 1$ when N is general); indeed every $u \in BV(\Omega, \{0,1\})$ can be approximated by a sequence $(u_n) \subset \mathcal{D}$ so that $\mathcal{H}^1(Su_n) \to \mathcal{H}^1(Su)$; this is an immediate consequence of a well-known approximation result for finite perimeter sets by smooth sets, see [Giu84], Theorem 1.24. (In fact another typical choice for $\mathcal{D}$ is given by the class of all u such that Su is a smooth curve with boundary included in $\partial\Omega$). Thus we take $u \in \mathcal{D}$, and given $\epsilon > 0$ we construct u_ϵ as follows: we cover Su with disjoint rectangles R_i with width $\epsilon^{2/3}$, up to a residual set with measure of order $\epsilon^{2/3}$.

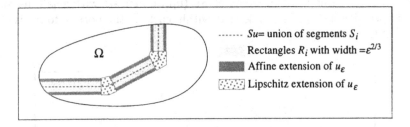

Fig. 4. Covering of Su

In each rectangle R_i (see Fig. 4 above) we set $u_\epsilon(x) := \gamma(x_i/\epsilon)$ where x_i is the oriented distance of the point x from the segment S_i (so that it is positive on the side of S_i where $u = 1$ and negative on the side where $u = 0$). In the darker rectangles we take an affine extension of u_ϵ which agrees with u on the sides which border the white region, and we choose the width of each rectangle so that the slope of u_ϵ is $1/\epsilon$ (therefore this width has order $o(\epsilon)$). In the white region we take u_ϵ equal to u. Finally, we define u_ϵ in the interior of the remaining dotted regions by taking any Lipschitz extension with the same Lipschitz constant as on the boundary, which has order $O(\epsilon^{-1})$. Hence, within each R_i (and in the corresponding darker rectangles) the function u_ϵ varies only in the direction ν_i normal to S_i, and $F_\epsilon(u_\epsilon, R_i) \leq \sigma\mathcal{H}^1(S_i)$; while the contributions of the other regions vanish as $\epsilon \to 0$. Thus $F_\epsilon(u_\epsilon) \leq \sigma\mathcal{H}^1(Su) + o(1)$, and the proof of statement (iii) is completed.

4   Final remarks

Theorem 1 was conjectured by E. De Giorgi and then proved by L. Modica and S. Mortola [MM77] in 1977, shortly after the definition of Γ-convergence was given in [GF75]; the connection with the Cahn-Hilliard model was established by L. Modica [Mod87] only in 1987. Since then several results were given which extend Theorem 1 in different directions (cf. paragraphs 4.1, 4.2 and 4.4). The idea of defining $\bar{\sigma}$ via a suitable minimum problem involving only non-scaled functionals is common knowledge, and I was not able to trace the source. The idea of proving compactness and lower bound inequality via Young measures (and Lemma 1) is essentially contained in [ABS94].

Remark 10. The minimum problem (3.5) leads to the Euler-Lagrange equation $2\ddot{u} - \dot{W}(u) = 0$. Hence solution of (3.5) are standing waves for the parabolic equation $u_t = 2u_{xx} - f(u)$ for certain choices of f. Standing and travelling waves for this equation have been widely studied in the literature, since connected with the asymptotic behaviour of the solutions of the scalings of the Allen-Cahn equation $u_t = \Delta u - f(u)$; we refer the reader to [Amb99] and the references therein.

Remark 11. Notice that throughout the whole proof of Theorem 1, what we really need is the positivity of $\bar{\sigma}$, while it is not strictly necessary that the infimum in (3.5) is attained (in fact a suitable modification of the proof of statement (iii) works even if no optimal profile is available). Nevertheless the existence of the optimal profile has a deeper meaning than it appears in the proof above. Indeed if (v_ϵ) is a sequence of minimizers of F_ϵ which converges to some $v \in BV(\Omega, \{0, 1\})$, then the upper bound inequality is verified, and we would naturally conclude that if we blow-up the functions v_ϵ at some fixed singular point $\bar{x}$ of v by taking the functions $\gamma_\epsilon(x) := v_\epsilon(\epsilon(x - \bar{x}))$, then γ_ϵ should more and more resemble an optimal profile. In other words we expect the optimal profiles to be the asymptotic shapes of the minimizers v_ϵ close to the discontinuity points of v. Yet a precise statement cannot be easily formulated in the current framework.

Remark 12. The existence of a solution of the optimal profile problem (3.5) cannot be deduced by standard semicontinuity and compactness results: indeed not only the functional $F(\cdot, \mathbf{R})$ is translation invariant, but also its natural domain is the class of all $u : \mathbf{R} \to [0, 1]$ such that $\nabla u \in L^2(\mathbf{R})$, and for such functions the limits at $\pm\infty$ are not always defined (take for instance $u(x) := \sin^2 \log(1 + x^2)$). An alternative way to find a solution of (3.5) is via rearrangement. Given a function $u : \mathbf{R} \to [0, 1]$ which tends to 1 at $+\infty$ and to 0 at $-\infty$, then each sublevel $E_t := \{x : u(x) > t\}$ with $t \in (0, 1)$ can be written as the disjoint union of a bounded A_t and an half line $(b_t, +\infty)$; we define the *increasing rearrangement* of u as the function u^* whose sublevels are the half-lines $E_t^* := (b_t - a_t, +\infty)$, where a_t is the measure of A_t. This rearrangement operator decreases the functional $F_1(\cdot, \mathbf{R})$ among others, that is, $F_1(u, \mathbf{R}) \geq F_1(u^*, \mathbf{R})$ (see [Kaw85]). Hence in (3.5) we can restrict to the subclass of increasing functions u such that $\nabla u \in L^2(\mathbf{R})$ and $u(0) = 1/2$ (we add this constraint to work out the translation invariance of the functional), which is compact with respect to the strong convergence; then the existence of an optimal profile follows from the (strong) semicontinuity of $F_1(\cdot, \mathbf{R})$.

4.1 The vectorial case

The mechanical model described at the beginning of section 2 applies to mixtures of two fluids only, but can be generalized to mixtures of an arbitrary number m of fluids. In this case every configuration of the macroscopic model can be described by a function $u : \Omega \to \{\alpha_1, \ldots, \alpha_m\}$ where $\alpha_1, \ldots, \alpha_m$ are arbitrarily chosen affinely independent points in $\mathbf{R}^{m-1}$ (each one corresponds to one fluid in the mixture), and the energy $F(u)$ in (2.1) must be rewritten as $F(u) := \sum_{i<j} \sigma_{ij} \mathcal{H}^2(S_{ij}u)$, where $S_{ij}u$ is the interface between the phases $\{u = \alpha_i\}$ and $\{u = \alpha_j\}$, and σ_{ij} is the corresponding surface tension. If V_i is the total volume of the phase α_i, then the admissible configurations are all u which satisfy the volume constraint $\int_\Omega u = \sum V_i \alpha_i$. Notice that the corresponding minimum problem is well-posed if the coefficients σ_{ij} satisfy the following *wetting conditions*: $\sigma_{ij} + \sigma_{jk} \geq \sigma_{ik}$ for all i, j, k. In fact, if any of these inequalities does not hold, then F is not lower semicontinuous on $BV(\Omega, \{\alpha_1, \ldots, \alpha_m\})$ (it is worth trying to understand why!)

In the continuous model u takes values in the convex hull T of $\{\alpha_1, \ldots, \alpha_m\}$, and the associated energy $E_\epsilon(u)$ is given as in (2.2) with W a continuous positive function on $\mathbf{R}^{m-1}$ which vanishes at $\alpha_1, \ldots, \alpha_m$. If we modify (2.3) and (2.4) accordingly, then Theorem 1 holds, provided we set $\sigma_{ij} := \mathrm{dist}(\alpha_i, \alpha_j)$ where dist is the geodesic distance on T associated with the metric $\sqrt{W(u)}\, du$ (this result was first proved in [Bal90], see also [BF94] for a more general result). The proof can be achieved by modifying the argument of section 3; the optimal profile problem (3.5) becomes now

$$\bar{\sigma}_{ij} := \inf \left\{ F_1(u, \mathbf{R}) : u : \mathbf{R} \to [0,1], \lim_{x \to -\infty} u(x) = \alpha_i, \lim_{x \to +\infty} u(x) = \alpha_j \right\}.$$

Notice that one of the main technical difficulties in this proof was due to the lack of a "nice" dense subclass for $BV(\Omega, \{\alpha_1, \ldots, \alpha_m\})$ (cf. the proof in paragraph 3.9).

4.2   The general anisotropic case

Anisotropic functionals of type (2.2) are obtained for instance when we replace the Dirichlet integral $\int |\nabla u|^2$ with the quadratic form $\int \langle A \nabla u; \nabla u \rangle$ where A is a symmetric positive definite $n \times n$ matrix. This is the quadratic case of a larger class of anisotropic functionals considered in [Bou90], [OS91] and then in [BF94]. Here the surface tension σ depends on the orientation of the interface, and the Γ-limit of F_ϵ is given by the integral over the interface Su of $\sigma(\nu)$ where ν is the normal to Su. For every direction e the value of $\sigma(e)$ is given by an N-dimensional version of the optimal profile problem (3.5). In some cases (including the quadratic case) it is still possible to apply some rearrangement theorem to prove that the optimal profile problem reduces to a one-dimensional minimization problem.

4.3   The Ginzburg-Landau functionals

An important variation of the Cahn-Hilliard functionals are the Ginzburg-Landau functionals, which are defined as in (2.2) for all $u : \Omega \to \mathbf{R}^2$, by taking W a continuous positive function which vanishes on the unit circle $S^1 := \{|u| = 1\}$.

In this case when ϵ tends to 0 the function u is forced to take values closer and closer to the unit circle, and then we expect that the Γ-limit of suitable rescalings of E_ϵ is finite on functions $u : \Omega \to S^1$ with singularities of co-dimension 2 (in particular when Ω has dimension 2 we expect point singularities). For this reason the asymptotic behaviour of these functionals differs deeply from what described in Theorem 1. However no Γ-convergence result is available so far, while a complete description of the asymptotic behaviour of the minimizing sequences (under some boundary constraint) was carried out when Ω has dimension 2 in [BBH94].

4.4 A different type of interaction energy

Another variation of the Cahn-Hilliard functionals is obtained by replacing the Dirichlet energy $\epsilon^2 \int |\nabla u|^2$ in (2.2) with suitable scalings of a non-local interaction energy $I_\epsilon(u)$. In [AB98a] and [ABCP96] is considered the case

$$I_\epsilon(u) := \iint J_\epsilon(x' - x) \left(u(x') - u(x) \right)^2 dx' dx \ ,$$

where $J_\epsilon(h) := \epsilon^{-N} J(h/\epsilon)$ and J is a positive interaction potential in $L^1(\mathbf{R}^N)$. These kind of functionals arises as scalings of the free energy of a continuum limit of spin systems on lattices, or Ising systems . Theorem 1 is still true, but now the surface tension σ is directly defined through the optimal profile problem (3.5), and cannot be computed explicitly in term of J and W; the existence of an optimal profile and the positivity of σ have been proved in this case via rearrangement, as described in Remark 12 (see [AB98b]).

4.5 The original proof of Theorem 1

The proof of statement (i) and (ii) of Theorem 1 given in [Mod87] is very simple and elegant, and works directly in the N-dimensional case; it is quite interesting to compare it with the proof given in section 3. Given $u : \Omega \to [0, 1]$, one applies the inequality $a^2 + b^2 \geq 2ab$ with $a := \sqrt{W(u)/\epsilon}$ and $b := \sqrt{\epsilon}|\nabla u|$ and obtains

$$\begin{aligned} F_\epsilon(u) = \int_\Omega \left(\epsilon|\nabla u|^2 + \epsilon^{-1}W(u) \right) dx \\ \geq 2 \int_\Omega \sqrt{W(u)} \, |\nabla u| \, dx = \int_\Omega |\nabla(H(u))| \, dx \end{aligned} \tag{4.1}$$

where $H : [0, 1] \to \mathbf{R}$ satisfies $\dot{H} = 2\sqrt{W}$. Take now a sequence (u_ϵ) such that $F_\epsilon(u_\epsilon) \leq C < +\infty$. Then (4.1) implies that the functions $H(u_\epsilon)$ are uniformly bounded in $BV(\Omega)$, and then pre-compact in $L^1(\Omega)$, and since H admits a continuous inverse, also (u_ϵ) is pre-compact in $L^1(\Omega)$, and every limit point must takes values 0 or 1 a.e. by the usual argument. Assume now that (u_ϵ) converge to some limit u. Since u takes values 0 and 1 only, then $H(u)$ takes values $H(0)$ and $H(1)$ only, and therefore the total variation of the measure derivative $D(H(u))$ is equal to the total variation of Du multiplied by a factor $H(1) - H(0)$, which is equal to σ (cf. Theorem 1). Hence (4.1) yields

$$\liminf_{\epsilon \to 0} F_\epsilon(u_\epsilon) \geq \liminf_{\epsilon \to 0} \int_\Omega |D(H(u_\epsilon))| \geq \|D(H(u))\| = \sigma \|Du\| = \sigma \mathcal{H}^{N-1}(Su)$$

(the total variation of the measure derivatives is lower semicontinuous with respect to the strong convergence of functions). This completes the proof of statements (i) and (ii), while the proof of statement (iii) is quite similar to the one sketched in paragraph 3.9.

Some aspects of De Giorgi's barriers for geometric evolutions

G. Bellettini and M. Novaga

1 Introduction

Motion by mean curvature has been the subject of several recent papers, and is considered an interesting example of geometric evolution [Ham89, Ilm95, Str96]. A family of sets $(E(t))_{t \in [0,T]}$ evolves smoothly by mean curvature if, letting

$$d(t,x) := \text{dist}(x, E(t)) - \text{dist}(x, \mathbf{R}^n \setminus E(t)), \qquad (t,x) \in [0,T] \times \mathbf{R}^n$$

the oriented distance from $\partial E(t)$ negative inside $E(t)$, there exists an open set $A \subseteq \mathbf{R}^n$ containing $\partial E(t)$ for any $t \in [0,T]$, such that $d \in C^\infty([0,T] \times A)$, and

$$\frac{\partial d}{\partial t} - \Delta d = 0 \qquad \text{on } \partial E(t), \quad t \in [0,T].$$

One of the main features of motion by mean curvature is its variational character, since it can be interpreted as the gradient flow associated with the area functional; if $(E(t))_{t \in [0,T]}$ flows smoothly by mean curvature, we have the following formula for the decrease of the area $\mathcal{H}^{n-1}(\partial E(t))$ of $\partial E(t)$:

$$\frac{d}{dt} \mathcal{H}^{n-1}(\partial E(t)) = - \int_{\partial E(t)} (\Delta d)^2 \, d\mathcal{H}^{n-1}.$$

From the viewpoint of applications, mean curvature flow describes simplified physical phenomena with tension effects, such as phase transitions in material sciences [Mul56, AC79]. Also more general geometric evolutions , possibly depending on the normal vector ∇d, the second fundamental form $\nabla^2 d$, and explicitly on space-time, such as

$$\frac{\partial d}{\partial t} + F(t, x, \nabla d, \nabla^2 d) = 0 \qquad \text{on } \partial E(t),$$

may have a geometric or a physical meaning [TCH93]; we recall, among others, motion by mean curvature with a forcing term (representing, for instance, the action of an exterior field [Spo93]) and the anisotropic mean curvature flow, which arises naturally in mathematical models of phenomena such as dendritic growth, crystal growth [Tay92, CT94], and also may be used to describe the propagation of an electric stimulus in the cardiac tissue [BFP97].

A crucial rôle in studying qualitative properties of (possibly weak) solutions to motion by mean curvature is played by the maximum principle, which is obtained imposing on F a monotonicity dependence (degenerate ellipticity) in the last matrix variable.

Short time existence of smooth solutions, starting from a smooth compact initial boundary, has been proved by Gage-Hamilton in [GH86] in a general setting (see also [Ham75] and [Hui84]). A different proof, based on the oriented distance function , has been given by Evans-Spruck in [ES92a], and generalized in [GG92a] (see also [Lun94] and [Ang90, Ang91] for related results). Qualitative properties of the evolution in the case of curves have been obtained, among others, by Gage [Gag83, Gag84], Gage-Hamilton [GH86], Grayson [Gra87, Gra89b], Angenent [Ang90, Ang91], Altschuler [Alt91], Altschuler-Grayson [AG92]. In [Gra87] Grayson proved that an initial smooth compact embedded plane curve flows smoothly by curvature up to the extinction time (an alternative proof of this result has been recently given in [Hui98]).

In [Hui84] Huisken proved that a smooth strictly convex compact hypersurface shrinks smoothly to a point (see also [ES91, And94a, And94b]). Ecker-Huisken [EH89, EH91] proved that if the initial smooth hypersurface is an entire graph then, under suitable assumptions on the slope and the curvature of the graph, the flow exists globally and it is smooth (see also [Gio90a]).

It is however well known that smooth compact hypersurfaces flowing by mean curvature can develop singularities at finite time, as proved by Grayson in [Gra89a]. Examples of singularities (before the extinction time) can be constructed, among others, by starting from suitable tori or dumbbell shaped sets [AAG95, Ang92, Gio90b, Gio91a, PV92, SS93]. An important tool in studying the asymptotic shape of singularities is the monotonicity formula, discovered by Huisken in [Hui90], which shows that singularity formation is modelled by self shrinking evolutions. Results concerning the description of various types of singularities have been obtained, among others, by Angenent [Ang91], Altschuler [Alt91], Ilmanen [Ilm92], Angenent-Ilmanen-Velasquez [AIVc, AIVb], White [Whi94, Whi] and Hamilton.

The presence of singularities justifies the necessity of introducing weak definitions of motion by mean curvature and, more generally, of geometric evolutions. Clearly, any weak evolution must agree with the classical one as long as the latter exists. Among the generalized methods to treat geometric evolutions past singularities we recall: the approach of Brakke [Bra78], which studies the mean curvature flow in the context of varifolds theory; the approach of Angenent [Ang90, Ang91], concerning curves shortening on surfaces; the approach of Evans-Spruck [ES91, ES92b, ES95], Chen-Giga-Goto [CGG91], Giga-Goto-Ishii-Sato [GGIS91], which consider the level set of the solutions, in the viscosity sense, of suitable parabolic fully non linear partial differential equations, where the notion of viscosity solution was introduced by Crandall- P.-L. Lions [CL83], P.-L. Lions [Lio83], Jensen [Jen88] (see also Jasnow-Kawasaki-Ohta [JKO82], Osher-Sethian [OS88], Ishii [Ish89], Soner [Son93b, Son93a], and Ilmanen [Ilm92]); the solutions that can be obtained as asymptotic limits of the scaled Allen-Cahn equation [BSS93, BK91, Gio90a, Gio91a, MS95, ESS92, Ilm93a, JSb, JSa, MM77] and of a nonlocal equation [MOPT93, MOPT94b, MOPT94a, MOPT96a, MOPT96b, KS94, KS95, KS97]; the variational approach of Almgren-Taylor-Wang [ATW93] (see also Luckhaus-Sturzenhecker [LS95]) and its possible generalizations by means of the minimizing movements of De Giorgi [Gio93a, Amb95a, FK95]; the elliptic regularization method of Ilmanen [Ilm94]; the method of set-theoretic subsolutions of Ilmanen [Ilm95] (see also [Whi95]); the semigroup approach of Bence-Merriman-Osher [BMO92] and Evans [Eva93]; the barriers approach of De Giorgi [Gio93b, Gio94b]; the penalization method on higher derivatives of De Giorgi [Gio96]; the approach of Barles-Souganidis [BS98]. We remark that the relations between all these approaches (after the onset of singularities) have not been completely clarified.

The barriers method, which is the argument we are concerned with in the present paper, provides a weak solution for a number of differential equations. In the geometric context, it gives a natural notion of weak evolution for a large class of flows, such as motion by mean curvature in arbitrary codimension (see [Gio94b, AS96, Amb99]). In this geometric framework, properties such as uniqueness of the weak evolution, the comparison principle , and the agreement of the minimal barrier with the classical flow as long as it exists, are immediate consequences of the definitions. Moreover the method is intrinsic, since it is mainly based on the distance function and on the inclusion relation between sets.

Let us briefly explain the concept of geometric minimal barrier in $\mathbf{R}^n$, and some of its properties. First we choose a nonempty family $\mathcal{F}$ of maps which take some time interval into the set $\mathcal{P}(\mathbf{R}^n)$ of all subsets of $\mathbf{R}^n$: for instance $\mathcal{F}$ can be the family of all smooth local evolutions with respect to a given geometric law. Then we define the class $\mathcal{B}(\mathcal{F})$ of all maps $\phi : [\,\bar{t}, +\infty[\to \mathcal{P}(\mathbf{R}^n)$ which are barriers for $\mathcal{F}$ in $[\,\bar{t}, +\infty[$ with respect to the inclusion of sets, that is, if $f : [a, b] \subseteq [\,\bar{t}, +\infty[\to \mathcal{P}(\mathbf{R}^n)$ belongs to $\mathcal{F}$ and $f(a) \subseteq \phi(a)$, then it must hold $f(b) \subseteq \phi(b)$ (here $\bar{t} \in \mathbf{R}$ is fixed). Finally, we define the minimal barrier $\mathcal{M}(E, \mathcal{F}, \bar{t})(t)$ with origin in the set $E \subseteq \mathbf{R}^n$, with respect to $\mathcal{F}$, at time $t \in [\,\bar{t}, +\infty[$, as

$$\mathcal{M}(E, \mathcal{F}, \bar{t})(t) := \bigcap \Big\{ \phi(t) : \quad \phi : [\,\bar{t}, +\infty[\to \mathcal{P}(\mathbf{R}^n), \phi \in \mathcal{B}(\mathcal{F}), \phi(\bar{t}) \supseteq E \Big\}. \quad (1.1)$$

We stress that the minimal barrier depends on $\mathcal{F}$ and is unique and globally defined, for an arbitrary initial set E; moreover, it enjoys the comparison principle, that is

$$E_1 \subseteq E_2 \Rightarrow \mathcal{M}(E_1, \mathcal{F}, \bar{t})(t) \subseteq \mathcal{M}(E_2, \mathcal{F}, \bar{t})(t), \qquad t \in [\,\bar{t}, +\infty[,$$

and, under minor assumptions on $\mathcal{F}$, the semigroup property.

If we choose $\mathcal{F} = \mathcal{F}_F$ as the family of all smooth local geometric (super) solutions of an equation of the form

$$\frac{\partial u}{\partial t} + F(\nabla u, \nabla^2 u) = 0, \quad (1.2)$$

then $\mathcal{M}(E, \mathcal{F}_F, \bar{t})$ is defined under no assumptions on F and, if $f : [a, b] \subseteq [\,\bar{t}, +\infty[\to \mathcal{P}(\mathbf{R}^n)$, $f \in \mathcal{F}_F$, then $\mathcal{M}(f(a), \mathcal{F}_F, a)(t) \supseteq f(t)$ for any $t \in [a, b]$. It is not difficult to verify that the equality holds true when F is degenerate elliptic (i.e., we have agreement with smooth flows) but, in general, it does not hold for a not degenerate elliptic function F, when it happens that the elements of $\mathcal{F}_F$ are not necessarily elements of $\mathcal{B}(\mathcal{F}_F)$. In this respect it turns out that, if F is lower semicontinuous and if $\mathcal{F}_F^{\geq}$ denotes the family of all strict local geometric supersolutions of (1.2), then

$$\mathcal{B}(\mathcal{F}_F^{\geq}) = \mathcal{B}(\mathcal{F}_{F^+}^{\geq}), \tag{1.3}$$

where F^+ is the smallest degenerate elliptic function greater than or equal to F, that is

$$F^+(p, X) := \sup\{F(p, Y) : Y \geq X\}.$$

Here $(p, X) \in (\mathbf{R}^n \setminus \{0\}) \times \text{Sym}(n) =: J_0$, where $\text{Sym}(n)$ denotes the space of all symmetric real $(n \times n)$-matrices.

In particular, from (1.3) we deduce that $\mathcal{M}(E, \mathcal{F}_F^{\geq}, \bar{t}) = \mathcal{M}(E, \mathcal{F}_{F^+}^{\geq}, \bar{t})$. This shows that, in presence of a non degenerate elliptic function F, the generalized evolution of any set by (1.2) is governed by the parabolic equation in which F is replaced by F^+. For instance, if we consider a geometric evolution of sets where the normal velocity of the interface is a function ζ of the mean curvature, where $\zeta(s)$ is increasing for s in some bounded interval, and decreasing for large $|s|$, then the resulting evolution is defined by means of the smallest decreasing function greater than or equal to ζ. In Proposition 2, under further assumptions on F (which may explicitly depend on (t, x)) we prove the analogue of (1.3) in the viscosity framework, namely that the family of all viscosity subsolutions of

$$\frac{\partial u}{\partial t} + F(t, x, \nabla u, \nabla^2 u) = 0 \tag{1.4}$$

coincides with the family of all viscosity subsolutions of

$$\frac{\partial u}{\partial t} + F^+(t, x, \nabla u, \nabla^2 u) = 0. \tag{1.5}$$

Starting from the minimal barrier, there is a natural way to construct two set-valued maps $\mathcal{M}_*(E, \mathcal{F}, \bar{t})$, $\mathcal{M}^*(E, \mathcal{F}, \bar{t})$ which play a crucial rôle both in the general theory and in the comparison between the barriers and other generalized evolutions. Precisely, given any set $E \subseteq \mathbf{R}^n$ and $\varrho > 0$, let

$$E_\varrho^- := \{x \in \mathbf{R}^n : \text{dist}(x, \mathbf{R}^n \setminus E) > \varrho\}, \tag{1.6}$$

$$E_\varrho^+ := \{x \in \mathbf{R}^n : \text{dist}(x, E) < \varrho\}, \tag{1.7}$$

and define the lower and upper regularizations

$$\mathcal{M}_*(E, \mathcal{F}, \bar{t}) := \bigcup_{\varrho > 0} \mathcal{M}(E_\varrho^-, \mathcal{F}, \bar{t}), \qquad \mathcal{M}^*(E, \mathcal{F}, \bar{t}) := \bigcap_{\varrho > 0} \mathcal{M}(E_\varrho^+, \mathcal{F}, \bar{t}).$$

The map $\mathcal{M}^*(E, \mathcal{F}, \bar{t})$ belongs to $\mathcal{B}(\mathcal{F})$, while $\mathcal{M}_*(E, \mathcal{F}, \bar{t})$ is an $\mathcal{F}$-barrier under minor assumptions on $\mathcal{F}$; moreover $\mathcal{M}_*(E, \mathcal{F}, \bar{t})$, $\mathcal{M}^*(E, \mathcal{F}, \bar{t})$ are stable with respect to topological interior part and closure respectively, and provide a lower and an upper bound for *any* generalized evolution of sets which extends the smooth evolutions, satisfies the comparison principle and the semigroup property. Also, these two maps allow to define the so called n-dimensional *fattening phenomenon*, which is a peculiar singularity of geometric evolutions, i.e., whenever, for some $t_1 \in [\bar{t}, +\infty[$,

$$\mathcal{H}^n\left(\mathcal{M}^*(E, \mathcal{F}, \bar{t})(t) \setminus \mathcal{M}_*(E, \mathcal{F}, \bar{t})(t)\right) = 0 \qquad \text{for } t \in [\bar{t}, t_1],$$

$$\mathcal{H}^n\left(\mathcal{M}^*(E, \mathcal{F}, \bar{t})(t) \setminus \mathcal{M}_*(E, \mathcal{F}, \bar{t})(t)\right) > 0 \qquad \text{for some } t \in]t_1, +\infty[,$$

where $\mathcal{H}^n$ denotes the n-dimensional Hausdorff measure.

Interesting properties of geometric minimal barriers are the disjoint sets property and the joint sets property with respect to $(\mathcal{F}, \mathcal{G})$, where $\mathcal{F}, \mathcal{G}$ are two arbitrary families of set-valued maps. Due to elementary counterexamples (to the joint sets property, for instance, in case of motion by curvature of plane curves) we introduce the regularized versions of these two properties, which read as follows:

$$
\begin{aligned}
E_1 \cap E_2 = \emptyset &\Rightarrow \mathcal{M}_*(E_1, \mathcal{F}, \bar{t})(t) \cap \mathcal{M}^*(E_2, \mathcal{G}, \bar{t})(t) = \emptyset, && t \geq \bar{t}, \\
E_1 \cup E_2 = \mathbf{R}^n &\Rightarrow \mathcal{M}_*(E_1, \mathcal{F}, \bar{t})(t) \cup \mathcal{M}^*(E_2, \mathcal{G}, \bar{t})(t) = \mathbf{R}^n, && t \geq \bar{t}.
\end{aligned}
\tag{1.8}
$$

When $\mathcal{F} = \mathcal{F}_F$ and $\mathcal{G} = \mathcal{F}_G$ for two functions $F, G : J_0 \to \mathbf{R}$, these two properties can be characterized in terms of F and G. In particular, if we let $F_c(p, X) := -F(-p, -X)$ for any $(p, X) \in J_0$, the following assertion holds. Assume that $F : J_0 \to \mathbf{R}$ is continuous, $F^+ < +\infty$ and F^+ is continuous. Then the regularized disjoint sets property and the regularized joint sets property with respect to $(\mathcal{F}_F, \mathcal{F}_{F_c})$ (resp. with respect to $(\mathcal{F}_F, \mathcal{F}_F)$) hold if and only if F is degenerate elliptic (resp. if and only if F^+ is odd).

Remarkably, for motion by mean curvature of boundaries, the regularized disjoint and joint sets properties hold, and therefore in this case the complement of the lower regularization starting at E coincide with the upper regularization starting at $\mathbf{R}^n \setminus E$. The disjoint sets property in general fails for motion by mean curvature with a constant forcing term, corresponding to the equation $\frac{\partial d}{\partial t} - \Delta d + c = 0$ on $\partial E(t)$, for $c \in \mathbf{R}$.

We observe that, in general, the assertions referring to the joint sets property are more difficult to prove that the corresponding ones concerning the disjoint sets property. We remark also that the disjoint and joint sets properties, and hence their characterization, are related to the fattening phenomenon.

Concerning the comparison between barriers and viscosity solutions, it turns out that the sublevel sets of a viscosity subsolution of (1.4) are barriers and, conversely, that a function whose sublevel sets are barriers is a viscosity subsolution of (1.4). Summarizing the comparison results whenever there exists a unique uniformly continuous viscosity solution of (1.4), one obtains the following theorem. Let $E \subseteq \mathbf{R}^n$ be a bounded set; denote by v the unique continuous viscosity solution of (1.4) with $v(\bar{t}, x) := (-1) \vee d_E(x) \wedge 1$, where d_E is the oriented distance function from ∂E negative inside E. Then for any $t \in [\, \overline{t}, +\infty[$ we have

$$
\begin{aligned}
\mathcal{M}_*(E, \mathcal{F}_F, \overline{t})(t) &= \{x \in \mathbf{R}^n : v(t, x) < 0\}, \\
\mathcal{M}^*(E, \mathcal{F}_F, \overline{t})(t) &= \{x \in \mathbf{R}^n : v(t, x) \leq 0\}.
\end{aligned} \tag{1.9}
$$

In particular

$$
\mathcal{M}^*(E, \mathcal{F}_F, \overline{t})(t) \setminus \mathcal{M}_*(E, \mathcal{F}_F, \overline{t})(t) = \{x \in \mathbf{R}^n : v(t, x) = 0\}. \tag{1.10}
$$

Equality (1.10) connects the fattening phenomenon defined through the barriers approach with the one defined through the level set approach. In case of nonuniqueness of viscosity solutions, the minimal barrier selects the maximal viscosit subsolution of (1.4).

The outline of the paper is the following. In Section 2 we give some notation. In Section 3 we recall the abstract definitions of barriers, local (in space) barriers and inner barriers, and some of their properties, such as the relations with the test family $\mathcal{F}$ (see (3.1) and Proposition 1), the semigroup property (see (3.3)), and a useful consequence of the translation invariance (see (3.5)). Using this latter property we prove that the function $\mathcal{M}_{u_0, \mathcal{F}} : [\, \overline{t}, +\infty[\times \mathbf{R}^n \to \mathbf{R}$ defined as

$$
\mathcal{M}_{u_0, \mathcal{F}}(t, x) := \inf \left\{ \lambda : \mathcal{M}(\{u_0 < \lambda\}, \mathcal{F}, \overline{t})(t) \ni x \right\}, \qquad (t, x) \in [\, \overline{t}, +\infty[\times \mathbf{R}^n,
$$

which is the weak evolution (as a function) of an *arbitrary* initial function u_0, preserves the Lipschitz constant (Proposition 3). In (3.9), (3.11), and (3.12) we list some properties of the regularizations and their connections with the minimal barrier. The section contains several examples showing the behaviour of the barriers (choice of convex test hypersurfaces in Example 3, inverse mean curvature flow in Example 4, backward mean curvature flow in Example 5, nonconvex anisotropic curvature flow in Example 6) and motivating the lower and upper regularizations (Example 2). Theorem 1 is concerned with the relations between barriers and local barriers, whereas in Theorem 2 and Proposition 2 we deepen the relations between $\mathcal{B}(\mathcal{F}_F)$ and $\mathcal{B}(\mathcal{F}_{F+})$. We conclude Section 3 by proving some results on the outer regularity of the minimal barrier (Proposition 4) and on the right continuity of the distance function between minimal barriers (Lemma 1 and Corollary 1). In Section 4 we introduce the notion of barrier solution (Definition 9) and we study existence and stability properties (Proposition 5 and 6); these two properties are reminiscent of the existence and stability of viscosity solutions [CGG91]. In Theorems 4 and 5 we recall the connections between the barriers and the level set flow; Theorem 6 is concerned with the characterization of the complement of regularized barriers, and Theorem 7 with the connections between barriers and inner barriers. The comparison results between barriers and level set flow are generalized in Lemma 2 of subsection 4.2, where we introduce the notion of comparison flow (by extending a similar definition given in [BP95b]). In Section 5 we recall some results on the disjoint and joint sets properties; in Proposition 7 we reinterpret the disjoint sets property by means of the distance function. In Section 6 we discuss some aspects of the fattening phenomenon. Fattening for geometric evolutions in two dimensions is discussed in subsection 6.1, and fattening in dimension $n \geq 3$ is discussed in subsection 6.2. In presence of fattening, the connections between different weak approaches have not been, to our knowledge, completely clarified, even in two dimensions (see Example 10 and below). In this section we include an explicit example of three-dimensional fattening for motion by mean curvature in codimension 2 (see Example 15). For completeness, in Appendix A we include the abstract definition of barrier and minimal barrier [41]. In Appendix B we list some assumptions used in the paper, following the notation of [GGIS91].

Most of the results discussed in the present paper are proved in [BN98, BN97b] (see also [Bel97, BP95b, Ilm93b]); we will prove in details only original statements not appearing in [BN98, BN97b].

2   Some notation

In the following we let $I := [t_0, +\infty[$, for a fixed $t_0 \in \mathbf{R}$. We denote by $\mathcal{P}(\mathbf{R}^n)$ (resp. $\mathcal{A}(\mathbf{R}^n)$, $\mathcal{C}(\mathbf{R}^n)$) the family of all (resp. open, closed) subsets of $\mathbf{R}^n$, $n \geq 1$, and by $\mathcal{H}^m$ the m-dimensional Hausdorff measure in $\mathbf{R}^n$, for $m \in [0, n]$.

Given a set $E \subseteq \mathbf{R}^n$, we denote by $\text{int}(E)$, $\overline{E}$ and ∂E the interior part, the closure and the boundary of E, respectively. We set $\text{dist}(\cdot, \emptyset) \equiv +\infty$, $d_E(x) := \text{dist}(x, E) - \text{dist}(x, \mathbf{R}^n \setminus E)$. Given a map $\phi : L \to \mathcal{P}(\mathbf{R}^n)$, where $L \subseteq \mathbf{R}$ is a convex set, we let $d_\phi : L \times \mathbf{R}^n \to \mathbf{R}$ be the function defined as

$$d_\phi(t, x) := \text{dist}\big(x, \phi(t)\big) - \text{dist}\big(x, \mathbf{R}^n \setminus \phi(t)\big) = d_{\phi(t)}(x). \qquad (2.1)$$

If $\phi_1, \phi_2 : L \to \mathcal{P}(\mathbf{R}^n)$, by $\phi_1 \subseteq \phi_2$ (resp. $\phi_1 = \phi_2$, $\phi_1 \cap \phi_2$) we mean $\phi_1(t) \subseteq \phi_2(t)$ (resp. $\phi_1(t) = \phi_2(t)$, $\phi_1(t) \cap \phi_2(t)$) for any $t \in L$.

Given a function $v : L \times \mathbf{R}^n \to \mathbf{R}$ we denote by v_* (resp. v^*) the lower (resp. upper) semicontinuous envelope of v.

For $x \in \mathbf{R}^n$ and $R > 0$ we set $B_R(x) := \{y \in \mathbf{R}^n : |y - x| < R\}$. We let $a \vee b := \max\{a, b\}$ and $a \wedge b := \min\{a, b\}$.

Given $p \in \mathbf{R}^n \setminus \{0\}$, we set $P_p := \text{Id} - p \otimes p/|p|^2$, and

$$J_0 := \big(\mathbf{R}^n \setminus \{0\}\big) \times \text{Sym}(n), \qquad J_1 := I \times \mathbf{R}^n \times \big(\mathbf{R}^n \setminus \{0\}\big) \times \text{Sym}(n).$$

Given a function $F : J_1 \to \mathbf{R}$ we denote by F_* (resp. F^*) the lower (resp. upper) semicontinuous envelope of F, defined on $\overline{J_1}$.

We recall that F is *geometric* [GG92b] if $F(t, x, \lambda p, \lambda X + \sigma p \otimes p) = \lambda F(t, x, p, X)$ for any $\lambda > 0$, $\sigma \in \mathbf{R}$, $(t, x, p, X) \in J_1$, and that F is *degenerate elliptic* if

$$F(t, x, p, X) \geq F(t, x, p, Y), \qquad (t, x, p, X) \in J_1, \ Y \in \text{Sym}(n), X \leq Y. \quad (2.2)$$

In the sequel we shall always assume that F is geometric.

We say that F is locally Lipschitz in X if for any $(t, x, p) \in I \times \mathbf{R}^n \times (\mathbf{R}^n \setminus \{0\})$ the function $F(t, x, p, \cdot)$ is locally Lipschitz.

We say that F is *bounded below* if, for any compact set $K \subset J_0$, there exists a constant $c_K \in \mathbf{R}$ such that

$$\inf \left\{ F(t, x, p, X) : t \in I, x \in \mathbf{R}^n, (p, X) \in K \right\} \geq c_K.$$

For any $(t, x, p, X) \in J_1$ we set

$$
\begin{aligned}
F_c(t, x, p, X) &:= -F(t, x, -p, -X), \\
F^+(t, x, p, X) &:= \sup \left\{ F(t, x, p, Y) : Y \geq X \right\}, \\
F^-(t, x, p, X) &:= \inf \left\{ F(t, x, p, Y) : Y \leq X \right\}.
\end{aligned}
\tag{2.3}
$$

Note that F^+ and F^- are always degenerate elliptic, and F_c is degenerate elliptic if and only if F is degenerate elliptic. Furthermore, $\left(F_c \right)_c = F$ and $\left(F^+ \right)_c = \left(F_c \right)^-$.

We give definitions similar to (2.3) if the function F is defined on J_0 (resp. $\overline{J_0}, \overline{J_1}$).

We say that $F : J_0 \to \mathbf{R}$ is *compatible from above* (resp. *from below*) if there exists an odd degenerate elliptic function $F_1 : J_0 \to \mathbf{R}$ such that $F_1 \geq F$ (resp. $F_1 \leq F$).

We notice that $F : J_0 \to \mathbf{R}$ is compatible from above (resp. below) if and only if $\left(F^+ \right)_c \geq F^+$ (resp. $\left(F^- \right)_c \leq F^-$).

Unless otherwise specified, when we deal with the viscosity theory we mean the one developed in [GGIS91] (see also [CIL92] and references therein).

3   Barriers, local barriers and inner barriers

Let us recall the definitions of geometric barriers and minimal barriers with respect to the inclusion $\subseteq$ between subsets of $\mathbf{R}^n$ and to a family $\mathcal{F}$ of set-valued maps; we refer to Appendix A and to the original papers [Gio93b, Gio94b] for the abstract definition of barrier and minimal barrier.

Definition 1. *Let $\mathcal{F}$ be a family of functions with the following property: for any $f \in \mathcal{F}$ there exist $a, b \in \mathbf{R}$, $a < b$, such that $f : [a, b] \to \mathcal{P}(\mathbf{R}^n)$. A function ϕ is a barrier with respect to $\mathcal{F}$ if and only if ϕ maps a convex set $L \subseteq I$ into $\mathcal{P}(\mathbf{R}^n)$ and the following property holds: if $f : [a, b] \subseteq L \to \mathcal{P}(\mathbf{R}^n)$ belongs to $\mathcal{F}$ and $f(a) \subseteq \phi(a)$ then $f(b) \subseteq \phi(b)$. Given such a map ϕ, we shall write $\phi \in \mathcal{B}(\mathcal{F}, L)$. When $L = I$, we simply write $\phi \in \mathcal{B}(\mathcal{F})$.*

Notice that if $\phi_i \in \mathcal{B}(\mathcal{F}, L)$ for every $i \in \Lambda$, Λ any family of indices, then $\bigcap_{i \in \Lambda} \phi_i \in \mathcal{B}(\mathcal{F}, L)$.

Definition 2. *Let $E \subseteq \mathbf{R}^n$ be a given set and let $\bar{t} \in I$. The minimal barrier $\mathcal{M}(E, \mathcal{F}, \bar{t}) : [\,\bar{t}, +\infty[\, \to \mathcal{P}(\mathbf{R}^n)$ (with origin in E at time $\bar{t}$) with respect to the family $\mathcal{F}$ at any time $t \geq \bar{t}$ is defined by*

$$\mathcal{M}(E, \mathcal{F}, \bar{t})(t) := \bigcap \left\{ \phi(t) : \phi : [\,\bar{t}, +\infty[\, \to \mathcal{P}(\mathbf{R}^n), \phi \in \mathcal{B}(\mathcal{F}, [\,\bar{t}, +\infty[), \phi(\bar{t}) \supseteq E \right\}.$$

Clearly
$$\mathcal{M}(E, \mathcal{F}, \bar{t}) \in \mathcal{B}(\mathcal{F}, [\bar{t}, +\infty[).$$

Moreover the following properties are immediate:

- comparison principle: $E_1 \subseteq E_2 \Rightarrow \mathcal{M}(E_1, \mathcal{F}, \bar{t}) \subseteq \mathcal{M}(E_2, \mathcal{F}, \bar{t})$;
- initial datum: $\mathcal{M}(E, \mathcal{F}, \bar{t})(\bar{t}) = E$;
- relaxation of the elements of $\mathcal{F}$: if $f : [a, b] \subseteq [\bar{t}, +\infty[\to \mathcal{P}(\mathbf{R}^n)$, $f \in \mathcal{F}$, then

$$f(t) \subseteq \mathcal{M}(f(a), \mathcal{F}, a)(t), \qquad t \in [a, b]; \tag{3.1}$$

- semigroup property: assume that the family $\mathcal{F}$ satisfies the following assumption:

$$\text{if } f : [a, b] \subseteq [\bar{t}, +\infty[\to \mathcal{P}(\mathbf{R}^n), \ f \in \mathcal{F}, \ t \in \ ]a, b[, \text{ then } f_{|[a,t]}, f_{|[t,b]} \in \mathcal{F}. \tag{3.2}$$

Then

$$\mathcal{M}(E, \mathcal{F}, \bar{t})(t_2) = \mathcal{M}(\mathcal{M}(E, \mathcal{F}, \bar{t})(t_1), \mathcal{F}, t_1)(t_2), \qquad \bar{t} \le t_1 \le t_2. \tag{3.3}$$

Moreover $\mathcal{F} \subseteq \mathcal{G} \Rightarrow \mathcal{B}(\mathcal{F}, [\bar{t}, +\infty[) \supseteq \mathcal{B}(\mathcal{G}, [\bar{t}, +\infty[)$, hence $\mathcal{M}(E, \mathcal{F}, \bar{t}) \subseteq \mathcal{M}(E, \mathcal{G}, \bar{t})$.

In the sequel, unless otherwise specified, we shall assume $\bar{t} = t_0$, and we often drop it in the notation; for instance, we write $\mathcal{M}(E, \mathcal{F})$ in place of $\mathcal{M}(E, \mathcal{F}, t_0)$, and similarly for all the other set valued maps (regularizations, local barriers, inner barriers).

We say that $\mathcal{F}$ is *translation invariant* (in space) if, for any $f \in \mathcal{F}$ and $y \in \mathbf{R}^n$, the map $f + y : t \to f(t) + y := \{x \in \mathbf{R}^n : (x - y) \in f(t)\}$ belongs to $\mathcal{F}$. If $\mathcal{F}$ is translation invariant one can check that

$$\phi \in \mathcal{B}(\mathcal{F}) \Rightarrow \text{int}(\phi) \in \mathcal{B}(\mathcal{F}), \tag{3.4}$$

which implies that the minimal barrier with origin in an open set remains open for any time.

Another useful property of minimal barriers, which implies (under a further assumption on $\mathcal{F}$) the preservation of the Lipschitz constant for $\mathcal{M}_{u_0, \mathcal{F}}$ (see Proposition 3), is the following:

- if $\mathcal{F}$ is translation invariant, for any $\varrho > 0$ and any $t \in I$ we have

$$\mathcal{M}(E_\varrho^+, \mathcal{F})(t) \supseteq \left(\mathcal{M}(E, \mathcal{F})(t)\right)_\varrho^+. \tag{3.5}$$

The following example clarifies the choice of $\mathcal{F}$ when dealing with geometric equations of the form (1.4).

Example 1. Let $F : J_1 \to \mathbf{R}$. We write $f \in \mathcal{F}_F$ (and we say that f is a smooth local geometric supersolution of (1.4)) if and only if there exist $a, b \in \mathbf{R}$, $a < b$, $[a, b] \subseteq I$ such that $f : [a, b] \to \mathcal{P}(\mathbf{R}^n)$, and the following conditions hold:

(i) $f(t)$ is closed and $\partial f(t)$ is compact for any $t \in [a, b]$;

(ii) there exists an open set $A \subseteq \mathbf{R}^n$ such that $d_f \in C^\infty([a,b] \times A)$ and $\partial f(t) \subseteq A$ for any $t \in [a,b]$;

(iii) the following inequality holds

$$\frac{\partial d_f}{\partial t}(t,x) + F\left(t, x, \nabla d_f(t,x), \nabla^2 d_f(t,x)\right) \geq 0, \qquad t \in [a,b], \; x \in \partial f(t).$$
(3.6)

We write $f \in \mathcal{F}_F^{>}$ (resp. $f \in \mathcal{F}_F^{<}$, $f \in \mathcal{F}_F^{=}$) if the strict inequality (resp. the inequality $\leq$, the equality) holds in (3.6).

The minimal barrier with respect to $\mathcal{F}_F$ starting from the set $E \subseteq \mathbf{R}^n$ will be from now on our definition of weak evolution of E, concerning equations of the form (1.4). Clearly, if F does not depend on x, then all families in Example 1 are translation invariant.

We recall that motion by mean curvature of hypersurfaces corresponds to the choice

$$F(t,x,p,X) = -\mathrm{tr}(P_p X P_p),$$
(3.7)

and that motion by mean curvature of manifolds of codimension $k \geq 1$ corresponds to the choice

$$F(p,X) = -\sum_{i=1}^{n-k} \lambda_i,$$
(3.8)

where $\lambda_1 \leq \ldots \leq \lambda_{n-1}$ are the eigenvalues of the matrix $P_p X P_p$ which correspond to eigenvectors orthogonal to p, see [Gio94b, AS96, BN].

We remark that, when dealing with the evolution of oriented *hypersurfaces*, we prefer to think of the evolution of the *solid* set E rather than of the evolution of its boundary ∂E (see Remark 2 and below).

The following example [BP95b] shows that, unless that E is open, $\mathcal{M}(E, \mathcal{F}_F)$ is very sensible to slight modifications of the original set E.

Example 2. Let $n = 2$, $E = \{x = (x_1, x_2) \in \mathbf{R}^2 : |x_1| \leq 1, |x_2| \leq 1\}$ and let F be as in (3.7). Then, as a consequence of the definitions and the strong maximum principle one has

$$\mathcal{M}(E, \mathcal{F}_F)(t) = \mathcal{M}\left(\mathrm{int}(E), \mathcal{F}_F\right)(t), \qquad t > t_0.$$

Similarly, if $E = \{(x_1, x_2) \in \mathbf{R}^2 : x_1^2 + x_2^2 \leq 1\}$ and $x^* \in \partial E$, then

$$\mathcal{M}(E \setminus \{x^*\}, \mathcal{F}_F)(t) = \mathcal{M}\left(\mathrm{int}(E), \mathcal{F}_F\right)(t), \qquad t > t_0.$$

In view of Example 2, the minimal barrier $\mathcal{M}(E, \mathcal{F})$ is not always "topologically stable"; on the other hand, the regularization maps $\mathcal{M}_*(E, \mathcal{F})$, $\mathcal{M}^*(E, \mathcal{F})$ defined in the Introduction, enjoy the following stability property:

- stability of the lower and upper regularizations with respect to interior part and closure: if $\mathcal{F}$ is translation invariant, then for any $t \in I$ we have

$$\mathcal{M}_*(E, \mathcal{F})(t) = \mathcal{M}_*\left(\mathrm{int}(E), \mathcal{F}\right)(t) \in \mathcal{A}(\mathbf{R}^n),$$
$$\mathcal{M}^*(E, \mathcal{F})(t) = \mathcal{M}^*\left(\overline{E}, \mathcal{F}\right)(t) \in \mathcal{C}(\mathbf{R}^n).$$
(3.9)

We have already observed that $\mathcal{M}^*(E,\mathcal{F})$ belongs to $\mathcal{B}(\mathcal{F})$. One can ask under which conditions on $\mathcal{F}$ it holds $\mathcal{M}_*(E,\mathcal{F}) \in \mathcal{B}(\mathcal{F})$. The following result holds:

- write $f \in \mathcal{F}^c$ if and only if

$$\exists a < b, [a,b] \subseteq I : \ f:[a,b] \to \mathcal{P}(\mathbf{R}^n), f \in \mathcal{F}, f(t) \text{ is compact } \forall t \in [a,b].$$
$$(3.10)$$

If $\mathcal{F}$ is translation invariant, then $\mathcal{M}_*(E,\mathcal{F}^c) \in \mathcal{B}(\mathcal{F}^c)$ and for any $t \in I$ there holds

$$E \in \mathcal{A}(\mathbf{R}^n) \Rightarrow \mathcal{M}_*(E,\mathcal{F}^c)(t) = \mathcal{M}(E,\mathcal{F}^c)(t) \in \mathcal{A}(\mathbf{R}^n). \qquad (3.11)$$

If additionally $\mathcal{F}$ satisfies (3.2), then $\mathcal{M}_*(E,\mathcal{F}^c)$ satisfies the semigroup property.

If $F : J_1 \to \mathbf{R}$ is bounded below, it turns out that $\mathcal{B}(\mathcal{F}_F) = \mathcal{B}((\mathcal{F}_F)^c)$, and therefore in this case we can ensure that $\mathcal{M}_*(E,\mathcal{F}_F) \in \mathcal{B}(\mathcal{F}_F)$.

We also note that, under mild conditions on F, and possibly regularizing the minimal barrier, we can interchange $\mathcal{F}_F$ with $\mathcal{F}_F^{\geq}$ when defining the minimal barrier. Indeed, the following property holds:

- assume that $F : J_0 \to \mathbf{R}$ is either lower semicontinuous and locally Lipschitz in X, or is continuous and degenerate elliptic. Then, for any $E \subseteq \mathbf{R}^n$ we have

$$\mathcal{M}_*(E,\mathcal{F}_F) = \mathcal{M}_*(E,\mathcal{F}_F^{\geq}), \qquad \mathcal{M}^*(E,\mathcal{F}_F) = \mathcal{M}^*(E,\mathcal{F}_F^{\geq}). \qquad (3.12)$$

The following examples show the rôle of the choice of $\mathcal{F}$ in the definition of the minimal barriers: Example 3 concerns motion by curvature whenever $\mathcal{F}$ consists of smooth *convex* evolutions, and Example 4 concerns the case of inverse mean curvature flow.

Example 3. Let $n = 2$, F be as in (3.7) and let

$$f \in \mathcal{C}_F \Longleftrightarrow \exists a,b \in \mathbf{R} :[a,b] \subseteq I, f :[a,b] \to \mathcal{P}(\mathbf{R}^2), f \in \mathcal{F}_F, f(t) \text{ convex}, t \in [a,b],$$
$$f \in \mathcal{D}_F \Longleftrightarrow \exists a,b \in \mathbf{R} :[a,b] \subseteq I, f : [a,b] \to \mathcal{P}(\mathbf{R}^2), f \in \mathcal{F}_F^{=}, f(a) \text{ is convex}.$$

Then, for any $E \subseteq \mathbf{R}^2$ we have

$$\begin{aligned} \mathcal{M}_*(E,\mathcal{C}_F) &= \mathcal{M}_*(E,\mathcal{D}_F) = \mathcal{M}_*(E,\mathcal{F}_{F \wedge 0}), \\ \mathcal{M}^*(E,\mathcal{C}_F) &= \mathcal{M}^*(E,\mathcal{D}_F) = \mathcal{M}^*(E,\mathcal{F}_{F \wedge 0}). \end{aligned} \qquad (3.13)$$

Example 4. Let us define the family $\mathcal{G}$ as follows. A function f belongs to $\mathcal{G}$ if and only if there exist $a < b$ such that $[a,b] \subseteq I$, $f : [a,b] \to \mathcal{P}(\mathbf{R}^n)$, $f(t)$ is compact for any $t \in [a,b]$, there exists an open set $A \subseteq \mathbf{R}^n$ such that $d_f \in C^\infty([a,b] \times A)$, $\partial f(t) \subseteq A$ for any $t \in [a,b]$, and

$$\Delta d_f > 0, \qquad \frac{\partial d_f}{\partial t} + \frac{1}{\Delta d_f} \geq 0 \qquad t \in [a,b], \quad x \in \partial f(t).$$

Then $\mathcal{M}(E,\mathcal{G})$ provides a definition of weak evolution of any set $E \subseteq \mathbf{R}^n$ by the inverse mean curvature [HI].

Inclusion (3.1) becomes an equality ("agreement with smooth flows") whenever $\mathcal{F} = \mathcal{F}_F$ for suitable functions F (in particular for motion by mean curvature of hypersurfaces).

Proposition 1. *Assume that $F : J_1 \to \mathbf{R}$ does not depend on x and is degenerate elliptic. Let $f : [a,b] \subseteq I \to \mathcal{P}(\mathbf{R}^n)$, $f \in \mathcal{F}_{\overline{F}}$. Then*

$$f(t) = \mathcal{M}\big(f(a), \mathcal{F}_F, a\big)(t), \qquad t \in [a,b]. \tag{3.14}$$

Proof. It is enough to show that $f(t) \supseteq \mathcal{M}(f(a), \mathcal{F}_F, a)(t)$ for any $t \in [a,b]$. Observe that the following geometric maximum principle holds: let $g, h : [a,b] \subseteq I \to \mathcal{P}(\mathbf{R}^n)$, $g \in \mathcal{F}_F$, $h \in \mathcal{F}_{\overline{F}}^{\leq}$. Then

$$g(a) \subseteq h(a) \Rightarrow g(b) \subseteq h(b).$$

It follows that $f \in \mathcal{B}(\mathcal{F}_F, [a,b])$, and (3.14) follows.   □

Notice that we have defined $\mathcal{M}(E, \mathcal{F}_F)$ under no assumptions on F (such as continuity and degenerate ellipticity); clearly, to end up with a nontrivial minimal barrier, we have to ensure that $\mathcal{B}(\mathcal{F}_F)$ is nonempty, which is true under minor assumptions, such as boundedness below of F. Even under these assumptions, it may happen that the minimal barrier becomes trivial for all times $t \in \,]t_0, +\infty[$. Indeed, let us consider the following example.

Example 5. Let $F : J_0 \to \mathbf{R}$ be the function giving motion by mean curvature with the "wrong" sign (corresponding to the backward heat equation for the signed distance function), i.e.,

$$F(p, X) := \mathrm{tr}(P_p X P_p),$$

and let $A \in \mathcal{A}(\mathbf{R}^n)$. Then, as a consequence of (3.16) below, we have

$$\mathcal{M}(A, \mathcal{F}_F)(t) = \mathbf{R}^n, \qquad t > t_0.$$

Barriers and their properties are a *global* concept, since they are defined through sets inclusions. In order to derive differential properties of the evolution, one can look for locality properties of barriers. Following [BN98], we introduce the local barriers (the localization is with respect to the space variable) and the inner barriers.

Definition 3. *Let $\mathcal{F}$ be as in Definition 1. A function ϕ is a local barrier with respect to $\mathcal{F}$ if and only if there exists a convex set $L \subseteq I$ such that $\phi : L \to \mathcal{P}(\mathbf{R}^n)$ and the following property holds: for any $x \in \mathbf{R}^n$ there exists $R > 0$ (depending on ϕ and x) so that if $f : [a,b] \subseteq L \to \mathcal{P}(\mathbf{R}^n)$ belongs to $\mathcal{F}$ and $f(a) \subseteq \phi(a) \cap B_R(x)$, then $f(b) \subseteq \phi(b)$. We denote by $\mathcal{B}_{loc}(\mathcal{F})$ the family of all local barriers ϕ such that $L = I$ (that is, local barriers on the whole of I).*

Definition 4. *Let $E \subseteq \mathbf{R}^n$ be a given set and let $\bar{t} \in I$. The local minimal barrier $\mathcal{M}_{loc}(E, \mathcal{F}, \bar{t}) : [\bar{t}, +\infty[\to \mathcal{P}(\mathbf{R}^n)$ (with origin in E at time $\bar{t}$) with respect to the family $\mathcal{F}$ at any time $t \geq \bar{t}$ is defined by*

$$\mathcal{M}_{loc}(E, \mathcal{F}, \bar{t})(t) := \bigcap \Big\{ \phi(t) : \phi : [\bar{t}, +\infty[\to \mathcal{P}(\mathbf{R}^n), \phi \in \mathcal{B}_{loc}(\mathcal{F}, [\bar{t}, +\infty[), \phi(\bar{t}) \supseteq E \Big\}.$$

The definitions of regularized local barriers can be given in the obvious way. Notice also that from Definition 4 it does not directly follow that the local minimal barrier is a local barrier, because of the dependence of R on ϕ.

The following theorem connects barriers with local barriers.

Theorem 1. *Assume that $F : J_0 \to \mathbf{R}$ is lower semicontinuous. Then*

$$\mathcal{B}_{loc}\big(\mathcal{F}_F^{\geq}\big) = \mathcal{B}\big(\mathcal{F}_F^{\geq}\big).$$

In particular, for any $E \subseteq \mathbf{R}^n$ we have $\mathcal{M}_{loc}(E, \mathcal{F}_F^{\geq}) = \mathcal{M}(E, \mathcal{F}_F^{\geq})$.

Considering the opposite sets inclusion in Definition 1, we can define the inner barriers.

Definition 5. *Let $\mathcal{F}$ be as in Definition 1. A function $\tilde{\phi}$ is an inner barrier with respect to $\mathcal{F}$ if and only if $\tilde{\phi}$ maps a convex set $L \subseteq I$ into $\mathcal{P}(\mathbf{R}^n)$ and the following property holds: if $f : [a, b] \subseteq L \to \mathcal{P}(\mathbf{R}^n)$ belongs to $\mathcal{F}$ and $\tilde{\phi}(a) \subseteq int(f(a))$ then $\tilde{\phi}(b) \subseteq int(f(b))$. Given such a map $\tilde{\phi}$, we shall write $\tilde{\phi} \in \widetilde{B}(\mathcal{F}, L)$. When $L = I$, we simply write $\tilde{\phi} \in \widetilde{B}(\mathcal{F})$.*

The definition of local inner barrier can be given in the obvious way.

Definition 6. *Let $E \subseteq \mathbf{R}^n$ be a given set and let $\bar{t} \in I$. The maximal inner barrier $\mathcal{N}(E, \mathcal{F}, \bar{t}) : [\bar{t}, +\infty[\to \mathcal{P}(\mathbf{R}^n)$ (with origin in E at time $\bar{t}$) with respect to the family $\mathcal{F}$ at any time $t \geq \bar{t}$ is defined by*

$$\mathcal{N}(E, \mathcal{F}, \bar{t})(t) := \bigcup \Big\{ \tilde{\phi}(t) : \tilde{\phi} : [\bar{t}, +\infty[\to \mathcal{P}(\mathbf{R}^n), \tilde{\phi} \in \widetilde{B}(\mathcal{F}, [\bar{t}, +\infty[), \tilde{\phi}(\bar{t}) \subseteq E \Big\}.$$

Its lower and upper regularization are defined by

$$\mathcal{N}_*(E, \mathcal{F}, \bar{t})(t) := \bigcup_{\varrho > 0} \mathcal{N}(E_\varrho^-, \mathcal{F}, \bar{t})(t), \quad \mathcal{N}^*(E, \mathcal{F}, \bar{t})(t) := \bigcap_{\varrho > 0} \mathcal{N}(E_\varrho^+, \mathcal{F}, \bar{t})(t).$$

Note that $\phi \in \mathcal{B}(\mathcal{F}_F)$ if and only if $\mathbf{R}^n \setminus \phi \in \widetilde{B}(\mathcal{F}_{F_c}^{\leq})$. Consequently, for any $E \subseteq \mathbf{R}^n$, $\mathbf{R}^n \setminus \mathcal{M}(E, \mathcal{F}_F) = \mathcal{N}(\mathbf{R}^n \setminus E, \mathcal{F}_{F_c}^{\leq})$, hence

$$\begin{aligned} \mathbf{R}^n \setminus \mathcal{M}_*(E, \mathcal{F}_F) &= \mathcal{N}^*(\mathbf{R}^n \setminus E, \mathcal{F}_{F_c}^{\leq}), \\ \mathbf{R}^n \setminus \mathcal{M}^*(E, \mathcal{F}_F) &= \mathcal{N}_*(\mathbf{R}^n \setminus E, \mathcal{F}_{F_c}^{\leq}). \end{aligned} \tag{3.15}$$

If the function F is not degenerate elliptic, then the minimal barriers do not agree, in general, with the smooth evolutions (whenever they exist). One can ask what the minimal barrier represents in this case. It turns out that the minimal barrier with respect to $\mathcal{F}_F$ coincides with the minimal barrier with respect to $\mathcal{F}_{F+}$, where F^+ is defined in (2.3) and is degenerate elliptic. More precisely, there holds the following representation result of the minimal barrier for a not degenerate elliptic function F, which is one of the main results on barriers.

Theorem 2. *Assume that $F : J_0 \to \mathbf{R}$ is lower semicontinuous. Then*

$$\mathcal{B}(\mathcal{F}_F^\geq) = \mathcal{B}(\mathcal{F}_{F+}^\geq).$$

In particular, for any $E \subseteq \mathbf{R}^n$ we have

$$\mathcal{M}(E, \mathcal{F}_F^\geq) = \mathcal{M}(E, \mathcal{F}_{F+}^\geq). \tag{3.16}$$

This theorem clarifies inclusion (3.1) (when $\mathcal{F} = \mathcal{F}_F$). Under further assumptions on F, we can prove a viscosity version of Theorem 2.

Proposition 2. *Let $F : J_1 \to \mathbf{R}$ and $u : I \times \mathbf{R}^n \to \mathbf{R}$ be given functions. Assume that F is lower (resp. upper) semicontinuous, $(F^+)_* < +\infty$ (resp. $(F^-)^* > -\infty$) on $\overline{J}_1$, and $(F^+)_*(t, x, 0, X) = (F_*)^+(t, x, 0, X)$ (resp. $(F^-)^*(t, x, 0, X) = (F^*)^-(t, x, 0, X)$) for any $t \in I$, $x \in \mathbf{R}^n$ and $X \in Sym(n)$. Then u is a viscosity subsolution (resp. supersolution) of (1.4) in $]t_0, +\infty[\times \mathbf{R}^n$ if and only if u is a viscosity subsolution (resp. supersolution) of (1.5) (resp. of $\frac{\partial u}{\partial t} + F^-(t, x, \nabla u, \nabla^2 u) = 0$) in $]t_0, +\infty[\times \mathbf{R}^n$.*

Proof. As $(F^+)_c = (F_c)^-$, it is enough to show the assertion for subsolutions. As $F^+ \geq F$, we only need to show that if u is a subsolution of (1.4) then u is a subsolution of (1.5). Observe that F^+ is lower semicontinuous on J_1, hence $(F^+)_* = (F_*)^+$ on $\overline{J}_1$. Let $(\bar{t}, \bar{x}) \in\ ]t_0, +\infty[\times \mathbf{R}^n$ and let ψ be a smooth function such that $(u^* - \psi)$ has a maximum at $(\bar{t}, \bar{x})$. Assume by contradiction that

$$\frac{\partial \psi}{\partial t} + (F^+)_*(\bar{t}, \bar{x}, \nabla \psi, \nabla^2 \psi) = 2c > 0 \quad \text{at } (\bar{t}, \bar{x}). \tag{3.17}$$

By definition of $(F^+)_* = (F_*)^+$ and since $(F^+)_* < +\infty$, there exists $X \in Sym(n)$, $X \geq \nabla^2 \psi(\bar{t}, \bar{x})$, such that

$$(F^+)_*(\bar{t}, \bar{x}, \nabla \psi, \nabla^2 \psi) \leq F_*(\bar{t}, \bar{x}, \nabla \psi, X) + c \quad \text{at } (\bar{t}, \bar{x}). \tag{3.18}$$

Define

$$\Psi(t, x) := \psi(t, x) + \frac{1}{2} \langle (x - \bar{x}), (X - \nabla^2 \psi(\bar{t}, \bar{x}))(x - \bar{x}) \rangle, \tag{3.19}$$

where $\langle \cdot, \cdot \rangle$ stands for the scalar product. Then $\nabla^2 \Psi(\bar{t}, \bar{x}) = X$ and $(u^* - \Psi)$ has a maximum at $(\bar{t}, \bar{x})$. Therefore, using the fact that u is a subsolution of (1.4), (3.18) and (3.17), at $(\bar{t}, \bar{x})$ we have

$$0 \geq \frac{\partial \Psi}{\partial t} + F_*(\bar{t}, \bar{x}, \nabla \Psi, \nabla^2 \Psi) = \frac{\partial \psi}{\partial t} + F_*(\bar{t}, \bar{x}, \nabla \psi, X)$$

$$\geq \frac{\partial \psi}{\partial t} + (F^+)_*(\bar{t}, \bar{x}, \nabla \psi, \nabla^2 \psi) - c = c > 0,$$

a contradiction. $\qquad \square$

We observe that, in general, $(F^+)_* \geq (F_*)^+$ on $\overline{J}_1$ and the equality holds if F is degenerate elliptic. If F is lower semicontinuous, then $(F^+)_* = (F_*)^+$ on J_1. Also, the equality $(F^+)_*(t,x,0,X) = (F_*)^+(t,x,0,X)$ holds for a geometric function $F : J_1 \to \mathbf{R}$ which coincides with F^+ outside a compact set K of J_1, and is bounded on K.

Example 6. Let $n = 2$ and consider the anisotropic motion by mean curvature given by

$$F(p,X) := -\mathrm{tr}(P_p X P_p)\psi(\theta)(\psi(\theta) + \psi''(\theta)), \qquad (3.20)$$

where $\psi : \mathbf{S}^1 \to \,]0,+\infty[$ is a smooth function and $p = (p_1,p_2) = (\cos\theta, \sin\theta)$ (see [BP96a]). Then, if $\psi + \psi'' \geq 0$ on $\mathbf{S}^1$ (i.e., convex anisotropy), we have $F^+ = F$. If the anisotropy is not convex, then there exists $\bar{\theta} \in \mathbf{S}^1$ such that $\psi(\bar{\theta}) + \psi''(\bar{\theta}) < 0$, which implies $F^+(\bar{p}, X) = +\infty$ for any $X \in \mathrm{Sym}(2)$, where $\bar{p} = (\cos\bar{\theta}, \sin\bar{\theta})$. Indeed, $F(\bar{p}, \cdot)$, being linear with the "wrong sign", behaves as the backward mean curvature flow, compare Example 5.

One application of the barrier approach is related to crystalline motion by mean curvature. Assume indeed that it is possible to define what is a "smooth" crystal and to define what is a "smooth" local crystalline mean curvature evolution of the crystal (this has been done for crystalline motion by curvature of plane polygonal curves [TCH93]). Then, if we define $\mathcal{F}$ as the family of all smooth crystalline flows, the corresponding minimal barrier provides a definition of weak crystalline evolution.

3.1   The function $\mathcal{M}_{u_0,\mathcal{F}}$

In this subsection, starting from the weak evolution defined on sets as in the previous sections, we recall the definition of the weak solution as a function, and we study some of its properties. The procedure we follow is the one in [BN98], and is the *opposite* with respect to the one used to define the level set flow. This kind of procedure has also been used by Evans in [Eva93] when considering the semigroup approach to motion by mean curvature.

We have seen that the minimal barrier starting from an arbitrary set E is unique and globally defined. Therefore, given any initial function $u_0 : \mathbf{R}^n \to \mathbf{R}$, there is a natural way to construct a unique global evolution function $\mathcal{M}_{u_0,\mathcal{F}}(t,x)$ (assuming u_0 as initial datum): it is indeed defined as that function which, for any $\lambda \in \mathbf{R}$, has $\mathcal{M}(\{u_0 < \lambda\}, \mathcal{F})(t)$ as λ-sublevel set at time $t \in I$.

Definition 7. *Let $u_0 : \mathbf{R}^n \to \mathbf{R}$ be a given function. The function $\mathcal{M}_{u_0,\mathcal{F}} : I \times \mathbf{R}^n \to \mathbf{R} \cup \{\pm\infty\}$ is defined by*

$$\mathcal{M}_{u_0,\mathcal{F}}(t,x) := \inf\{\lambda \in \mathbf{R} : \mathcal{M}(\{u_0 < \lambda\}, \mathcal{F})(t) \ni x\}. \qquad (3.21)$$

If $\mathcal{B}(\mathcal{F}) = \mathcal{B}(\mathcal{F}^c)$ (see (3.10)), if $\mathcal{M}(A,\mathcal{F})(t) \in \mathcal{A}(\mathbf{R}^n)$ for any $A \in \mathcal{A}(\mathbf{R}^n)$, and if $u_0 : \mathbf{R}^n \to \mathbf{R}$ is upper semicontinuous, then for any $\lambda \in \mathbf{R}^n \cup \{\pm\infty\}$ there holds

$$\mathcal{M}_{u_0,\mathcal{F}}(t,x) = \inf\{\lambda \in \mathbf{R} : \mathcal{M}_*(\{u_0 < \lambda\}, \mathcal{F})(t) \ni x\},$$

and

$$\{x \in \mathbf{R}^n : \mathcal{M}_{u_0, \mathcal{F}}(t, x) < \lambda\} = \mathcal{M}(\{u_0 < \lambda\}, \mathcal{F})(t), \qquad t \in I. \qquad (3.22)$$

Hence, under these assumptions, $\mathcal{M}_{u_0, \mathcal{F}}(t, \cdot)$ is upper semicontinuous; in addition, the Lipschitz constant is preserved, as the following proposition shows.

Proposition 3. *Assume that $\mathcal{F}$ is translation invariant and that $\mathcal{B}(\mathcal{F}) = \mathcal{B}(\mathcal{F}^c)$. Let $u_0 : \mathbf{R}^n \to \mathbf{R}$ be a Lipschitz function, and let $k > 0$ be its Lipschitz constant. Then*

$$|\mathcal{M}_{u_0, \mathcal{F}}(t, x) - \mathcal{M}_{u_0, \mathcal{F}}(t, y)| \le k|x - y|, \qquad x, y \in \mathbf{R}^n, \ t \in I,$$

where we assume that the left hand side is zero if $\mathcal{M}_{u_0, \mathcal{F}}(t, x)$ and $\mathcal{M}_{u_0, \mathcal{F}}(t, y)$ are both equal to $+\infty$ or $-\infty$.

Proof. Let $\lambda, \mu \in \mathbf{R} \cup \{\pm\infty\}$ be such that $\mu \ge \lambda$, and set $E_\lambda := \{u_0 < \lambda\}$, $E_\mu := \{u_0 < \mu\}$. Then we have

$$E_\mu \supseteq (E_\lambda)^+_{\frac{\mu-\lambda}{k}}. \qquad (3.23)$$

Indeed, the inclusion is obvious if $|\lambda|$ or $|\mu|$ is equal to $+\infty$; otherwise, if $z \in (E_\lambda)^+_{\frac{\mu-\lambda}{k}}$ then $z = x + \frac{\mu-\lambda}{k}q$ for some $x \in E_\lambda$ and some $q \in \mathbf{R}^n$ with $|q| < 1$. Hence, as u_0 is k-Lipschitz,

$$u_0(z) = u_0\left(x + \frac{\mu-\lambda}{k}q\right) < u_0(x) + \frac{\mu-\lambda}{k}k < \mu,$$

so that $z \in E_\mu$. Then, by (3.22), (3.23) and (3.5) we find

$$\{x : \mathcal{M}_{u_0, \mathcal{F}}(t, x) < \mu\} = \mathcal{M}(E_\mu, \mathcal{F})(t) \supseteq \mathcal{M}\left((E_\lambda)^+_{\frac{\mu-\lambda}{k}}, \mathcal{F}\right)(t)$$
$$\supseteq \left(\mathcal{M}(E_\lambda, \mathcal{F})(t)\right)^+_{\frac{\mu-\lambda}{k}} = \left(\{x : \mathcal{M}_{u_0, \mathcal{F}}(t, x) < \lambda\}\right)^+_{\frac{\mu-\lambda}{k}}. \qquad (3.24)$$

Let $y, w \in \mathbf{R}^n$ and set $\lambda := \mathcal{M}_{u_0, \mathcal{F}}(t, y)$, $\mu := \mathcal{M}_{u_0, \mathcal{F}}(t, w)$; let us prove that

$$|\mu - \lambda| \le k|y - w|. \qquad (3.25)$$

If $\lambda = \mu$ there is nothing to prove. Without loss of generality, we can assume $\lambda < \mu$. Let $\epsilon > 0$ be such that $\lambda + \epsilon < \mu$. As $w \in \{\mathcal{M}_{u_0, \mathcal{F}}(t, \cdot) \ge \mu\}$, by (3.24) we have $w \notin (\{\mathcal{M}_{u_0, \mathcal{F}}(t, \cdot) < \lambda + \epsilon\})^+_{\frac{\mu-\lambda-\epsilon}{k}}$. As $y \in \{\mathcal{M}_{u_0, \mathcal{F}}(t, \cdot) < \lambda + \epsilon\}$, we then obtain $|y - w| \ge \frac{\mu-\lambda-\epsilon}{k}$. Letting $\epsilon \downarrow 0$ we get (3.25). In particular, it follows that either $\mathcal{M}_{u_0, \mathcal{F}}(t, \cdot) \equiv +\infty$, or $\mathcal{M}_{u_0, \mathcal{F}}(t, \cdot) \equiv -\infty$, or $\mathcal{M}_{u_0, \mathcal{F}}(t, \cdot) : \mathbf{R}^n \to \mathbf{R}$ is a Lipschitz function with Lipschitz constant less than or equal to k. □

If v_0 is smooth, bounded, constant outside a bounded subset of $\mathbf{R}^n$, and if F is as in (3.7), Evans-Spruck [ES92b] showed that

$$\sup_{t \in I} \int_{\mathbf{R}^n} |\nabla \mathcal{M}_{v_0, \mathcal{F}_F}| \, dx \leq \int_{\mathbf{R}^n} |\nabla v_0| \, dx, \qquad (3.26)$$

where we have used the fact that $\mathcal{M}_{v_0, \mathcal{F}_F}$ coincides with the (Lipschitz continuous) viscosity solution v assuming v_0 as initial datum, see Theorem 4 below. As we have seen, we can define $\mathcal{M}_{v_0, \mathcal{F}}$ under no restriction on v_0; we do not know wether inequality (3.26) still holds for an initial datum $v_0 : \mathbf{R}^n \to \mathbf{R}$ which is upper semicontinuous and with bounded variation on $\mathbf{R}^n$ (interpreting the integrals as total variations).

3.2 Outer regularity of the minimal barrier and right continuity of the distance between minimal barriers

In this subsection we show how a suitable notion of outer regularity of $\mathcal{F}$ reflects on the outer regularity of the minimal barrier, and we study some continuity properties of the distance function between barriers.

Definition 8. *We say that $\mathcal{F}$ is outer regular if for any $f : [a, b] \subseteq I \to \mathcal{P}(\mathbf{R}^n)$, $f \in \mathcal{F}$, we have $f(t) = \overline{int(f(t))}$ for any $t \in [a, b]$.*

Given $E \subseteq \mathbf{R}^n$, we set
$$E^r := E \cap \overline{int(E)}.$$

If $E = E^r$, we say that the set E is outer regular.

The following proposition shows that the minimal barrier is outer regular, provided that $\mathcal{F}$ is outer regular.

Proposition 4. *Assume that $\mathcal{F}$ is outer regular. Let $\phi : I \to \mathcal{P}(\mathbf{R}^n)$, $\phi \in \mathcal{B}(\mathcal{F})$, and let $\phi^r : I \to \mathcal{P}(\mathbf{R}^n)$ be the map defined by $\phi^r(t) := \phi(t)^r$ for any $t \in I$. Then $\phi^r \in \mathcal{B}(\mathcal{F})$. Moreover, for any $E \subseteq \mathbf{R}^n$, we have*

$$\mathcal{M}(E, \mathcal{F})(t) = \mathcal{M}(E, \mathcal{F})(t)^r = \mathcal{M}(E^r, \mathcal{F})(t), \qquad t > t_0. \qquad (3.27)$$

In particular $\mathcal{M}(E, \mathcal{F})(t) \subseteq \mathcal{M}(\overline{int(E)}, \mathcal{F})(t)$ for any $t > t_0$, and if E is closed then

$$\mathcal{M}(E, \mathcal{F})(t) = \mathcal{M}(\overline{int(E)}, \mathcal{F})(t), \qquad t > t_0. \qquad (3.28)$$

Proof. Let $f : [a, b] \subseteq I \to \mathcal{P}(\mathbf{R}^n)$, $f \in \mathcal{F}$, $f(a) \subseteq \phi^r(a) \subseteq \phi(a)$. As $\phi \in \mathcal{B}(\mathcal{F})$ we have $f(b) \subseteq \phi(b)$, and then $\overline{int(f(b))} = f(b) \subseteq \overline{int(\phi(b))}$. Hence $f(b) \subseteq \phi^r(b)$.

The inclusion $\mathcal{M}(E, \mathcal{F}) \supseteq \mathcal{M}(E, \mathcal{F})^r$ is immediate. To prove the opposite inequality, it is enough to show that the map $\phi : I \to \mathcal{P}(\mathbf{R}^n)$ defined by

$$\phi(t) := \begin{cases} E & \text{if } t = t_0; \\ \mathcal{M}(E, \mathcal{F})(t)^r & \text{if } t > t_0 \end{cases}$$

belongs to $\mathcal{B}(\mathcal{F})$. Let $f : [a, b] \subseteq I \to \mathcal{P}(\mathbf{R}^n)$, $f \in \mathcal{F}$, $f(a) \subseteq \phi(a)$. If $a > t_0$ then $f(b) \subseteq \phi(b)$ by the previous assertion. If $a = t_0$, then $f(a) = \mathrm{int}(f(a)) \subseteq \mathrm{int}(E)$, hence $f(a) \subseteq E^r = \mathcal{M}(E, \mathcal{F})(t_0)^r$, so that $f(b) \subseteq \phi(b)$.

In addition $\mathcal{M}(E^r, \mathcal{F}) \subseteq \mathcal{M}(E, \mathcal{F})$, and the opposite inclusion follows by observing that the map $\psi : I \to \mathcal{P}(\mathbf{R}^n)$ defined by

$$\psi(t) := \begin{cases} E & \text{if } t = t_0, \\ \mathcal{M}(E^r, \mathcal{F})(t) & \text{if } t > t_0 \end{cases}$$

belongs to $\mathcal{B}(\mathcal{F})$. $\qquad\qquad\qquad\qquad\qquad\qquad\qquad\qquad\qquad\qquad \square$

Equality (3.27) does not hold in general for $\mathcal{M}^*(E, \mathcal{F})$; take for instance $n = 2$, $E := \{(x_1, x_2) \in \mathbf{R}^2 : x_2 = 0\}$ and F as in (3.7). Then $\mathcal{M}^*(E, \mathcal{F}_F)(t) = E$ for any $t \geq t_0$, while $\mathcal{M}^*(E, \mathcal{F}_F)(t)^r = \emptyset$ for any $t \geq t_0$.

Notice that, if $\mathcal{F}$ is outer regular and E has empty interior, then from (3.28) we deduce that $\mathcal{M}(E, \mathcal{F})(t) = \emptyset$ for any $t > t_0$.

The following lemma will be useful to prove a continuity property of the distance between minimal barriers .

Lemma 1. *Let $F_1, F_2 : J_0 \to \mathbf{R}$ be bounded below and let $\phi \in \mathcal{B}(\mathcal{F}_{F_1})$, $\psi \in \mathcal{B}(\mathcal{F}_{F_2})$. Set*

$$\eta(t) := \mathrm{dist}(\mathbf{R}^n \setminus \phi(t), \mathbf{R}^n \setminus \psi(t)), \qquad t \in I. \qquad (3.29)$$

Then

$$\eta(t_0) \leq \liminf_{s \downarrow t_0} \eta(s), \qquad \limsup_{\sigma \uparrow t} \eta(\sigma) \leq \eta(t) \leq \liminf_{s \downarrow t} \eta(s), \qquad t > t_0. \quad (3.30)$$

Assume in addition that $\mathbf{R}^n \setminus \phi \in \mathcal{B}(\mathcal{F}_{F_3})$ and $\mathbf{R}^n \setminus \psi \in \mathcal{B}(\mathcal{F}_{F_4})$ for two suitable functions $F_3, F_4 : J_0 \to \mathbf{R}$ bounded below. If $\mathbf{R}^n \setminus \phi = (\mathbf{R}^n \setminus \phi)^r$ and $\mathbf{R}^n \setminus \psi = (\mathbf{R}^n \setminus \psi)^r$, then

$$\eta(t_0) = \lim_{s \downarrow t_0} \eta(s), \qquad \limsup_{\sigma \uparrow t} \eta(\sigma) \leq \eta(t) = \lim_{s \downarrow t} \eta(s), \qquad t > t_0. \quad (3.31)$$

Proof. Given any $F : J_0 \to \mathbf{R}$ bounded below, there exists [BN98] a strictly increasing function $\varrho_F : [0, +\infty[\to [0, +\infty[$, $\varrho_F \in \mathcal{C}([0, +\infty[) \cap \mathcal{C}^\infty(]0, +\infty[)$, $\varrho_F(0) = 0$, such that if we take any $t_0 \leq a < b$, $\epsilon > 0$ and $x \in \mathbf{R}^n$, we have that the map $t \in [a, b] \to \overline{B}_{\varrho_F(\epsilon + b - t_0 - t)}(x)$ belongs to $\mathcal{F}_F^{\geq}$.

It follows that, if $\chi \in \mathcal{B}(\mathcal{F}_F)$ and $t \in I$, then

$$\{x \in \mathbf{R}^n : \mathrm{dist}(x, \mathbf{R}^n \setminus \chi(t)) > \varrho_F(s - t)\} \subseteq \mathrm{int}(\chi(s)), \qquad s > t. \quad (3.32)$$

Indeed, let $s > t$ and $x \in \chi(t)$ be such that $\mathrm{dist}(x, \mathbf{R}^n \setminus \chi(t)) > \overline{\varrho} > \varrho_F(s - t)$. Let us evolve the ball $\overline{B}_{\overline{\varrho}}(x)$ as explained above on $[t, s]$, and denote this evolution by $\sigma \in [t, s] \to \overline{B}(\sigma)$; since it belongs to $\mathcal{F}_F$, we have $x \in B(s) \subseteq \mathrm{int}(\chi(s))$ and (3.32) is proved.

Consequently, for any $s > t$ we have

$$\mathbf{R}^n \setminus \text{int}(\phi(s)) \subseteq \{x \in \mathbf{R}^n : \text{dist}(x, \mathbf{R}^n \setminus \phi(t)) \leq \varrho_{F_1}(s-t)\}$$
$$\mathbf{R}^n \setminus \text{int}(\psi(s)) \subseteq \{x \in \mathbf{R}^n : \text{dist}(x, \mathbf{R}^n \setminus \psi(t)) \leq \varrho_{F_2}(s-t)\}. \tag{3.33}$$

For any $\epsilon > 0$ let $y \in \mathbf{R}^n \setminus \text{int}(\phi(s))$, $z \in \mathbf{R}^n \setminus \text{int}(\psi(s))$ be such that $|y - z| \leq \eta(s) + \epsilon$. By (3.33) we have

$$\text{dist}(y, \mathbf{R}^n \setminus \phi(t)) \leq \varrho_{F_1}(s-t), \qquad \text{dist}(z, \mathbf{R}^n \setminus \psi(t)) \leq \varrho_{F_2}(s-t).$$

Using the triangular property of the distance and setting $\varrho := \varrho_{F_1} + \varrho_{F_2}$, we have

$$\eta(t) \leq \eta(s) + \epsilon + \text{dist}(y, \mathbf{R}^n \setminus \phi(t)) + \text{dist}(z, \mathbf{R}^n \setminus \psi(t)) \leq \eta(s) + \varrho(s-t) + \epsilon.$$

Letting $\epsilon \to 0^+$ we get $\eta(s) \geq \eta(t) - \varrho(s-t)$, which implies (3.30).

Let us now prove (3.31). Let $t \in I$, $\epsilon > 0$, $x_\epsilon \in \mathbf{R}^n \setminus \text{int}(\phi(t))$, $y_\epsilon \in \mathbf{R}^n \setminus \text{int}(\psi(t))$ be such that $|x_\epsilon - y_\epsilon| \leq \eta(t) + \epsilon$. We can assume that $x_\epsilon \in \partial(\mathbf{R}^n \setminus \phi(t))$ and $y_\epsilon \in \partial(\mathbf{R}^n \setminus \psi(t))$. In particular, as $\mathbf{R}^n \setminus \phi = (\mathbf{R}^n \setminus \phi)^r$, $\mathbf{R}^n \setminus \psi = (\mathbf{R}^n \setminus \psi)^r$, we have $x_\epsilon \in \text{int}(\mathbf{R}^n \setminus \phi(t))$, $y_\epsilon \in \text{int}(\mathbf{R}^n \setminus \psi(t))$. Hence, for any $\varrho > 0$ the set $B_\varrho(x_\epsilon) \cap \text{int}(\mathbf{R}^n \setminus \phi(t))$ (resp. the set $B_\varrho(y_\epsilon) \cap \text{int}(\mathbf{R}^n \setminus \psi(t))$) contains a closed ball D_1 (resp. D_2). As $\mathbf{R}^n \setminus \phi \in \mathcal{B}(\mathcal{F}_{F_3})$ and $\mathbf{R}^n \setminus \psi \in \mathcal{B}(\mathcal{F}_{F_4})$, we can find $s = s(\epsilon, x_\epsilon, y_\epsilon) > t$ so that a suitable evolution $D_1(\sigma)$ (resp. $D_2(\sigma)$) of D_1 (resp. of D_2) belongs to $\mathcal{F}_{F_3}$ (resp. to $\mathcal{F}_{F_4}$) and it is contained in $\mathbf{R}^n \setminus \phi(\sigma)$ (resp. in $\mathbf{R}^n \setminus \psi(\sigma)$) for any $\sigma \in [t, s]$.

Let $z_\sigma \in D_1(\sigma)$ and $w_\sigma \in D_2(\sigma)$. By the triangular property of η we have, for any $\sigma \in [t, s]$,

$$\eta(\sigma) \leq |z_\sigma - w_\sigma| \leq |z_\sigma - x_\epsilon| + \eta(t) + \epsilon + |y_\epsilon - w_\sigma| \leq \eta(t) + \epsilon + 2\varrho.$$

Letting $\varrho, \epsilon \to 0$, we have

$$\eta(t) \geq \limsup_{s \downarrow t} \eta(s) \qquad t \in I, \tag{3.34}$$

and (3.31) follows. $\square$

The following result is concerned with the right continuity of the distance between minimal barriers.

Corollary 1. *Assume that $F : J_0 \to \mathbf{R}$ is lower semicontinuous and $F^+ : J_0 \to \mathbf{R}$ is upper semicontinuous. Given $A, B \subseteq \mathbf{R}^n$, let $d : I \to \mathbf{R} \cup \{+\infty\}$ be the function defined as*

$$d(t) := \text{dist}(\mathcal{M}(A, \mathcal{F}_F)(t), \mathcal{M}(B, \mathcal{F}_F)(t)), \qquad t \in I.$$

Then d is right continuous on $]t_0, +\infty[$.

Proof. By Theorem 8 below (applied with $G = (F^+)_c$) we have

$$\mathbf{R}^n \setminus \mathcal{M}(E, \mathcal{F}_F) \in \mathcal{B}(\mathcal{F}_{(F^+)_c})$$

for any $E \subseteq \mathbf{R}^n$ (see (5.6)).

Moreover, by Proposition 4 the minimal barrier is outer regular for $t > t_0$. We now apply Lemma 1 with $\phi = \mathbf{R}^n \setminus \mathcal{M}(A, \mathcal{F}_F)$, $\psi = \mathbf{R}^n \setminus \mathcal{M}(B, \mathcal{F}_F)$, $F_1 = F_2 = (F^+)_c$, $F_3 = F_4 = F$, and we get the thesis. $\square$

4   Barrier solutions, level set flow and comparison flows

In this section we study some properties of barriers, and some relations with other generalized flows. Let us introduce the following definition of barriers solution.

Definition 9. *Given a function $F : J_1 \to \mathbf{R}$, we define the family $S(F)$ of all barrier solutions of equation (1.4) as*

$$S(F) := \mathcal{B}(\mathcal{F}_F) \cap \tilde{\mathcal{B}}(\mathcal{F}_{\bar{F}}^{\leq}). \tag{4.1}$$

By Proposition 1, if F does not depend on x and is degenerate elliptic, then the barrier solutions coincide with the smooth evolutions, whenever the latter exist.

The next proposition shows that, under some monotonicity assumptions on F, there always exists a barrier solution of (1.4); this result is reminiscent of the proposition asserting the existence of viscosity solutions. Note that we do not still have an uniqueness result (see Theorem 7 below).

Proposition 5. *Let $F : J_0 \to \mathbf{R}$ be degenerate elliptic and let $\phi, \tilde{\phi} : I \to \mathcal{P}(\mathbf{R}^n)$, $\phi \in \mathcal{B}(\mathcal{F}_F)$, $\tilde{\phi} \in \tilde{\mathcal{B}}(\mathcal{F}_{\bar{F}}^{\leq})$ and $\tilde{\phi} \subseteq \phi$. Then there exists $\psi \in S(F)$ such that $\tilde{\phi} \subseteq \psi \subseteq \phi$.*

Proof. Let $\psi : I \to \mathcal{P}(\mathbf{R}^n)$ be defined as

$$\psi(t) := \bigcap \Big\{ \chi(t) : \ \chi : I \to \mathcal{P}(\mathbf{R}^n), \ \chi \in \mathcal{B}(\mathcal{F}_F), \ \tilde{\phi} \subseteq \chi \Big\}, \qquad t \in I.$$

Clearly $\psi \in \mathcal{B}(\mathcal{F}_F)$ and $\psi \supseteq \tilde{\phi}$. Let us show that $\psi \in \tilde{\mathcal{B}}(\mathcal{F}_{\bar{F}}^{\leq})$. Assume by contradiction that there exists $f : [a,b] \subseteq I \to \mathcal{P}(\mathbf{R}^n)$, $f \in \mathcal{F}_{\bar{F}}^{\leq}$, such that $\psi(a) \subseteq \mathrm{int}(f(a))$ and $\psi(b)$ is not contained in $\mathrm{int}(f(b))$. Define

$$\psi_1(t) := \begin{cases} \psi(t), & t \in I \setminus [a,b] \\[2mm] \psi(t) \cap \mathrm{int}\big(f(t)\big), & t \in [a,b]. \end{cases}$$

As F is degenerate elliptic and independent of x, one can check that $\mathrm{int}(f) \in \mathcal{B}(\mathcal{F}_F, [a,b])$. Hence $\psi_1 \in \mathcal{B}(\mathcal{F}_F)$; since $\mathrm{int}(f(a)) \supseteq \psi(a) \supseteq \tilde{\phi}(a)$ and $f \in \mathcal{F}_{\bar{F}}^{\leq}$, $\tilde{\phi} \in \tilde{\mathcal{B}}(\mathcal{F}_{\bar{F}}^{\leq})$, we have $\mathrm{int}(f(b)) \supseteq \tilde{\phi}(b)$, and we conclude that $\psi_1 \supseteq \tilde{\phi}$, therefore $\psi_1 \supseteq \psi$. However $\psi_1(b)$ is strictly contained in $\psi(b)$, a contradiction. Therefore $\psi \in \tilde{\mathcal{B}}(\mathcal{F}_{\bar{F}}^{\leq})$ so that $\psi \in S(F)$. Since $\psi \subseteq \phi$, the proof is concluded. $\square$

The following Proposition is reminiscent of the stability of viscosity subsolutions.

Proposition 6. *Let $F : J_0 \to \mathbf{R}$ be bounded below. Let $F_m : J_0 \to \mathbf{R}$ be such that $\liminf\limits_{m \to +\infty} \inf\limits_{K} (F_m - F) \geq 0$, for any compact set $K \subset J_0$. For any $m \in \mathbf{N}$, let $\phi_m \in \mathcal{B}(\mathcal{F}_{F_m}^{\geq})$ and set $\phi := \bigcup\limits_{h \in \mathbf{N}} int\Big(\bigcap\limits_{m \geq h} \phi_m \Big)$. Then $\phi \in \mathcal{B}(\mathcal{F}_F^{\geq})$.*

Proof. Let $f : [a, b] \subseteq I \to \mathcal{P}(\mathbf{R}^n)$, $f \in \mathcal{F}_F^>$, $f(a) \subseteq \phi(a)$; we have to prove that $f(b) \subseteq \phi(b)$. As we have already observed, we can assume that $f(t)$ is compact for any $t \in [a, b]$. As $f \in \mathcal{F}_F^>$, there exists a constant $0 < c < +\infty$ such that

$$\frac{\partial d_f}{\partial t}(t, x) + F\left(\nabla d_f(t, x), \nabla^2 d_f(t, x)\right) \geq 2c, \qquad x \in \partial f(t), \ t \in [a, b].$$

Set $K_f := \left\{\left(\nabla d_f(t, x), \nabla^2 d_f(t, x)\right) : x \in \partial f(t), t \in [a, b]\right\}$ which is a compact set, and let $\overline{m} \in \mathbf{N}$ be such that $\inf_{K_f}\left(F_m - F\right) \geq -c$ for any $m \geq \overline{m}$. Then for any $t \in [a, b]$, $x \in \partial f(t)$, $m \geq \overline{m}$, we have

$$\frac{\partial d_f}{\partial t}(t, x) + F_m\left(\nabla d_f(t, x), \nabla^2 d_f(t, x)\right)$$

$$\geq \frac{\partial d_f}{\partial t}(t, x) + F\left(\nabla d_f(t, x), \nabla^2 d_f(t, x)\right) - c \geq c,$$

which implies $f \in \mathcal{F}_{F_m}^>$. Given $h \in \mathbf{N}$, we set $\psi_h := \text{int}\left(\bigcap_{m \geq h} \phi_m\right)$. As $f(a) \subseteq \bigcup_h \psi_h(a)$ and $f(a)$ is compact, there exists $\overline{h}$ such that $f(a) \subseteq \psi_{\overline{h}}(a)$, which implies $f(a) \subseteq \left[\phi_m(a)\right]_\varrho^-$ for some $\varrho > 0$ and for any $m \geq \overline{h}$. Taking $N \geq \overline{m} \vee \overline{h}$, we have $f \in \mathcal{F}_{F_m}^>$ and $f(a) \subseteq \left[\phi_m(a)\right]_\varrho^-$ for any $m \geq N$, therefore, as $\mathcal{F}_{F_m}^>$ is translation invariant, $f(b) \subseteq \left[\phi_m(b)\right]_\varrho^-$. This implies

$$f(b) \subseteq \bigcap_{m \geq N} \left[\phi_m(b)\right]_\varrho^- \subseteq \left[\bigcap_{m \geq N} \phi_m(b)\right]_{\frac{\varrho}{2}}^- \subseteq \psi_N(b) \subseteq \phi(b).$$

This concludes the proof.   □

4.1   Barriers and viscosity subsolutions

The following theorem is proved in [GGIS91], Theorem 4.9 (see Appendix B for the notation).

Theorem 3. *Assume that $F : J_1 \to \mathbf{R}$ satisfies either (F1)-(F4), (F8), or (F1), (F3), (F4), (F9), (F10). Let $v_0 : \mathbf{R}^n \to \mathbf{R}$ be a continuous function which is constant outside a bounded subset of $\mathbf{R}^n$. Then there exists a unique continuous viscosity solution (constant outside a bounded subset of $\mathbf{R}^n$) of (1.4) with $v(t_0, x) = v_0(x)$.*

Given a bounded open set $E \subseteq \mathbf{R}^n$ we define the viscosity evolutions $V(E)(t)$, $\Gamma(t)$ of $\text{int}(E)$, ∂E respectively (the so-called level set flow) as

$$V(E)(t) := \{x \in \mathbf{R}^n : v(t, x) < 0\}, \qquad \Gamma(t) := \{x \in \mathbf{R}^n : v(t, x) = 0\}, \quad (4.2)$$

where v is as in Theorem 3 with $v_0(x) := (-1) \vee d_E(x) \wedge 1$.

The following results [BN98] show the connection between the minimal barrier and the viscosity solution (notice that it applies, in particular, to the case of motion by mean curvature in arbitrary codimension).

Theorem 4. *Assume that $F : J_1 \to \mathbf{R}$ satisfies (F1), (F3), (F4), (F6'), (F7), (F9), (F10). Let $E \subseteq \mathbf{R}^n$ be a bounded set. Then for any $t \in I$ we have*

$$\mathcal{M}_*(E, \mathcal{F}_F^{\geq})(t) = \mathcal{M}_*(E, \mathcal{F}_F)(t) = V(E)(t), \tag{4.3}$$

$$\mathcal{M}^*(E, \mathcal{F}_F^{\geq})(t) = \mathcal{M}^*(E, \mathcal{F}_F)(t) = V(E)(t) \cup \Gamma(t). \tag{4.4}$$

In particular $\mathcal{M}_{v_0, \mathcal{F}_F} = v$.

The difficult part of the proof of Theorem 4 relies in showing that, given a bounded open set $A \subseteq \mathbf{R}^n$, there holds $\mathcal{M}(A, \mathcal{F}_F^{\geq}) \supseteq V(A)$. To prove this, the idea is to show that the function χ, defined by

$$\chi(t, x) := -\chi_{\mathcal{M}(A, \mathcal{F}_F^{\geq})(t)}(x), \qquad (t, x) \in I \times \mathbf{R}^n,$$

(where $\chi_C(x) := 1$ if $x \in C$ and $\chi_C(x) := 0$ if $x \notin C$) is a viscosity subsolution of (1.4) in $]t_0, +\infty[\times \mathbf{R}^n$. The use of characteristic functions is needed because of the explicit dependence on x of the function F; when F does not depend on x, one can equivalently reason by using the distance function.

Theorem 4 in the case of driven motion by mean curvature of hypersurfaces has been proved in [BP95b], where the minimal barriers are compared with any generalized evolution of sets satisfying the semigroup property, the comparison principle, and the extension of smooth evolutions (see Corollary 2 below). Notice that, as a consequence of Theorem 4, it follows (under the same assumptions on F) that $\mathcal{M}_*(E, \mathcal{F}_F)$ and $\mathcal{M}^*(E, \mathcal{F}_F)$ verify the semigroup property.

The proof of Theorem 4 is based on the facts that the sublevel sets of a viscosity subsolution of (1.4) are barriers and, conversely, that a function whose sublevel sets are barriers is a viscosity subsolution of (1.4). Using these observations, one can show that the minimal barrier selects the maximal viscosity subsolution.

Theorem 5. *Assume that $F : J_1 \to \mathbf{R}$ satisfies (F1), (F3), (F4), (F6'), (F7), (F9), (F10). Let $u_0 : \mathbf{R}^n \to \mathbf{R}$ be a given upper semicontinuous function. Define S_{u_0} as the family of all viscosity subsolutions v of (1.4) in $]t_0, +\infty[\times \mathbf{R}^n$ such that $v^*(t_0, x) = u_0(x)$. Then*

$$\mathcal{M}_{u_0, \mathcal{F}_F} = \mathcal{M}_{u_0, \mathcal{F}_F^{\geq}} = \sup\{v : v \in S_{u_0}\}. \tag{4.5}$$

Remark 1. A similar assertion of Theorem 5 can be given for supersolutions, see [BN98].

The following results characterizes the complement of regularized barriers, and does not cover the case where E is unbounded and $\mathbf{R}^n \setminus E$ is unbounded.

Theorem 6. *Assume that* $F : J_1 \to \mathbf{R}$ *satisfies (F1), (F3), (F4), (F6'), (F7), (F9), (F10). Then, for any bounded set* $E \subseteq \mathbf{R}^n$ *we have*

$$
\begin{aligned}
\mathcal{M}_*(E, \mathcal{F}_F) &= \mathbf{R}^n \setminus \mathcal{M}^*(\mathbf{R}^n \setminus E, \mathcal{F}_{F_c}), \\
\mathcal{M}^*(E, \mathcal{F}_F) &= \mathbf{R}^n \setminus \mathcal{M}_*(\mathbf{R}^n \setminus E, \mathcal{F}_{F_c}).
\end{aligned}
\tag{4.6}
$$

Moreover, if $F = F_c$ *then*

$$
\mathcal{M}^*(E, \mathcal{F}_F) \setminus \mathcal{M}_*(E, \mathcal{F}_F) \in \mathcal{B}(\mathcal{F}_F).
\tag{4.7}
$$

Concerning the connections between the minimal barriers and the viscosity evolutions without growth conditions on F (see [IS95]) and for unbounded sets E with unbounded complement, there holds the following result. Assume that $F : J_0 \to \mathbf{R}$ is continuous and degenerate elliptic. Given any $E \subseteq \mathbf{R}^n$, let $v : I \times \mathbf{R}^n \to \mathbf{R}$ be the unique continuous viscosity solution of (1.4), in the sense of [IS95], with $v(t_0, x) = d_E(x)$. Then, for any $t \in I$, (4.3), (4.4), (4.6) and (4.7) hold. In particular $\mathcal{M}^*(E, \mathcal{F}_F)(t) \setminus \mathcal{M}_*(E, \mathcal{F}_F)(t) = \{x \in \mathbf{R}^n : v(t, x) = 0\}$ and $\mathcal{M}_{d_E, \mathcal{F}_F} = v$.

The next theorem shows the connection between the minimal barrier and the maximal inner barrier. This property is reminiscent of the uniqueness theorem for viscosity solutions.

Theorem 7. *Assume that* $F : J_1 \to \mathbf{R}$ *satisfies (F1), (F3), (F4), (F6'), (F7), (F9), (F10). Then, for any bounded set* $E \subseteq \mathbf{R}^n$ *we have*

$$
\mathcal{N}_*(E, \mathcal{F}_{\overline{F}}^{\leq}) = \mathcal{M}_*(E, \mathcal{F}_F), \qquad \mathcal{N}^*(E, \mathcal{F}_{\overline{F}}^{\leq}) = \mathcal{M}^*(E, \mathcal{F}_F).
\tag{4.8}
$$

Moreover, if $F : J_0 \to \mathbf{R}$ *is continuous and degenerate elliptic, then (4.8) holds for any* $E \subseteq \mathbf{R}^n$.

We remark that, to prove Theorem 7, we need to pass through the viscosity theory, and we miss a self-contained proof based only on barriers.

4.2 Comparison flows

In this subsection we generalize the comparison results discussed above; indeed, we compare the minimal barrier with an abstract comparison flow, which is defined as follows.

Definition 10. *Let* $F : J_1 \to \mathbf{R}$ *be a given function. Let* $\mathcal{Q}$ *be a family of sets containing the open and the close subsets of* $\mathbf{R}^n$. *We say that a map* $\mathcal{R}$ *is a comparison flow for (1.4) if and only if, for any* $E \in \mathcal{Q}$ *and* $\bar{t} \in I$, $\mathcal{R} = \mathcal{R}(E, \bar{t})$ *maps* $[\bar{t}, +\infty[$ *into* $\mathcal{Q}$, $\mathcal{R}(E, \bar{t})(\bar{t}) = E$, *and the following properties hold:*
(i) (semigroup property) for any $E \in \mathcal{Q}$ *we have*

$$
\mathcal{R}(E, t_1)(t) = \mathcal{R}\big(\mathcal{R}(E, t_1)(t_2), t_2\big)(t), \qquad t_1 \leq t_2 \leq t;
$$

(ii) (relaxation of the elements of $\mathcal{F}_F$ and $\mathcal{F}_{\bar{F}}^{\leq}$) for any $f : [a,b] \subseteq I \to \mathcal{P}(\mathbf{R}^n)$, $f \in \mathcal{F}_F$, $g : [c,d] \subseteq I \to \mathcal{P}(\mathbf{R}^n)$, $g \in \mathcal{F}_{\bar{F}}^{\leq}$, we have

$$f(t) \subseteq \mathcal{R}(f(a),a)(t), \qquad t \in [a,b],$$
$$int(g(t)) \supseteq \mathcal{R}(int(g(c)),c)(t), \qquad t \in [c,d];$$

(iii) (comparison principle) for any $A, B \in \mathcal{Q}$, $A \subseteq B$, and any $\bar{t} \in I$ we have

$$\mathcal{R}(A,\bar{t})(t) \subseteq \mathcal{R}(B,\bar{t})(t), \qquad t \geq \bar{t}.$$

If $\bar{t} = t_0$ we simply write $\mathcal{R}(E)$ instead of $\mathcal{R}(E,t_0)$; moreover, we define the lower and upper regularizations of $\mathcal{R}$ as

$$\mathcal{R}_*(E,\bar{t}) := \bigcup_{\varrho > 0} \mathcal{R}(E_\varrho^-,\bar{t}), \qquad \mathcal{R}^*(E,\bar{t}) := \bigcap_{\varrho > 0} \mathcal{R}(E_\varrho^+,\bar{t}),$$

and we note that they are defined on the whole of $\mathcal{P}(\mathbf{R}^n)$.

Lemma 2. *Let $\mathcal{R}$ be a comparison flow and let $E \in \mathcal{Q}$. Then $\mathcal{R}(E) \in \mathcal{S}(F)$ (see (4.1)), which implies*

$$\mathcal{M}(E,\mathcal{F}_F) \subseteq \mathcal{R}(E) \subseteq \mathcal{N}(E,\mathcal{F}_{\bar{F}}^{\leq}). \tag{4.9}$$

Proof. Let $f : [a,b] \subseteq I \to \mathcal{P}(\mathbf{R}^n)$, $f \in \mathcal{F}_F$, $f(a) \subseteq \mathcal{R}(E)(a)$. By property (ii) of Definition 10, we have $f(t) \subseteq \mathcal{R}(f(a),a)(t)$ for any $t \in [a,b]$. Therefore, using properties (iii) and (i) we get

$$f(b) \subseteq \mathcal{R}(f(a),a)(b) \subseteq \mathcal{R}(\mathcal{R}(E)(a),a)(b) = \mathcal{R}(E)(b).$$

Hence $\mathcal{R}(E) \in \mathcal{B}(\mathcal{F}_F)$. Reasoning in a similar way, one can check that $\mathcal{R}(E) \in \tilde{\mathcal{B}}(\mathcal{F}_{\bar{F}}^{\leq})$, and the thesis follows. □

We are now in a position to prove that regularized minimal barriers are essentially the *only* regularized comparison flows for (1.4).

Corollary 2. *Assume that the function $F : J_1 \to \mathbf{R}$ satisfies (F1), (F3), (F4), (F6'), (F7), (F9), (F10). Then, for any bounded set $E \subseteq \mathbf{R}^n$, we have*

$$\mathcal{M}_*(E,\mathcal{F}_F) = \mathcal{R}_*(E), \qquad \mathcal{M}^*(E,\mathcal{F}_F) = \mathcal{R}^*(E). \tag{4.10}$$

Moreover, if $F : J_0 \to \mathbf{R}$ is continuous and degenerate elliptic, then (4.10) holds for any $E \subseteq \mathbf{R}^n$.

Proof. The assertions follow from (4.9) and Theorem 7. □

Problem. Implement the barrier method for geometric evolutions on an open set Ω, with suitable boundary conditions, and compare it with other generalized approaches.

5   The disjoint and the joint sets properties

In this section we recall the notions of disjoint sets property and joint sets property [BN97b]. We remark that these properties are close to uniqueness of barrier solutions and are related to the fattening phenomenon.

Definition 11. *We say that the disjoint sets property (resp. the regularized disjoint sets property) with respect to $(\mathcal{F}, \mathcal{G})$ holds if for any $E_1, E_2 \subseteq \mathbf{R}^n$ and $\bar{t} \in I$*

$$E_1 \cap E_2 = \emptyset \;\Rightarrow\; \mathcal{M}(E_1, \mathcal{F}, \bar{t}) \cap \mathcal{M}(E_2, \mathcal{G}, \bar{t}) = \emptyset \tag{5.1}$$

$$\left(resp. \; E_1 \cap E_2 = \emptyset \;\Rightarrow\; \mathcal{M}_*(E_1, \mathcal{F}, \bar{t}) \cap \mathcal{M}^*(E_2, \mathcal{G}, \bar{t}) = \emptyset \right). \tag{5.2}$$

We say that the joint sets property (resp. the regularized joint sets property) with respect to $(\mathcal{F}, \mathcal{G})$ holds if for any $E_1, E_2 \subseteq \mathbf{R}^n$ and $\bar{t} \in I$

$$E_1 \cup E_2 = \mathbf{R}^n \;\Rightarrow\; \mathcal{M}(E_1, \mathcal{F}, \bar{t}) \cup \mathcal{M}(E_2, \mathcal{G}, \bar{t}) = \mathbf{R}^n, \tag{5.3}$$

$$\left(resp. \; E_1 \cup E_2 = \mathbf{R}^n \;\Rightarrow\; \mathcal{M}_*(E_1, \mathcal{F}, \bar{t}) \cup \mathcal{M}^*(E_2, \mathcal{G}, \bar{t}) = \mathbf{R}^n \right). \tag{5.4}$$

Example 7. As proved in [BP95b] motion by mean curvature enjoys both the regularized disjoint sets property and the regularized joint sets property with respect to $(\mathcal{F}_F, \mathcal{F}_F)$. Notice that in this case $F = F^+$ and F is odd.

Example 8. Let

$$F(t, x, p, X) := -\operatorname{tr}(P_p X P_p) + g(t, x)|p| \tag{5.5}$$

(i.e., motion by mean curvature with a forcing term g). Then, in general the disjoint sets property and the regularized disjoint sets property with respect to $(\mathcal{F}_F, \mathcal{F}_F)$ fail, compare Example 10. Notice that, in this case, $F = F^+$ and F is not odd.

Example 9. Let $n = 2$, $E = \{x = (x_1, x_2) \in \mathbf{R}^2 : |x_1| \leq 1, |x_2| \leq 1\}$ and let F be as in (3.7). Then, recalling Example 2, (3.11) and (3.9) we have, for $t > \bar{t}$,

$$\mathcal{M}(E, \mathcal{F}_F, \bar{t})(t) = \mathcal{M}(\operatorname{int}(E), \mathcal{F}_F, \bar{t})(t) \in \mathcal{A}(\mathbf{R}^n),$$

$$\mathcal{M}(\mathbf{R}^n \setminus E, \mathcal{F}_F, \bar{t})(t) \in \mathcal{A}(\mathbf{R}^n),$$

and the joint sets property with respect to $(\mathcal{F}_F, \mathcal{F}_F)$ does not hold ("we instantly loose ∂E").

Notice that if (5.2) holds then $\mathcal{M}^*(E_1, \mathcal{F}, \bar{t}) \cap \mathcal{M}_*(E_2, \mathcal{G}, \bar{t}) = \emptyset$, and conversely. Similarly, if (5.4) holds then $\mathcal{M}^*(E_1, \mathcal{F}, \bar{t}) \cup \mathcal{M}_*(E_2, \mathcal{G}, \bar{t}) = \mathbf{R}^n$, and conversely.

If $\mathcal{F}$ satisfies (3.2), then the disjoint sets property with respect to $(\mathcal{F}, \mathcal{G})$ is equivalent to the assertion

for any $E \subseteq \mathbf{R}^n$ there holds $\mathbf{R}^n \setminus \mathcal{M}(E, \mathcal{F}, \bar{t}) \in \mathcal{B}(\mathcal{G}, [\, \bar{t}, +\infty[). \tag{5.6}$

Moreover, the disjoint (resp. joint) sets property with respect to $(\mathcal{F}, \mathcal{G})$ implies the regularized disjoint (resp. joint) sets property with respect to $(\mathcal{F}, \mathcal{G})$.

The following theorems characterize the disjoint and joint sets property in terms of the functions F, G describing the evolution.

Theorem 8. *Assume that $F, G : J_0 \to \mathbf{R}$ are lower semicontinuous. Then the disjoint sets property (equivalently, the regularized disjoint sets property) with respect to $(\mathcal{F}_F, \mathcal{F}_G)$ holds if and only if $(F^+)_c \geq G^+$. In particular*
(i) if $F_c = G$, then the disjoint sets property with respect to $(\mathcal{F}_F, \mathcal{F}_{F_c})$ holds if and only if F is degenerate elliptic;
(ii) if $F = G$ then the disjoint sets property with respect to $(\mathcal{F}_F, \mathcal{F}_F)$ holds if and only if F is compatible from above.

The disjoint sets property, under suitable assumptions on an abstract family $\mathcal{F}$, can be restated by means of the distance function.

Proposition 7. *Assume that $\mathcal{F}$ is translation invariant and satisfies (3.2). Let $E \subseteq \mathbf{R}^n$ and $\phi \in \mathcal{B}(\mathcal{F})$. Then the function*

$$t \in I \to \eta(t) := dist\big(\mathcal{M}(E, \mathcal{F})(t), \mathbf{R}^n \setminus \phi(t)\big)$$

is nondecreasing.
Moreover, assume that $F : J_0 \to \mathbf{R}$ is lower semicontinuous and compatible from above. Let $A, B \subseteq \mathbf{R}^n$. Then the function

$$t \in I \to dist\big(\mathcal{M}(A, \mathcal{F}_F)(t), \mathcal{M}(B, \mathcal{F}_F)(t)\big) \tag{5.7}$$

is nondecreasing.

Proof. Let $t_2 > t_1 \geq t_0$; we have to prove that $\eta(t_2) \geq \eta(t_1)$. We can assume that $\eta(t_1) = \delta > 0$. Notice that for any $B \subseteq \mathbf{R}^n$

$$B \subseteq \phi(t_1) \Rightarrow \mathcal{M}(B, \mathcal{F}, t_1)(t_2) \subseteq \phi(t_2). \tag{5.8}$$

By (3.3), (3.5) and (5.8) we have

$$\big(\mathcal{M}(E, \mathcal{F})(t_2)\big)_\delta^+ = \big(\mathcal{M}(\mathcal{M}(E, \mathcal{F})(t_1), \mathcal{F}, t_1)(t_2)\big)_\delta^+$$
$$\subseteq \mathcal{M}\big((\mathcal{M}(E, \mathcal{F})(t_1))_\delta^+, \mathcal{F}, t_1\big)(t_2) \subseteq \phi(t_2),$$

which proves the monotonicity of η.
Let us prove (5.7). Setting $\phi := \mathbf{R}^n \setminus \mathcal{M}(B, \mathcal{F}_F)$, we have $\phi \in \mathcal{B}(\mathcal{F}_F)$ by Theorem 8, hence (5.7) follows from the previous assertion.   □

Theorem 9. *Assume that $F, G : J_0 \to \mathbf{R}$ are continuous, $F^+ < +\infty$, $G^+ < +\infty$ and F^+, G^+ are continuous. Then the regularized joint sets property with respect to $(\mathcal{F}_F, \mathcal{F}_G)$ holds if and only if $(F^+)_c \leq G^+$. In particular*
(i) if $F_c = G$, then the regularized joint sets property with respect to $(\mathcal{F}_F, \mathcal{F}_{F_c})$ holds for any function F satisfying the hypotheses;
(ii) if $F = G$ then the regularized joint sets property with respect to $(\mathcal{F}_F, \mathcal{F}_F)$ holds if and only if F^+ is compatible from below.

We remark that the proof of Theorem 9 passes through the viscosity theory, and we miss a self-contained proof based only on barriers.

Remark 2. Assume that $F : J_0 \to \mathbf{R}$ is continuous, odd and degenerate elliptic. Then for any $E \subseteq \mathbf{R}^n$ we have

$$\mathcal{M}^*(\partial E, \mathcal{F}_F) = \mathcal{M}^*(E, \mathcal{F}_F) \setminus \mathcal{M}_*(E, \mathcal{F}_F). \tag{5.9}$$

In [Ilm93b] Ilmanen introduced a notion of weak evolution, for motion by mean curvature of hypersurfaces, called set-theoretic subsolution, which essentially coincides with $\mathbf{R}^n \setminus \mathcal{M}_*(\mathbf{R}^n \setminus \partial E, \mathcal{F}_F)$, F as in (3.7); hence, thanks to (4.6) of Theorem 6 and the fact that F is *odd*, the set-theoretic subsolution of Ilmanen is $\mathcal{M}^*(\partial E, \mathcal{F}_F)$ (or, more generally, $\mathcal{M}^*(E, \mathcal{F}_F)$, E any given closed set which is not necessarily a boundary). In his paper Ilmanen proved that the set-theoretic subsolution coincides with the level set flow, which is consistent with (5.9), (4.3) and (4.4), for F as in (3.7). Also, a comparison result between barriers and the level set flow for sets E with compact boundary when F is as in (5.5) has been proved in [BP95b]; notice that in this case F is no more odd. The results of [BP95b, Ilm93b] are based on Ilmanen's interposition lemma and on Huisken's estimates [Hui84] of the existence time for the evolution of a smooth compact hypersurface in dependence on the L^∞ norm of its second fundamental form, without requiring bounds on further derivatives of the curvatures. The above results of Ilmanen and Huisken apply basically to the case of motion by mean curvature; it seems difficult to recover the time estimates of [Hui84] for a general evolution law of the form (1.4). This is the main reason for which the proof of Theorem 4 follows a completely different approach.

Solving the next problem (which asks, basically, which conditions we need to impose on a smooth elliptic function F, in order to let evolve $C^{1,1}$ compact hypersurfaces) would allow, following the arguments of Ilmanen in [Ilm93b], to give an alternative proof (with respect to [BN98]) of the comparison results between barriers and level set flows, for a class of evolutions including driven motion by mean curvature. Assume that F does not depend on x, it is smooth and uniformly elliptic. Let $\{E_\epsilon\}$ be a sequence of sets so that there exists a bounded open set $A \subseteq \mathbf{R}^n$ such that $\partial E_\epsilon \subseteq A$ and $d_{E_\epsilon} \in C^\infty(A)$ for any ϵ, and $\sup_\epsilon \sup_{x \in \partial E_\epsilon} |\nabla^2 d_{E_\epsilon}(x)| < +\infty$. Let $E_\epsilon(t)$ be the unique smooth evolution [GG92a] of the set E_ϵ under (1.4) for small times $t \in [t_0, t_0 + \tau_\epsilon[$, that is

$$\frac{\partial d_{E_\epsilon(t)}}{\partial t} + F\big(t, \nabla d_{E_\epsilon(t)}, \nabla^2 d_{E_\epsilon(t)}\big) = 0 \qquad \text{on } \partial E_\epsilon(t), \; t \in [t_0, t_0 + \tau_\epsilon[\, .$$

Which further conditions on F are needed in such a way that τ_ϵ can be chosen *independently* of ϵ?

We conclude this section by noticing that an abstract evolution of sets could be defined starting from the joint and the disjoint sets properties; since these properties are global, one should then show that they can be suitably localized, in order to obtain some geometric evolution law depending pointwise on the normal and the second fundamental form of the front.

6   The fattening phenomenon

In this section we discuss some aspects and examples concerning the fattening phenomenon, which is considered an interesting kind of singularity in geometric evolutions.

Definition 12. *Let $E \subseteq \mathbf{R}^n$. We say that the set E develops m-dimensional fattening with respect to $\mathcal{F}$ at time $t_1 \in I$ if*

$$\mathcal{H}^m \Big(\mathcal{M}^*(E, \mathcal{F}, \bar{t})(t) \setminus \mathcal{M}_*(E, \mathcal{F}, \bar{t})(t) \Big) = 0 \quad \text{for } t \in [\,\bar{t}, t_1],$$
$$\mathcal{H}^m \Big(\mathcal{M}^*(E, \mathcal{F}, \bar{t})(t) \setminus \mathcal{M}_*(E, \mathcal{F}, \bar{t})(t) \Big) > 0 \quad \text{for some } t \in \,]t_1, +\infty[\,, \tag{6.1}$$

where $m \in \,]0, n]$.

Concerning motion by mean curvature of manifolds of arbitrary codimension, F has the expression in (3.8), and $\mathcal{M}_*(E, \mathcal{F}_F)$ is usually empty; Definition 12 then reduces to

$$\mathcal{H}^m \Big(\mathcal{M}^*(E, \mathcal{F}, \bar{t})(t) \Big) = 0 \quad \text{for } t \in [\,\bar{t}, t_1],$$
$$\mathcal{H}^m \Big(\mathcal{M}^*(E, \mathcal{F}, \bar{t})(t) \Big) > 0 \quad \text{for some } t \in \,]t_1, +\infty[\,. \tag{6.2}$$

Unless otherwise specified, throughout this section we will consider n-dimensional fattening, i.e., $m = n$.

Fattening was defined [BSS93, ES91] by means of the viscosity solution as follows. Let v be the unique viscosity solution of (1.4) with $v(t_0, x) = (-1) \vee d_E(x) \wedge 1$, see Theorem 3; then fattening occurs if $\{x : v(t, x) = 0\}$ has nonempty interior part.

Equalities (4.3) and (4.4) shows that definition (6.1) is consistent with the definition of fattening given by means of the (unique) viscosity solution, see [BP95b]. Adopting (6.1), fattening can be defined also in case of nonuniqueness of viscosity solutions.

Given a function F, the main issue could be to characterize those subsets E of $\mathbf{R}^n$ which fatten under (1.4); the complete characterization is clearly a difficult problem, which is still open even for motion by mean curvature.

6.1   Fattening in two dimensions

Examples of fattening in two dimensions for curvature flow can be given in the following two cases:

- (i) if the initial set E is not required to be smooth (Evans and Spruck [ES91] provided the example of the inside of the figure eight curve);
- (ii) if the boundary ∂E is smooth but not compact, see Example 12, which is due to Ilmanen.

On the other hand, if $E \subseteq \mathbf{R}^2$ has compact smooth boundary, fattening does not take place under motion by curvature, as a consequence of a theorem of Grayson [Gra87]. However, if one modifies the evolution law, for instance by adding a forcing term, the situation is completely different. Barles-Soner-Souganidis [BSS93] have given an example with a time-dependent forcing term (see [NPV96] for numerical evidence). Even more, one can choose the forcing term to be constant, as the following example proposed in [BP94] shows.

Example 10. Assume that we can exhibit a smooth bounded Lipschitz function $g : I \times \mathbf{R}^2 \to \mathbf{R}$ and an initial smooth compact set $E \subseteq \mathbf{R}^2$, with

$$E := L \cup R, \qquad \overline{L} \cap \overline{R} = \emptyset,$$

where L and R are homeomorphic to a ball, with the following properties: if we denote by $L(t)$ (resp. $R(t)$) the evolution of $L = L(t_0)$ (resp. of $R = R(t_0)$) under the law (1.4) with the choice of F as in (5.5), then there exist $t^* > t_0$ and $x^* \in \mathbf{R}^2$ such that:
(i) $L(t)$ and $R(t)$ are smooth for $t \in [t_0, t^* + \delta]$, for some $\delta > 0$;
(ii) $\overline{L(t)} \cap \overline{R(t)} = \emptyset$ for any $t \in [t_0, t^*[$;
(iii) $\partial L(t^*) \cap \partial R(t^*) = \{x^*\}$;
(iv) $\partial L(t^*)$ and $\partial R(t^*)$ meet at x^* with zero relative velocity;
(v) recalling that we are considering the evolution of L and R as *independent*, $L(t)$ and $R(t)$ would smoothly "bounce back" after the collision.
    Then, under the previous assumptions, fattening takes place. Notice that F is not odd and that L and R violate the disjoint sets property with respect to $(\mathcal{F}_F, \mathcal{F}_F)$. We remark that one can rearrange things in such a way that g and E can be chosen as follows:

$$g \equiv 1, \qquad L := B_{r_1}(z), \quad R := B_{r_2}(w), \tag{6.3}$$

for suitable $r_1, r_2 > 0$ and $z, w \in \mathbf{R}^2$, with $r_1 + r_2 < |z - w|$.

Consider the example in the case (6.3). The heuristic idea is the following. Given a small $\varrho > 0$, the set E_ϱ^- consists of two disjoint balls which, by comparison arguments, flow smoothly remaining disjoint in $[t_0, t^*]$. Moreover, the construction is such that they flow smoothly remaining distant, independently of ϱ, after some time bigger than t^*.

On the other hand, given any small $\varrho > 0$, the evolving set starting from E_ϱ^+ becomes connected and has the shape of a "bean". The main point is to prove the following assertion: there exist a time interval $[\alpha, \beta] \subset \,]t^*, +\infty[$ and an open set A, *independent of* $\varrho > 0$, such that

$$A \subset \mathcal{M}(E_\varrho^+, \mathcal{F}_F)(t) \qquad \text{for all } \varrho \text{ sufficiently small and all } t \in [\alpha, \beta].$$

Notice that α is strictly larger than t^*, since the fat region increases "continuously" in time after t^*. Notice also that, being the curvature very high near the collision point, we can, heuristically, drop out the forcing term (which is bounded) in the evolution. The crucial tool to prove the above assertions are the comparison principle and a Sturmian theorem of Angenent (see [Ang91], Theorem 3) which estimates the number of intersections of two curves flowing independently by curvature (without forcing term), which reads as follows.

Theorem 10. *Let be given two families of smooth curves which evolve (independently) by their curvature for $t \in [t_0, t_0 + T]$ of which at least one is compact. Then for any $t \in \,]t_0, t_0 + T]$ the number of intersections of the two curves at time t is finite, and this number does not increase with time; moreover, it decreases whenever the two curves are not tranverse.*

Problem. Is nonfattening "generic", for instance with respect to E (or with respect to g) in Example 10?

Problem. Let us consider F as in (5.5), choose g and E as in (6.3), and let $\Lambda(t)$ be the Almgren-Taylor-Wang [ATW93] evolution starting from E, constructed iteratively by minimizing the energy functional

$$P(B) + \frac{1}{\tau} \int_{(E \setminus B) \cup (B \setminus E)} \text{dist}(x, \partial E) \, dx - |B|, \tag{6.4}$$

where $\tau > 0$ is the time step, $B \subseteq \mathbf{R}^2$, and $P(B)$ and $|B|$ are the perimeter and the Lebesgue measure of B, respectively. Is it true that, after the time collision t^*, there holds $\Lambda(t) = \partial M^*(E, \mathcal{F}_F)(t)$? Moreover, replace in (6.4) the quantity $|B|$ with $(1 - \epsilon(\tau))|B|$; is it possible to choose $\epsilon(\tau)$ decreasing to zero as $\tau \to 0$ in such a way that the corresponding evolution in the sense of Almgren-Taylor-Wang coincides, just after the collision time, with $\partial M_*(E, \mathcal{F}_F)(t)$?

Problems of this type, in a different context, have been considered by Gobbino in [Gob].

Denote by u_ϵ be the solution of the reaction-diffusion equation $\frac{\partial u_\epsilon}{\partial t} = \Delta u_\epsilon - \frac{2}{\epsilon^2}(u_\epsilon^3 - u_\epsilon) + \frac{2}{3\epsilon}$, assuming an initial datum u_ϵ^0 which depends on ϵ, approximates, as $\epsilon \to 0$, the function defined as $+1$ on $E := L \cup R$ (L and R as in (6.3)) and -1 outside, and is such that $\sup_{\epsilon > 0} \int_{\mathbf{R}^n} \epsilon |\nabla u_\epsilon^0|^2 + \epsilon^{-1}(u_\epsilon^{0^2} - 1)^2 \, dx < +\infty$. The resulting evolution obtained as the limit of u_ϵ as $\epsilon \to 0$, or as the limit of some of its subsequences, depends on the choice of u_ϵ^0. Let $t \in [\alpha, \beta]$ and A be an open set contained in the fat region $M^*(E, \mathcal{F}_F)(t) \setminus M_*(E, \mathcal{F}_F)(t)$, where $F(p, X) := -\text{tr}(P_p X P_p) + |p|$. Can we find a sequence $\{u_\epsilon^0\}_\epsilon$ of initial data such that $\{u_\epsilon(0, \cdot) = 0\} \cap A \neq \emptyset$?

The following example shows that fattening can occur in two dimensions if the function $F(t, x, p, \cdot)$ is not Lipschitz (the dependence on (t, x) is irrelevant here).

Example 11. Let $n = 2$, $\zeta : \mathbf{R} \to \mathbf{R}$ be defined as

$$\zeta(s) := \begin{cases} 0 & \text{if } s \in [0, 1], \\ \sqrt{1 - s^{-1}} & \text{if } s > 1, \\ -\zeta(-s) & \text{if } s \leq 0. \end{cases}$$

Let $F(p, X) := -\zeta(\text{tr}(P_p X P_p))$, and $E := \{(x_1, x_2) \in \mathbf{R}^2 : x_1^2 + x_2^2 \leq 1\}$. For any $\varrho \in \,]0, 1[$ the set E_ϱ^+ stands still, while E_ϱ^- shrinks to a point at finite time $T_\varrho \leq 2$; hence E develops fattening.

The following example is a particular case of an example due to Ilmanen [Ilm92], and concerns the case of motion by curvature of an initial smooth set with non compact boundary.

Example 12. Let $n = 2$, F be as in (3.7), and

$$v_0(x_1, x_2) := x_2^2(1 + x_1^2)^2.$$

For any $\lambda > 0$ the set $E_\lambda := \{v_0 \leq \lambda\}$ is smooth, has non compact boundary and finite Lebesgue measure. Notice that the oriented distance d_{E_λ} is not smooth on a ρ-tubular neighbourhood of ∂E_λ, for any $\rho > 0$. It turns out that E_λ develops fattening instantly. Intuitively, since in two dimensions the shrinking time of a connected closed smooth bounded curve flowing by curvature depends on the enclosed area and since E_λ has finite Lebesgue measure, the set $\mathcal{M}_*(E_\lambda, \mathcal{F}_F)(t)$ becomes bounded for times arbitrarily close to t_0 (note that, for any $\varrho > 0$, $(E_\lambda)_\varrho^-$ is bounded). On the other hand, for any $\varrho > 0$, the boundary of each set $(E_\lambda)_\varrho^+$ is composed by two entire graphs, that smoothly evolve by curvature remaining graphs for all times [EH89, EH91].

Clearly v_0 is not uniformly continuous and Ilmanen proved nonuniqueness of continuous viscosity solutions of (1.4) with $v(t_0, x) = v_0(x)$; we point out that Ilmanen selected a special viscosity solution for this evolution, see [Ilm92], Definition 7.1]. Notice that $\mathcal{M}_{v_0, \mathcal{F}_F}$ is, by Theorem 5, the maximal viscosity (sub) solution. One can check, following [Ilm92], that there exist $t \in I$ and $x \in \mathbf{R}^n$ such that $\mathcal{M}_{v_0, \mathcal{F}_F}(t, x) > -\mathcal{M}_{-v_0, \mathcal{F}_F}(t, x)$, where $-\mathcal{M}_{-v_0, \mathcal{F}_F}$ represents the minimal viscosity (super) solution. We conclude the discussion of this example with two further observations. Assume that we are interested in the evolution of a special $E_{\overline{\lambda}}$: then, if we choose $v(t_0, x) := (-1) \vee d_{E_{\overline{\lambda}}}(x) \wedge 1$ as (Lipschitz continuous) initial datum, equation (1.4) has a unique viscosity solution; nevertheless, $E_{\overline{\lambda}}$ develops fattening. Finally, we remark that $\mathcal{M}^*(E_\lambda, \mathcal{F}_F)$ does not coincide with $\bigcap \{\mathcal{M}(A, \mathcal{F}_F) : A \in \mathcal{A}(\mathbf{R}^n), A \supseteq E_\lambda\}$.

To conclude this subsection, we recall that, as a consequence of a theorem of Angenent (see [Ang91], Theorem 8.1 for a precise statement) it results that if $F(p, X)$ is odd, uniformly elliptic and of class $C^{2,1}$ (with suitable growth) then any smooth compact set E evolving by (1.2) does not develop fattening.

Finally, an example of fattening for anisotropic motion by curvature, with a nonsymmetric anisotropy, has been recently proposed in [Pao97].

6.2   The n-dimensional case

In $n \geq 3$ dimensions the situation is much more complicated than in two dimensions. First of all, as a consequence of a result of Huisken [Hui84], a smooth bounded strictly convex set $E \subseteq \mathbf{R}^n$ flowing by mean curvature does not develop fattening, and the same holds if the set is bounded and convex [ES92a].

A few years ago De Giorgi [Gio91a, Gio90b] asked wether a torus of the form

$$\left\{ (x_1, \ldots, x_n) \in \mathbf{R}^n : \left(\left(\sum_{i=1}^{n-1} x_i^2 \right)^{1/2} - 1 \right)^2 + x_n^2 \leq \lambda \right\} \tag{6.5}$$

flowing by mean curvature develops fattening, for a suitable choice of the parameter λ. Soner-Souganidis in [SS93] showed the following result, confirmed by the numerical simulations of Paolini-Verdi [PV92] (we refer also to the paper [AAG95] of Altschuler-Angenent-Giga, concerning singularities of a smooth, compact, rotationally symmetric hypersurface). See also the paper of Evans-Spruck [ES91] for further discussions.

Theorem 11. *Let $n \geq 3$. Then the torus defined in (6.5) does not fatten under motion by mean curvature. Moreover, up to a parabolic scaling, at the singularity the torus converges to a cylinder.*

As already remarked, the complete characterization of those sets which fatten is still open. In this respect, Barles-Soner-Souganidis [BSS93], Theorem 4.3, gave the following sufficient condition for an initial set E of class C^2 to not develop fattening.

Theorem 12. *Suppose that $F : J_0 \to \mathbf{R}$ satisfies the assumptions of Theorem 3, and moreover*

$$F(\mu Q^t p, \mu^2 Q^t X Q) = \mu^2 F(p, X),$$

for all $\mu > 0$, $p \in \mathbf{R}^n \setminus \{0\}$, $X \in Sym(n)$, and any orthogonal $(n \times n)$-matrix Q. Assume that there exist nonnegative constants c_1, c_2, c_3, a skewsymmetric matrix M and $x_0 \in \mathbf{R}^n$ such that

$$c_1(x - x_0) \cdot \nabla d_E + c_2 M(x - x_0) \cdot \nabla d_E - c_3 F(\nabla d_E, \nabla^2 d_E) \neq 0 \qquad on \; \partial E.$$

Then E does not develop fattening.

A particular case of such a geometric condition corresponds to surfaces of positive mean curvature everywhere. However, this condition does not cover general rotationally symmetric hypersurfaces even in three dimensions.

To our knowledge, in three dimensions there are no examples of smooth *compact* sets which develop fattening at finite time under mean curvature flow. Also, there is no rigorous proof of the existence of a smooth set with non compact boundary developping fattening at finite time; in this direction, Angenent-Chopp-Ilmanen [ACI95] have exhibited an example which we briefly recall (the construction is not completely rigorous; it is however supported by numerical evidence, see also [Cho94]).

Example 13. Let $n = 3$ and F be as in (3.7). In [ACI95] it is numerically computed a complete, smooth, non compact surface ∂E_0 of genus three (invariant under certain symmetries) asymptotic to a suitable double cone at infinity, which shrinks self-similarly and, at a certain time, becomes a (not rotationally symmetric) double cone with a unique singularity at the origin, and aperture of approximately 72.3°. Using the evolutions of rotationally symmetric cones barriers (see Theorem 13), it is then rigorously proved that the evolving set develops fattening.

Still in the three-dimensional case, we recall the following example, which is studied by White in [Whi92] and answers to some questions raised by De Giorgi (see [Gio92] for further conjectures related to this example and to the fattening phenomenon in dimension $n \geq 3$).

Example 14. Let $n = 3$, F be as in (3.7),

$$v_0(x_1, x_2, x_3) := \sin x_1 + \sin x_2 + \sin x_3,$$

and for any $\lambda \in [-3, 0]$ set $E_\lambda := \{(x_1, x_2, x_3) \in \mathbf{R}^3 : v_0(x_1, x_2, x_3) \leq \lambda\}$. Notice that E_{-3} consists of isolated points, E_{-1} is not smooth, and for any $\lambda \in \,]-3, -1[\,\cup\,]-1, 0]$, the set E_λ is smooth. Notice also that $E_0 \cap \{(x_1, x_2, x_3) \in \mathbf{R}^3 : \sin x_1 = \sin x_2 = 1\} = \emptyset$. It turns out that, if $\lambda \in \,]-3, 0[\,$, then there exists a time $T(\lambda) \in \,]t_0, +\infty[$ such that $\mathcal{M}(E_\lambda, \mathcal{F}_F)(t) = \emptyset$ for any $t > T(\lambda)$. Moreover $\sup_{\lambda \in [-3, 0[} T(\lambda) = +\infty$. Finally, E_0 evolves smoothly by mean curvature for any $t \geq t_0$ and converges smoothly as $t \to +\infty$ to a triply periodic minimal surface; in particular E_0 does not fatten.

The following result is proved in [ACI95], Theorem 4, and generalizes the behaviour of the two-dimensional cross under motion by curvature.

Theorem 13. Let $n \geq 3$, F be as in (3.7), and let E_α be the double rotationally symmetric cone of aperture $\alpha \in \,]0, \pi/2[$. Then there exists $\alpha(n) \in \,]0, \pi/2[$ such that E_α develops fattening if and only if $\alpha \in [\alpha(n), \pi/2[$

In [AIVc, AIVb] Angenent-Ilmanen-Velasquez provided examples of fattening for non compact smooth hypersurfaces in dimension $n \geq 4$.

In [FP96] Fierro-Paolini showed numerical evidence of fattening for mean curvature flow of the smooth initial torus in $\mathbf{R}^4$

$$\left\{ (x_1, x_2, x_3, x_4) \in \mathbf{R}^4 : \left(\left(x_1^2 + x_2^2 \right)^{1/2} - 1 \right)^2 + x_3^2 + x_4^2 \leq \lambda \right\}.$$

It seems that in this case there are two critical choices $\lambda_* < \lambda_{**}$ of the parameter λ, corresponding to different singularities. If $\lambda = \lambda_{**}$, the singularity is similar to that of the dumbbell, and fattening is not expected. On the other hand, the shape of the singularity corresponding to λ_* seems to be the one of a cone with the proper aperture, in such a way that fattening is expected.

If the following question has a positive answer, if fattening occurs, it occurs at the same time everywhere in the connected component. Let $E \subseteq \mathbf{R}^n$ and F be as in (3.7). Assume that $\mathcal{M}^*(E, \mathcal{F}_F)$ is connected with nonempty interior in $[t_0, t_0 + \tau[$, for some $\tau > 0$. Can we say that $\mathcal{M}^*(E, \mathcal{F}_F)$ is outer regular in $]t_0, t_0 + \tau[$?

We conclude this section with some remarks on motion by mean curvature in codimension higher than one.

The following is the first explicit example, to our knowledge, of three-dimensional fattening for motion by mean curvature in codimension 2.

Example 15. Let $n = 3$ and F be as in (3.8). Then the set E which is the union of the three coordinate axes develops 3-dimensional fattening for any t arbitrarily close to t_0.

Proof. As $\mathcal{M}_*(E, \mathcal{F}_F) = \emptyset$, the thesis reduces to check that

$$\mathcal{H}^3\big(\mathcal{M}^*(E, \mathcal{F}_F)(t)\big) > 0. \tag{6.6}$$

In particular, it is enough to prove that for any $R > 0$ there exists $T(R) > t_0$, with $T(R) \downarrow t_0$ as $R \downarrow 0$, such that $\mathcal{M}^*(E, \mathcal{F}_F)(T) \supseteq B_R(0)$.

Fix $R > 0$. We recall that there exists $T_1 > t_0$ such that the generalized evolution by curvature of the two-dimensional cross $\{(x, y) \in \mathbf{R}^2 : xy = 0\}$ contains the ball $B_R(0) \cap \{z = 0\}$ at time T_1. This result implies that $\mathcal{M}^*(E, \mathcal{F}_F)(T_1)$ contains the boundary of any triangle with sides lying on the coordinate planes and which is contained in $B_R(0)$. We recall now that, if an initial curve is contained in a plane, then its evolution (as a space curve) coincides with the usual evolution by curvature in that plane. Hence, the evolution of the boundary of the above triangles can be regarded, after the initial time, as a classical curvature flow [ES92a] of codimension one in the plane containing the triangle, and this evolution exists for a time controlled by $R^2/2$. Following [AS96], Remark 6.2, it follows that, for any $\varrho > 0$ we have that $\mathcal{M}(E_\varrho^+, \mathcal{F}_F)$ is a barrier for such flows, and the same is true for $\mathcal{M}^*(E, \mathcal{F}_F)$. Considering now the evolutions of the boundaries of equilateral triangles, one obtains that $\mathcal{M}^*(E, \mathcal{F}_F)(t) \supseteq B_{\frac{R}{\sqrt{3}}}(0)$, for $t > T_1 + \frac{R^2}{3}$, which implies (6.6). $\qquad\square$

In [Gio94b, Gio94a] De Giorgi suggested to consider the evolution of the two knotted circles in $\mathbf{R}^3$

$$E := \big\{(x_1, x_2, x_3) : x_3 = 0, x_1^2 + x_2^2 = 1\big\} \cup \big\{(x_1, x_2, x_3) : x_1 = 0, (x_2 - 1)^2 + x_3^2 = 1\big\},$$

with F as in (3.8), and to study the behaviour of the minimal barrier $\mathcal{M}^*(E, \mathcal{F}_F)$ after the collision time. It has been recently proved in [BNP98] that E develops three dimensional fattening.

Appendix A

The general definitions of barrier and minimal barrier read as follows [Gio94b].

Definition 13. *Let S be a set and let $r \subseteq S^2$. Assume that $S = \bigcap\{E \subseteq S : r \subseteq E^2\}$. Let $\mathcal{F}$ be a family of functions of one real variable which satisfy the following property: for any $f \in \mathcal{F}$ there exist $a, b \in \mathbf{R}$ such that $a < b$ and $f : [a, b] \to S$. We say that a function ϕ is a barrier associated with the couple $(r, \mathcal{F})$, and we shall write $\phi \in \mathcal{B}(r, \mathcal{F})$, if there exists a convex set $J \subseteq I$ such that $\phi : J \to S$ and, whenever a, b, f satisfy the condition*

$$[a, b] \subseteq J, \quad f : [a, b] \to S, \quad f \in \mathcal{F}, \quad (f(a), \phi(a)) \in r,$$

then

$$(f(b), \phi(b)) \in r.$$

In Definition 13 r has the meaning of graph of a binary relation. The condition $S = \bigcap\{E : r \subseteq E^2\}$, that is equivalent to say that S is the ambient of r, can be rewritten as follows: each point of S is either first or second element of a couple belonging to r, and therefore each point of S is in relation with some elements of S.

Notice that if $f : [a, b] \to S$ belongs to $\mathcal{F}$, then f is not necessarily a barrier on $[a, b]$.

For any $T \subseteq S$ set

$$
\begin{aligned}
\mathcal{M}^-(r, T) &:= \{\eta \in S : (\eta, \chi) \in r \quad \forall \chi \in T\}, \\
\mathcal{M}^+(r, T) &:= \{\eta \in S : (\chi, \eta) \in r \quad \forall \chi \in T\}, \\
M^-(r, T) &:= T \cap \mathcal{M}^-(r, T), \qquad M^+(r, T) := T \cap \mathcal{M}^+(r, T), \\
\mathcal{I}(r, T) &= M^+(r, \mathcal{M}^-(r, T)), \qquad \mathcal{S}(r, T) = M^-(r, \mathcal{M}^+(r, T)).
\end{aligned}
$$

If the set $\mathcal{I}(r, T)$ (respectively $\mathcal{S}(r, T)$) consists of only one element of S this element is indicated with r–$\inf T$ (respectively with r–$\sup T$):

$$
\begin{aligned}
r\text{--}\inf T = \eta &\iff \{\eta\} = \mathcal{I}(r, T), \\
r\text{--}\sup T = \eta &\iff \{\eta\} = \mathcal{S}(r, T).
\end{aligned}
$$

Let us define the minimal barrier.

Definition 14. *Let $\eta \in S$; if there exists a function $\sigma : I = [t_0, +\infty[\to S$ defined, for any $t \in I$, by the formula*

$$
\sigma(t) = r\text{--}\inf\{\phi(t) : \quad \phi : I \to S, \ \phi \in \mathcal{B}(r, \mathcal{F}), \ (\eta, \phi(t_0)) \in r\},
$$

we shall say that σ is the minimal barrier associated with η, r, $\mathcal{F}$, and we shall write $\sigma = \mathcal{M}(\eta, r, \mathcal{F}, I)$.

Notice that, after the choice of $S, r, \mathcal{F}$, to have the minimal barrier one has first to prove that for any $t \in I$ the set r–$\inf\{\phi(t) : \quad \phi : I \to S, \ \phi \in \mathcal{B}(r, \mathcal{F}), \ (\eta, \phi(t_0)) \in r\}$ consists of only one element of S.

Motion by mean curvature of boundaries is obtained with the choice

$$
\begin{aligned}
r &:= \{(E, L) : E \subseteq L \subseteq \mathbf{R}^n\}, \qquad S := \mathcal{P}(\mathbf{R}^n), \\
\mathcal{F} &:= \mathcal{F}_F, \qquad F \text{ as in (3.7)}.
\end{aligned}
$$

Motion by mean curvature of manifolds of codimension $n - h$ in $\mathbf{R}^n$ (h an integer, with $1 \leq h \leq n - 1$) is obtained with the choice of r, S as

$$
r := \{(E, L) : E \subseteq L \subseteq \mathbf{R}^n\}, \qquad S := \mathcal{P}(\mathbf{R}^n),
$$

and of $\mathcal{F}$ as follows. For any $a < b$, $[a, b] \subseteq [0, +\infty[$, the elements of $\mathcal{F}$ are the functions $f : [a, b] \to \mathcal{P}(\mathbf{R}^n)$ satisfying the following properties:

(i) the set $\{(t, x) : a \leq t \leq b, x \in f(t)\}$ is compact;

(ii) setting $\eta(t, x) := \text{dist}(x, f(t))^2/2$, there exists an open set $A \subseteq \mathbf{R}^n$ such that
$\eta \in C^\infty([a, b] \times \mathbf{R}^n)$, $A \supseteq f(t)$ for any $t \in [a, b]$, and for any $t \in [a, b]$ and
$x \in f(t)$ the matrix $\nabla^2 \eta(t, x)$ has rank $n - h$;

(iii) for any $t \in [a, b]$ and $x \in f(t)$ the following system of equations hold:

$$\frac{\partial^2 \eta}{\partial t \partial x_i} = \Delta \frac{\partial \eta}{\partial x_i}, \qquad i = 1, \ldots, n.$$

We conclude this appendix by noticing that the application of the barrier method to other situations (besides geometric evolutions) deserves further investigation.

Appendix B

We list here some assumptions used in this paper. We follow the notation of [GGIS91], pp. 462-463; we omit those properties in [GGIS91] which are not useful in our context.

(F1) $F : J_1 \to \mathbf{R}$ is continuous;

(F2) F is degenerate elliptic;

(F3) $-\infty < F_*(t, x, 0, 0) = F^*(t, x, 0, 0) < +\infty$ for all $t \in I$, $x \in \mathbf{R}^n$;

(F4) for every $R > 0$, $\sup\{|F(t, x, p, X)| : |p|, |X| \leq R, (t, x, p, X) \in J_1\} < +\infty$;

(F6') for every $R > \varrho > 0$ there is a constant $c = c_{R,\varrho}$ such that

$$|F(t, x, p, X) - F(t, x, q, X)| \leq c|p - q|$$

for any $t \in I$, $x \in \mathbf{R}^n$, $\varrho \leq |p|, |q| \leq R$, $|X| \leq R$;

(F7) there are $\varrho_0 > 0$ and a modulus σ_1 such that

$$F^*(t, x, p, X) - F^*(t, x, 0, 0) \leq \sigma_1(|p| + |X|),$$
$$F_*(t, x, p, X) - F_*(t, x, 0, 0) \geq -\sigma_1(|p| + |X|),$$

provided $t \in I$, $x \in \mathbf{R}^n$, $|p|, |X| \leq \varrho_0$.

The following example shows that, if $F(t, \cdot, p, X)$ is not Lipschitz (the dependence on X is irrelevant here), then the viscosity solution of (1.4) is not necessarily continuous, and motivates assumption (F8).

Example 16. Let $n = 2$, $F(x, p) := -g(x)|p|$, where

$$g(x) := \begin{cases} 0 & \text{if } |x| \leq 1, \\ \sqrt{|x| - 1} & \text{if } |x| > 1. \end{cases}$$

Then g is uniformly continous and is not Lipschitz. Let $v_0(x) := \min(1, |x| - 1)$ and $E_\lambda := \{v_0 \leq \lambda\}$ for $\lambda \in \mathbf{R}$. Then for any $\lambda \in \,]-1, 0]$ the set E_λ stands still, while, if $\lambda \in \,]0, 1[$, E_λ shrinks to E_0 at time $T_\lambda = 2\sqrt{\lambda}$. The function v having as sublevels the evolution of all sets E_λ (corresponding to a viscosity solution of (1.4) with $v(t_0, x) = v_0(x)$) is therefore not continuous.

(F8) There is a modulus σ_2 such that

$$|F(t, x, p, X) - F(t, y, p, X)| \leq |x - y||p|\sigma_2(1 + |x - y|)$$

for $y \in \mathbf{R}^n$, $(t, x, p, X) \in J_1$;

(F9) there is a modulus σ_2 such that $F_*(t, x, 0, 0) - F^*(t, y, 0, 0) \geq -\sigma_2(|x - y|)$
for any $t \in I$, $x, y \in \mathbf{R}^n$;

(F10) suppose that

$$-\mu \begin{pmatrix} \text{Id} & 0 \\ 0 & \text{Id} \end{pmatrix} \leq \begin{pmatrix} X & 0 \\ 0 & Y \end{pmatrix} \leq \nu \begin{pmatrix} \text{Id} & \text{-Id} \\ \text{-Id} & \text{Id} \end{pmatrix}$$

with $\mu, \nu \geq 0$. Let $R \geq 2\nu \vee \mu$ and let $\varrho > 0$; then

$$F_*(t, x, p, X) - F^*(t, y, p, -Y) \geq -|x - y||p|\bar{\sigma}(1 + |x - y| + \nu|x - y|^2)$$

for $(t, x) \in I \times \mathbf{R}^n$, $\varrho \leq |p| \leq R$, with some modulus $\bar{\sigma} = \bar{\sigma}_{R,\varrho}$ independent of t, x, y, X, Y, μ, ν.

Acknowledgements

We wish to thank Ennio De Giorgi for the encouragement he gave us and for many useful suggestions and advices.

Partial Regularity for Minimizers of Free Discontinuity Problems with p-th Growth

A. Leaci

1 Introduction

In the last ten years many problems have been investigated which were denoted by Ennio De Giorgi in [Gio91b] as free discontinuity problems. In such variational problems we want to minimize a functional which depends on a closed set K and on a function u suitably smooth outside of K. A functional widely studied in this field was proposed by Mumford and Shah [MS89b] (see also [BZ87]) for a variational approach to the segmentation problem in computer vision theory. Other functionals have been considered in connection with fracture mechanics, liquid crystals theory, immiscible fluids, elastic-plastic plates (see [AFF93], [Amb90], [ACM97], [Car95], [CLT92], [CLT94], [CLT96], [CT91] and the references therein). In this paper we prove the existence of a minimizing pair for the functional

$$G(K, u) = \int_{\Omega \setminus K} (|Du|^p + \mu|u - g|^q) \, dy + \lambda \mathcal{H}^{n-1}(K \cap \Omega), \qquad (1.1)$$

over the class of the admissible pairs

$$\mathcal{A} = \{(K, u); \ K \subset \mathbf{R}^n \text{ closed set}, \ u \in C^1(\Omega \setminus K)\},$$

where $\Omega \subset \mathbf{R}^n$ is an open set, $p > 1$, $q \geq 1$, $g \in L^q(\Omega) \cap L^\infty(\Omega)$, $\lambda, \mu > 0$, and $\mathcal{H}^{n-1}$ is the $(n-1)$ dimensional Hausdorff measure. In the case $p = 2$ this result was proved in [GCL89], and in [MMS92] for $n = 2$. For $n = p = q = 2$ the functional G is the Mumford–Shah functional. In the context of Computer Vision Theory, Ω is a rectangle in $\mathbf{R}^2$ (the screen on which images appear), and $g(x)$ is the intensity of the light (or grey-level) at the point x of some given image. The Mumford-Shah model (1985) defines the segmentation problem as a joint smoothing (cancelling the discontinuities due to noise) and edges detection problem: namely, given an image g, one seeks simultaneously for a "piecewise smoothed image" u with a set K of abrupt discontinuities, the "edges" of g. Then, the "best" segmentation of a given image g is obtained by minimizing the functional

$$\int_{\Omega \setminus K} (|Du|^2 + \mu|u - g|^2) \, dy + \lambda \, \text{length}(K \cap \Omega).$$

In this framework $\mu > 0$ is a scale parameter and $\lambda > 0$ is a contrast parameter and a measure of immunity to noise. From the mathematical viewpoint we could assume $\mu = \lambda = 1$ due to the different behaviour of the three terms under rescaling and multiplication of the functions u and g by a constant. We want to prove the following theorem:

Theorem 1. *Let $n \in \mathbf{N}$, $n \geq 2$, let $\Omega \subset \mathbf{R}^n$ be an open set, $g \in L^q(\Omega) \cap L^\infty(\Omega)$, $\lambda, \mu > 0$; then there exists at least one admissible pair $(K_0, u_0) \in \mathcal{A}$ such that*

$$G(K_0, u_0) = \min_{\mathcal{A}} G(K, u).$$

Moreover $\|u_0\|_\infty \leq \|g\|_\infty$.

The main difficulty in the minimization problem for G is to find the unknown optimal set K; indeed, given K, the optimal function u is the unique solution of a Neumann problem. In particular, for $p = q = 2$ the function u solves the problem

$$\begin{cases} \Delta u = \mu(u - g) & \text{on } \Omega \setminus K \\ \frac{\partial u}{\partial n} = 0 & \text{on } \partial\Omega \cup K. \end{cases}$$

Since there is not a topology on $\mathcal{A}$ that ensures compactness of the minimizing sequences and lower semicontinuity of the functional G, we prove the existence of a minimizing pair for the functional G following a classical argument of the Direct Methods in the Calculus of Variations.

Step 1. Weak formulation of the minimum problem for a related functional $\mathcal{G}$ in a suitable space and proof of the existence of a minimizer for $\mathcal{G}$ by using compactness and lower semicontinuity theorems.

Step 2. Study of the properties of a weak solutions, in order to obtain a minimizing pair for the functional G.

The plan of the exposition is the following. In section 2 we recall the definition of the space $SBV(\Omega)$ and the existence theorem of a weak solution. In section 3 we prove a Poincaré–Wirtinger type inequality and a compactness result. In section 4 we study the behaviour of a sequence of minimizers of $\mathcal{G}$ and we prove the existence of a minimizing pair for the functional G and some properties of an optimal pair.

2   Weak formulation

The general framework for the free discontinuity problems is a new function space, named the space of the Special functions of Bounded Variation in Ω, $SBV(\Omega)$, whose elements admit essential discontinuities across sets of codimension one. This is a subspace of the classical space $BV(\Omega)$ of functions with bounded variation in Ω, and it was introduced by De Giorgi and Ambrosio in [GA88]. In order to define this function space, we give some preliminaries and notations.

For a given set $E \subset \mathbf{R}^n$ we denote by $\overline{E}$ its topological closure, by ∂E its topological boundary, and by $|E|$ its Lebesgue outer measure. If A, B are subsets of $\mathbf{R}^n$ with $A \Subset B$ we mean that $\overline{A}$ is compact and contained in the interior of B. We indicate by $B_\rho(x)$ the ball $\{y \in \mathbf{R}^n; |y - x| < \rho\}$ and we set $B_\rho = B_\rho(0)$, and $\omega_n = |B_1|$. By $(\mathbf{e}_i)$ we denote the canonical basis of $\mathbf{R}^n$. Finally, for every $a, \beta \in \mathbf{R}$ we set $\alpha \wedge \beta = \min\{\alpha, \beta\}$ and $\alpha \vee \beta = \max\{\alpha, \beta\}$.

We denote by $\mathcal{H}^k(E)$ (k integer, $k \geq 1$) the k-dimensional Hausdorff measure of E, defined by

$$\mathcal{H}^k(E) = \lim_{\epsilon \downarrow 0} \mathcal{H}^k_\epsilon(E) = \sup_{\epsilon > 0} \mathcal{H}^k_\epsilon(E),$$

where

$$\mathcal{H}^k_\epsilon(E) = \inf \left\{ \sum_{i=1}^\infty \omega_k \left(\frac{\mathrm{diam} E_i}{2} \right)^k ; \ E \subset \bigcup_{i=1}^\infty E_i, \ \mathrm{diam} E_i < \epsilon \right\},$$

and we set

$$\mathcal{H}^0(E) = \begin{cases} \text{number of elements of } E & \text{if } E \text{ is finite} \\ +\infty & \text{otherwise.} \end{cases}$$

Since we will be mainly concerned with the $(n-1)$ dimensional measure, we put $m = n - 1$ in order to simplify the notation.

For every $v \in L^1_{\mathrm{loc}}(\Omega)$ we define the total variation of v

$$\int_\Omega |Dv| = \sup \left\{ \int_\Omega v \, \mathrm{div} \phi \, dx; \ \phi \in C^1_0(\Omega; \mathbf{R}^n), |\phi| \leq 1 \right\}.$$

We denote by $BV(\Omega)$ the space of all functions v of $L^1(\Omega)$ with $\int_\Omega |Dv| < +\infty$ and it can be proved that $v \in BV(\Omega)$ if and only if there exists a vector measure $Dv = (D_1 v, \ldots, D_n v)$ representing its distributional derivative, i.e.

$$\int_\Omega v \frac{\partial \phi}{\partial x_i} dx = - \int_\Omega \phi \, dD_i v \qquad \forall \phi \in C^\infty_0(\Omega).$$

We denote by ∇v the density of the absolutely continuous part of Dv with respect to the Lebesgue measure and by S_v the discontinuity set of v, *i.e.* the set of all points in Ω which are not Lebesgue points of v.

If $v = \chi_E$ is the characteristic function of a set E, then $v \in BV(\Omega)$ if and only if E is a set with finite measure and finite perimeter in Ω and its perimeter is given by

$$P(E, \Omega) = \int_\Omega |D\chi_E|.$$

For every $v \in BV(\Omega)$ the following properties hold (see 4.5.9 in [Fed69]):

- the coarea formula

$$\int_\Omega |Dv| = \int_{-\infty}^{+\infty} P(\{v > t\}, \Omega) \, dt;$$

- S_v is countably $(\mathcal{H}^m, m)$ rectifiable, *i.e.* there exists a sequence of C^1 hypersurfaces (Γ_h) such that $\mathcal{H}^m \left(S_v \setminus \bigcup_{h=1}^\infty \Gamma_h \right) = 0;$

- for $\mathcal{H}^m$ almost all $x \in S_v$ there exist $\nu = \nu_v(x) \in \partial B_1$, $v^+(x) \in \mathbf{R}$ and $v^-(x) \in \mathbf{R}$ (outer and inner trace, respectively, of v at x in the direction ν) such that, setting $B_\rho^+ = \{y \in B_\rho(x) : (y-x) \cdot \nu > 0\}$ and $B_\rho^- = \{y \in B_\rho(x) : (y-x) \cdot \nu < 0\}$, then

$$\lim_{\rho \to 0} \rho^{-n} \left(\int_{B_\rho^+} |v(y) - v^+(x)|^{\frac{n}{m}} dy + \int_{B_\rho^-} |v(y) - v^-(x)|^{\frac{n}{m}} dy \right) = 0,$$

and

$$\int_\Omega |Dv| \geq \int_\Omega |\nabla v| \, dy + \int_{S_v} |v^+ - v^-| d\mathcal{H}^m.$$

Following [GA88] (see also [Amb89]), we define a class of special functions of bounded variation which are characterized by a property stronger than the inequality above.

Definition 1. $SBV(\Omega)$ *denotes the proper closed subspace of* $BV(\Omega)$ *whose members are the functions v such that*

$$\int_\Omega |Dv| = \int_\Omega |\nabla v| \, dy + \int_{S_v} |v^+ - v^-| d\mathcal{H}^m.$$

$SBV_{loc}(\Omega)$ *denotes the space of all functions which belong to $SBV(A)$ for every open subset A with $A \Subset \Omega$.*

We remark that the well-known Cantor-Vitali function has bounded variation, but it does not satisfy the previous equality, because the absolutely continuous part of its derivative is zero, the set S_v is empty, but Dv is a non zero measure whose support is the Cantor middle third set.

Denoting by $W^{1,p}(\Omega)$ $(p \geq 1)$ the Sobolev space of functions $v \in L^p(\Omega)$ such that $Dv \in (L^p(\Omega))^n$, we remark that, for $v \in SBV(\Omega)$,

$$v \in W^{1,p}(\Omega) \quad \text{iff} \quad \mathcal{H}^m(S_v) = 0 \text{ and } \int_\Omega \left(|v|^p + |\nabla v|^p \right) dy < +\infty$$

(see e.g. 4.5.9(30) in [Fed69]). We may define on the space $SBV(\Omega)$ a new functional $\mathcal{G} : SBV(\Omega) \to [0, +\infty]$ related with our original functional G. Precisely, for every $v \in SBV(\Omega)$, we set

$$\mathcal{G}(v) = \int_\Omega \left(|\nabla v|^p + \mu |v - g|^q \right) dy + \lambda \mathcal{H}^m(S_v); \tag{2.1}$$

in this formulation the only variable is v, the discontinuity set being deduced from the set S_v of its jumps. This set has null Lebesgue measure, hence we can integrate on the whole of Ω. From the next lemma, every admissible pair $(K, u) \in \mathcal{A}$ such that u is bounded and $G(K, u) < +\infty$ is related to a function in $SBV_{loc}(\Omega)$.

Lemma 1. *Let $\Omega \subset \mathbf{R}^n$ be open, and $u \in L^\infty_{loc}(\Omega)$. Let $K \subset \mathbf{R}^n$ be closed, and assume $u \in C^1(\Omega \setminus K)$ and*

$$\int_{A \setminus K} |Du|\, dx + \mathcal{H}^m(K \cap A) < +\infty$$

for every open set $A \Subset \Omega$. Then $u \in SBV_{loc}(\Omega)$ and $S_u \subset K$.

Proof. In the case $n = 1$ the set K is locally finite and the proof is very simple. We outline the proof of the lemma in the case $n > 1$ by a slicing argument. We can assume that $A = (0,1)^n$ and for every $x \in \hat{A} := (0,1)^m \times \{0\}$ and $t \in (0,1)$ we set $u_x(t) = u(x + te_n)$. We also define $K_x = \{t \in (0,1)\,; x + te_n \in K\}$. Then (by [Fed69], 3.2.22 and Fubini's Theorem)

$$\int_{\hat{A}} \mathcal{H}^0(K_x)\, d\mathcal{H}^m(x) \le \mathcal{H}^m(K \cap A) < +\infty,$$

$$\int_{\hat{A}} \int_{[0,1] \setminus K_x} |u'_x(t)|\, dt\, d\mathcal{H}^m(x) \le \int_{A \setminus K} |Du| dx < +\infty.$$

Hence by the one dimensional case $u_x \in SBV((0,1))$ for $\mathcal{H}^m$ almost every $x \in \hat{A}$. By the slicing theorem in $BV(A)$ (see [Zie88], theorem 5.3.5)

$$\int_A |D_n u| = \int_{\hat{A}} \left(\int_0^1 |Du_x| \right) d\mathcal{H}^m(x) \le \int_{A \setminus K} |Du| dx + 2\|u\|_{L^\infty} \mathcal{H}^m(K \cap A).$$

Analogous results hold for each distributional partial derivative of u, hence $u \in BV(A)$. By the slicing theorem in $SBV(A)$ (see [Amb89], theorem 3.3) we achieve the proof. $\square$

We remark that the space $BV(\Omega)$ is not suitable to study the functional $\mathcal{G}$, because it is not coercive on $BV(\Omega)$; moreover if e.g. we assume $n = 1$, $\Omega = (0,1)$, and if g is the Cantor–Vitali function, we have $0 = \mathcal{G}(g) < \inf_A G(K, u)$.

By a simple truncation argument we see that

$$\inf \{\mathcal{G}(v)\,; \ v \in SBV(\Omega) \} = \inf \{\mathcal{G}(v)\,; \ v \in SBV(\Omega),\ \|v\|_{L^\infty} \le \|g\|_{L^\infty}\} \quad (2.2)$$

and also

$$\inf \{G(K, u)\,; \ (K, u) \in A \} = \inf \{G(K, u)\,; \ (K, u) \in A,\ \|u\|_{L^\infty} \le \|g\|_{L^\infty}\}.$$

Thanks to the above L^∞ bounds, the existence of a weak solution for the minimum problem for $\mathcal{G}$ on $SBV_{loc}(\Omega)$ is established by a compactness and lower semicontinuity result in the space $SBV_{loc}(\Omega)$ with respect to $L^1_{loc}(\Omega)$ topology (see the lecture by A. Braides in this book or Theorem 2.1 in [Amb89]). Moreover, by Lemma 1, we get

$$\min \{\mathcal{G}(v)\,; \ v \in SBV_{loc}(\Omega) \} \le \inf_A G(K, u).$$

In the following we shall prove that if v is a minimizer for $\mathcal{G}$, then

$$\mathcal{H}^m(\overline{S}_v \cap \Omega) = \mathcal{H}^m(S_v), \tag{2.3}$$

v coincide a.e. with a function $\tilde{v} \in C^1(\Omega \setminus \overline{S}_v)$, hence

$$(\overline{S}_v, \tilde{v}) \in \mathcal{A}, \qquad G(\overline{S}_v, \tilde{v}) = \mathcal{G}(v) = \min_{\mathcal{A}} G(K, u).$$

Notice that a function $v \in SBV(\Omega)$ may have a singular set S_v which is not closed and which may be even dense in Ω (see an example in [Lea92], Example 4.2). In the problem we are dealing with, we show that for a minimizer of $\mathcal{G}$ this cannot happen and we shall prove (2.3) estimating the lower $(n-1)$ dimensional density of S_v.

3 A Poincaré–Wirtinger type inequality and a compactness theorem

Assume that B is an open ball in $\mathbf{R}^n$. It is well-known that in the space $BV(B)$ the following estimate holds (see [Fed69], 4.5.9):

$$\left(\int_B |v - \mathrm{med}(v, B)|^{\frac{n}{m}} \, dy \right)^{\frac{m}{n}} \leq \gamma_n \int_B |Dv| \tag{3.1}$$

where $\mathrm{med}(v, B)$ denotes the least median value of v in B (see the definition below) and γ_n is the isoperimetric constant relative to the balls of $\mathbf{R}^n$, $i.e.$ for every measurable set E

$$\min\{|E \cap B|^{\frac{m}{n}}, |B \setminus E|^{\frac{m}{n}}\} \leq \gamma_n P(E, B).$$

Arguing on functions v such that $\mathcal{G}(v)$ is finite, we do not have an estimate of the right hand side of the inequality (3.1) and, on the other hand, we have summability of the p-th power of ∇v. Hence we prove a suitable Poincaré–Wirtinger type inequality in the class $SBV(B)$ (see Theorem 3.1 in [GCL89]) and we can obtain, as a consequence, a useful compactness theorem. For every measurable function $v : B \to \mathbf{R}$ we define the least median of v in B as

$$\mathrm{med}(v, B) = \inf \left\{ t \in \mathbf{R}; \ |\{v < t\} \cap B| \geq \frac{1}{2}|B| \right\}.$$

We emphasize that $\mathrm{med}(\cdot, B)$ is a non linear operator and in general it has no relationship with $\fint_B \cdot \, dy$. Moreover for every $v \in SBV(B)$ such that $(2\gamma_n \mathcal{H}^m(S_v))^{\frac{n}{m}} < \frac{1}{2}|B|$, we set

$$\tau' = \tau'(v, B) = \inf \left\{ t \in \mathbf{R}; \ |\{v < t\}| \geq (2\gamma_n \mathcal{H}^m(S_v))^{\frac{n}{m}} \right\},$$

$$\tau'' = \tau''(v, B) = \inf \left\{ t \in \mathbf{R}; \ |\{v \geq t\}| \leq (2\gamma_n \mathcal{H}^m(S_v))^{\frac{n}{m}} \right\}.$$

Now we define the truncated function

$$\overline{v} = \tau'(v, B) \vee v \wedge \tau''(v, B) \tag{3.2}$$

and we prove an integral inequality for $\overline{v}$.

Theorem 2. *Let* $B \subset \mathbf{R}^n$ *be an open ball,* $n \geq 2$, $1 \leq p < n$ *and* $p^* = \frac{np}{n-p}$. *Let* $v \in SBV(B)$ *with* $(2\gamma_n \mathcal{H}^m(S_v))^{\frac{n}{m}} < \frac{1}{2}|B|$, *and let* $\bar{v}$ *be as in (3.2). Then*

$$\left(\int_B |\bar{v} - \mathrm{med}(v, B)|^{p^*} \, dy \right)^{\frac{1}{p^*}} \leq \frac{2\gamma_n mp}{n-p} \left(\int_B |\nabla \bar{v}|^p \, dy \right)^{\frac{1}{p}}.$$

Proof. We may assume that the right hand side is finite. If $\mathcal{H}^m(S_v) = 0$ then $v \in W^{1,p}(B)$ and the thesis is a version of the Sobolev inequality. In the general case, since $\bar{v} \in SBV(B)$ we have

$$\begin{aligned}
\int_B |D\bar{v}| &= \int_B |\nabla \bar{v}| \, dy + \int_{S_{\bar{v}}} |\bar{v}^+ - \bar{v}^-| d\mathcal{H}^m \\
&\leq \int_B |\nabla \bar{v}| \, dy + (\tau'' - \tau')\mathcal{H}^m(S_v).
\end{aligned} \tag{3.3}$$

By the coarea formula and the isoperimetric inequality we have

$$\int_B |D\bar{v}| = \int_{\tau'}^{\tau''} P(\{\bar{v} < t\}, B) \, dt$$

$$\geq \frac{1}{\gamma_n} \int_{\tau'}^{\mathrm{med}(v,B)} |\{\bar{v} < t\}|^{\frac{m}{n}} \, dt + \frac{1}{\gamma_n} \int_{\mathrm{med}(v,B)}^{\tau''} |\{\bar{v} \geq t\}|^{\frac{m}{n}} \, dt.$$

Since for $\tau' < t < \mathrm{med}(v, B)$ we have

$$\frac{1}{\gamma_n}|\{\bar{v} < t\}|^{\frac{m}{n}} \geq 2\mathcal{H}^m(S_v)$$

and for $\mathrm{med}(v, B) < t < \tau''$ we have

$$\frac{1}{\gamma_n}|\{\bar{v} \geq t\}|^{\frac{m}{n}} \geq 2\mathcal{H}^m(S_v),$$

then we obtain

$$\int_B |D\bar{v}| \geq 2(\tau'' - \tau')\mathcal{H}^m(S_v). \tag{3.4}$$

By comparison between (3.3) and (3.4), we get $(\tau'' - \tau')\mathcal{H}^m(S_v) \leq \int_B |\nabla \bar{v}| \, dy$ and then, by (3.3),

$$\int_B |D\bar{v}| \leq 2 \int_B |\nabla \bar{v}| \, dy.$$

By the inequality (3.1) we achieve the proof in the case $p = 1$. For $1 < p < n$ we can apply the previous inequality to the function $w = |\bar{v}|^{\frac{p^*}{1^*}-1}\bar{v}$, taking into account that $\bar{w} = w$. $\qquad\square$

Remark 1. With the notations and under the assumptions of the previous theorem, we get

(1) $|\{v \neq \overline{v}\}| \leq 2(2\gamma_n \mathcal{H}^m(S_v))^{\frac{n}{m}}$,

(2) $\int_B |D\overline{v}| \leq 2\int_B |\nabla \overline{v}| \, dy \leq 2|B|^{\frac{p-1}{p}} \left(\int_B |\nabla \overline{v}|^p \, dy\right)^{\frac{1}{p}}$,

(3) $\text{med}(\overline{v}, B) = \text{med}(v, B)$ and $\overline{\overline{v}} = \overline{v}$.

Remark 2. In the case $p \geq n$ we can estimate $\int_B |\overline{v} - \text{med}(v, B)|^s \, dy$ for every $s \geq 1^*$. Namely, choosing $\sigma < n$ such that $\sigma^* = s$, we apply Theorem 2 with σ in place of p and then, by Hölder inequality, we obtain

$$\left(\int_B |\overline{v} - \text{med}(v, B)|^s \, dy\right)^{\frac{1}{s}} \leq \frac{2\gamma_n m s}{n}|B|^{\frac{1}{s}+\frac{1}{n}-\frac{1}{p}}\left(\int_B |\nabla \overline{v}|^p \, dy\right)^{\frac{1}{p}}.$$

Now we prove a compactness theorem for sequences in $SBV(B)$ without L^∞ estimates and with vanishing discontinuity sets.

Theorem 3 (Compactness and lower semicontinuity). *Let $B \subset \mathbf{R}^n$ be an open ball, $(v_h) \subset SBV(B)$ and let $p > 1$. Assume*

$$\sup_h \int_B |\nabla v_h|^p \, dy < +\infty, \qquad \lim_h \mathcal{H}^m(S_{v_h}) = 0.$$

Then there exist a subsequence (v_{h_k}) and $w \in W^{1,p}(B)$ such that for every $s \in [1, p^*)$ ($s \geq 1$ if $p \geq n$)

$$\lim_h \left(\overline{v}_{h_k} - \text{med}(v_{h_k}, B)\right) = w \quad \text{strongly in } L^s(B),$$

$$\lim_k \left(v_{h_k} - \text{med}(v_{h_k}, B)\right) = w \quad \text{a.e. on } B,$$

$$\int_B |Dw|^p \, dy \leq \liminf_k \int_B |\nabla \overline{v}_{h_k}|^p \, dy,$$

where $\overline{v}_{h_k}$ is defined as in (3.2).

Proof. We assume $p < n$. Then we have by Theorem 2 and remark 1(2)

$$\sup_h \left(\int_B |\overline{v}_h - \text{med}(v_h, B)|^{p^*} \, dy + \int_B |D\overline{v}_h|\right) < +\infty.$$

Therefore the sequence $(\overline{v}_h - \text{med}(v_h, B))$ is bounded in $BV(B)$, hence there is $w \in BV(B)$ and a subsequence $(\overline{v}_{h_k} - \text{med}(v_{h_k}, B))$ converging to w in $L^1(B)$. By interpolation (see e.g. [Bre92], Theorem 18 in Appendix) we obtain that this subsequence converges strongly in $L^s(B)$ for every $s < p^*$. By remark 1(1), up to a subsequence we obtain convergence a.e. of $(v_{h_k} - \text{med}(v_{h_k}, B))$. By Theorem 2.1 of [Amb89] we have that $w \in SBV(B)$. By the assumptions and still by Theorem 2.1 of [Amb89] we get $w \in W^{1,p}(B)$ and the lower semicontinuity inequality. In the case $p \geq n$ the thesis follows by the previous case and by using a diagonalization argument. □

4   A Limit Equation and a Decay Theorem

To study the behaviour of the minimizers of $\mathcal{G}$, we introduce a new functional F obtained by localizing the two leading terms of $\mathcal{G}$ and a functional Ψ that measures how much a function is far from minimizing locally the functional F. We consider for every $v \in SBV_{loc}(\Omega)$, $c > 0$, and for every Borel set $B \subset \Omega$, the functionals

$$F(v, c, B) = \int_B |\nabla v|^p \, dy + c\mathcal{H}^m(S_v \cap B),$$

$$\Phi(v, c, B) = \inf \{F(w, c, B); w \in SBV_{loc}(\Omega), \ w = v \text{ a.e. on } \Omega \setminus B\}$$

and, in the case $\Phi(v, c, B) < +\infty$, we set

$$\Psi(v, c, B) = F(v, c, B) - \Phi(v, c, B).$$

The functional Ψ is usually called "deviation from minimality" or "excess" of v in B. We shall argue on the asymptotic behaviour, as $\rho \to 0$, of the functional $\rho^{-m} F(v, c, B_\rho(x))$. Here we recall a property, proved in [GCL89] with the inessential restriction $c = 1$.

Lemma 2. *Let* $v \in SBV(\Omega)$ *such that* $F(v, c, A) < +\infty$ *for every open set* $A \Subset \Omega$. *Then*

$$\lim_{\rho \to 0} \rho^{-m} F(v, c, B_\rho(x)) = 0$$

for $\mathcal{H}^m$ *almost all* $x \in \Omega \setminus S_v$.

Proof. Fix $t > 0$ and $A \Subset \Omega$. Define

$$A_t = \{x \in A \setminus S_v \ : \ \limsup_{\rho \to 0} \rho^{-m} F(v, c, B_\rho(x)) \geq \omega_m t \ \}.$$

From Theorem 2.2.15 in [Amb97] we deduce

$$t\mathcal{H}^m(A_t) \leq F(v, c, A_t) = \int_{A_t} |\nabla v|^p \, dy < +\infty.$$

The finiteness of $\mathcal{H}^m(A_t)$ implies $|A_t| = 0$, hence by the previous inequality $\mathcal{H}^m(A_t) = 0$. By the arbitrariness of t and A we achieve the proof. $\quad\square$

For a minimizer of $\mathcal{G}$ some upper density estimates can be easily proved.

Lemma 3 (Excess estimate). *Assume that* v_0 *is a minimizer in* $SBV_{loc}(\Omega)$ *of the functional* $\mathcal{G}$ *with* $g \in L^\infty(\Omega) \cap L^q(\Omega)$. *Then for every* $\overline{B}_r(x) \subset \Omega$ *we have*

$$\Phi(v_0, \lambda, \overline{B}_r(x)) \leq F(v_0, \lambda, \overline{B}_r(x)) \leq \lambda n \omega_n r^m + \mu \int_{B_r(x)} |g|^q \, dy,$$

$$\Psi(v_0, \lambda, \overline{B}_r(x)) \leq (2\|g\|_\infty)^q \mu \omega_n r^n.$$

Proof. The previous inequalities follow immediately by using the comparison function $v = (1 - \chi_{\overline{B}_r(x)}) v_0$ and taking into account the "maximum principle" (2.2). $\quad\square$

In the following lemmas we state some properties of the functionals just definied. Without loss of generality we consider only balls centered at 0.

Lemma 4. *Let $v \in SBV(B_r)$. For every $c > 0$ the functions*

$$\rho \to F(v, c, B_\rho) \qquad \rho \to \Psi(v, c, B_\rho)$$

are nondecreasing in $(0, r)$.

Lemma 5 (Scaling). *Let $v \in SBV(B_r)$ and $0 < \rho < r$. For $\sigma > 0$ and for every $x \in B_1$ set*

$$v_\rho(x) = \sigma^{-1} \rho^{\frac{1-p}{p}} v(\rho x).$$

Then $v_\rho \in SBV(B_1)$ and

$$F(v, c, B_\rho) = \sigma^p \rho^m F(v_\rho, \frac{c}{\sigma^p}, B_1)$$

$$\Phi(v, c, B_\rho) = \sigma^p \rho^m \Phi(v_\rho, \frac{c}{\sigma^p}, B_1).$$

Proof. By the change of variables $y = \rho x$ we get

$$\int_{B_\rho} |\nabla v(y)|^p dy = \rho^n \int_{B_1} |\nabla v(\rho x)|^p dx = \sigma^p \rho^m \int_{B_1} |\nabla v_\rho(x)|^p dx,$$

$$\mathcal{H}^m(S_v \cap B_\rho) = \rho^m \mathcal{H}^m(S_{v_\rho} \cap B_1)$$

and the thesis follows.   □

Lemma 6 (Matching). *Let $u, v \in SBV(B_r)$, $c > 0$ and $0 < \rho < r$. Suppose*

$$\mathcal{H}^m(S_u \cap \partial B_\rho) = \mathcal{H}^m(S_v \cap \partial B_\rho) = 0.$$

Then

$$|\Phi(u, c, \overline{B}_\rho) - \Phi(v, c, \overline{B}_\rho)| \le c \mathcal{H}^m(\{\tilde{u} \ne \tilde{v}\} \cap \partial B_\rho).$$

Proof. Fix $\epsilon > 0$ and let $w \in SBV(B_r)$ be such that $w = u$ in $B_r \setminus \overline{B}_\rho$ and

$$F(w, c, \overline{B}_\rho) \le \Phi(u, c, \overline{B}_\rho) + \epsilon.$$

Setting

$$z(x) = \begin{cases} w(x) & \text{for } x \in \overline{B}_\rho \\ v(x) & \text{for } x \in B_r \setminus \overline{B}_\rho, \end{cases}$$

we get

$$\Phi(v, c, \overline{B}_\rho) \le F(z, c, \overline{B}_\rho) \le F(w, c, \overline{B}_\rho) + c \mathcal{H}^m(\{\tilde{u} \ne \tilde{v}\} \cap \partial B_\rho)$$

$$\le \Phi(u, c, \overline{B}_\rho) + \epsilon + c \mathcal{H}^m(\{\tilde{u} \ne \tilde{v}\} \cap \partial B_\rho).$$

By the arbitrariness of ϵ and by interchanging the role of u and v we achieve the proof.   □

Lemma 7 (Joining). *Let $u, v \in SBV(B_r)$, $c > 0$, $0 < \rho' < \rho'' < r$ and set $d = \rho'' - \rho'$. For every $0 < \delta < 1$ there exists $N = N_{p,\delta} \in \mathbf{N}$ such that*

$$\Phi(u, c, \overline{B}_{\rho''}) \leq (1 + \delta)\left(F(v, c, \overline{B}_{\rho''}) + F(u, c, \overline{B}_{\rho''} \setminus B_{\rho'})\right)$$

$$+ \delta \left(\frac{2N}{d}\right)^p \int_{B_{\rho''} \setminus B_{\rho'}} |u - v|^p \, dy.$$

Proof. Fix $\delta > 0$ and let $N \in \mathbf{N}$ be the smallest integer such that $N > \frac{2^{p-1}}{\delta}$. Let $r_j = \rho' + j \frac{d}{N}$ $(j = 0, \ldots, N)$ and let $\varphi_j \in C_0^\infty(\mathbf{R}^n)$ $(j = 1, \ldots, N)$ with

$$0 \leq \varphi_j \leq 1, \ \operatorname{spt} \varphi_j \Subset B_{r_j}, \ \varphi_j = 1 \text{ in a neighborhood of } \overline{B}_{r_{j-1}}, \ |\nabla \varphi_j| \leq \frac{2N}{d}.$$

Setting $w_j = u + (v - u)\varphi_j$, we have $w_j \in SBV(B_r)$, $w_j = u$ on $B_r \setminus B_{r_j}$,

$$S_{w_j} \subset (S_v \cap B_{r_j}) \cup (S_u \cap (B_{r_j} \setminus B_{r_{j-1}})),$$

and, by the inequality $|a + b|^p \leq 2^{p-1}(|a|^p + |b|^p)$, we get

$$\int_{B_{\rho''}} |\nabla w_j|^p \, dy = \int_{B_{\rho''}} |\nabla((1 - \varphi_j)u + \varphi_j v)|^p \, dy$$

$$\leq \int_{B_{\rho''}} |\nabla v|^p \, dy + \int_{B_{\rho''} \setminus B_{\rho'}} |\nabla u|^p \, dy + \int_{B_{r_j} \setminus B_{r_{j-1}}} |\nabla((1 - \varphi_j)u + \varphi_j v)|^p \, dy$$

$$\leq \int_{B_{\rho''}} |\nabla v|^p \, dy + \int_{B_{\rho''} \setminus B_{\rho'}} |\nabla u|^p \, dy + 2^{p-1} \int_{B_{r_j} \setminus B_{r_{j-1}}} (|\nabla u|^p + |\nabla v|^p) \, dy$$

$$+ 2^{p-1} \left(\frac{2N}{d}\right)^p \int_{B_j \setminus B_{j-1}} |u - v|^p \, dy \ .$$

Since the sets $\left(B_{r_j} \setminus B_{r_{j-1}}\right)$ are disjoint, then it follows that

$$\min_{1 \leq j \leq N} \int_{B_{\rho''}} |\nabla w_j|^p \, dy \leq \int_{B_{\rho''}} |\nabla v|^p \, dy + \int_{B_{\rho''} \setminus B_{\rho'}} |\nabla u|^p \, dy$$

$$+ \frac{2^{p-1}}{N} \int_{B_{\rho''} \setminus B_{\rho'}} (|\nabla u|^p + |\nabla v|^p) \, dy + \frac{2^{p-1}}{N} \left(\frac{2N}{d}\right)^p \int_{B_{\rho''} \setminus B_{\rho'}} |u - v|^p \, dy$$

$$\leq (1 + \delta) \left(\int_{B_{\rho''}} |\nabla v|^p \, dy + \int_{B_{\rho''} \setminus B_{\rho'}} |\nabla u|^p \, dy\right) + \delta \left(\frac{2N}{d}\right)^p \int_{B_{\rho''} \setminus B_{\rho'}} |u - v|^p \, dy.$$

Therefore an index j exists such that, setting $w = w_j$, we have

$$F(w, c, \overline{B}_{\rho''}) = \int_{\overline{B}_{\rho''}} |\nabla w|^p \, dy + c\mathcal{H}^m(S_w \cap \overline{B}_{\rho''})$$

$$\leq (1 + \delta) \left(F(v, c, \overline{B}_{\rho''}) + F(u, c, \overline{B}_{\rho''} \setminus B_{\rho'}) \right) + \delta \left(\frac{2N}{d} \right)^p \int_{B_{\rho''} \setminus B_{\rho'}} |u - v|^p \, dy,$$

and the thesis follows immediately. $\square$

Now we can characterize the asymptotic behaviour of a sequence of "quasi minimizers" of F (see assumption (i) below) with vanishing discontinuity set.

Theorem 4 (Blow up equation). *Let $B_r \subset \Omega$ be an open ball, $(v_h) \subset SBV(\Omega)$, let (c_h) be a sequence of positive numbers and let $w \in W^{1,p}(B_r)$. Assume that*

(i) $\lim_h \Psi(v_h, c_h, B_r) = 0$,

(ii) $\lim_h \mathcal{H}^m(S_{v_h} \cap B_r) = 0$,

(iii) $\lim_h F(v_h, c_h, B_\rho) = \omega(\rho) < +\infty$ *for almost all $\rho < r$,*

(iv) $\lim_h (v_h - \text{med}(v_h, B_r)) = w$ *a.e. on B_r.*

Then w is p-harmonic in B_r (i.e. it is a local minimizer in $W^{1,p}(B_r)$ for the functional $\int_{B_r} |Du|^p dy$) and

$$\omega(\rho) = \int_{B_\rho} |Dw|^p \, dy \qquad \text{for almost all } \rho < r.$$

Proof. Without loss of generality we assume $\text{med}(v_h, B_r) = 0$ for every $h \in \mathbf{N}$. By Theorem 3 we know that for a.e. $\rho < r$

$$\int_{B_\rho} |Dw|^p \, dy \leq \liminf_h F(v_h, c_h, B_\rho) = \omega(\rho).$$

Hence we complete the proof by showing that for a.e. $\rho < r$ and for every $v \in W^{1,p}(B_r)$ such that $v = w$ a.e. on $B_r \setminus \overline{B}_\rho$ we have

$$\omega(\rho) \leq \int_{B_\rho} |Dv|^p \, dy. \tag{4.1}$$

Let $\overline{v}_h$ be as in Theorem 3, hence we can assume, up to a subsequence, that $\overline{v}_h \to w$ in $L^p(B_r)$. By assumption (iii) we have that $\left(c_h \mathcal{H}^m(S_{v_h}) \right)$ is bounded. By integration in polar coordinates and by remark 1(1) we get

$$c_h \int_0^r \mathcal{H}^m(\{\widetilde{\overline{v}_h} \neq \widetilde{v}_h\} \cap \partial B_\rho) \, d\rho = c_h |\{\overline{v}_h \neq v_h\} \cap B_r| \leq 2c_h \left(2\gamma_n \mathcal{H}^m(S_{v_h} \cap B_r) \right)^{\frac{n}{m}}.$$

Hence, up to a subsequence that we do not relabel, for a.e. $\rho < r$ we have

$$\lim_h c_h \mathcal{H}^m(\{\widetilde{\overline{v}_h} \neq \widetilde{v}_h\} \cap \partial B_\rho) = 0. \tag{4.2}$$

By Lemma 6 and assumption (i), for a.e. $\rho < r$ we get

$$\lim_h \Phi(\overline{v}_h, c_h, B_\rho) = \lim_h F(\overline{v}_h, c_h, B_\rho) = \omega(\rho). \tag{4.3}$$

Now we fix ρ' such that ω is continuous and (4.2), (4.3) hold at ρ' (these properties are true a.e. in $(0, r)$). Since we want to prove (4.1) by contradiction, we assume that there exist $\epsilon > 0$ and $v \in W^{1,p}(B_r)$ such that $v = w$ a.e. on $B_r \setminus \overline{B}_{\rho'}$ and

$$\int_{B_{\rho'}} |Dv|^p \, dy < \omega(\rho') - \epsilon. \tag{4.4}$$

Let $\rho' < \rho'' < r$ such that (4.3) holds at ρ'' and also

$$\omega(\rho'') - \omega(\rho') < \frac{\epsilon}{4}, \qquad \int_{B_{\rho''} \setminus B_{\rho'}} |Dv|^p \, dy < \frac{\epsilon}{4}. \tag{4.5}$$

By Lemma 7 with $\overline{v}_h$ replacing u, for every $0 < \delta < 1$ we obtain

$$\Phi(\overline{v}_h, c_h, \overline{B}_{\rho''}) \leq (1 + \delta) \Big(F(v, c_h, \overline{B}_{\rho''}) + F(\overline{v}_h, c_h, \overline{B}_{\rho''} \setminus B_{\rho'}) \Big)$$
$$+ \frac{\delta(2N)^p}{(\rho'' - \rho')^p} \int_{B_{\rho''} \setminus B_{\rho'}} |\overline{v}_h - v|^p \, dy$$

Since $\overline{v}_h \to v$ in $L^p(B_{\rho''} \setminus B_{\rho'})$, passing to the limit as $h \to +\infty$, by (4.3)-(4.5) we obtain

$$\omega(\rho'') \leq (1 + \delta) \Big(\omega(\rho') - \epsilon + \frac{\epsilon}{4} + \frac{\epsilon}{4} \Big);$$

hence, by the arbitrariness of δ, we have $\omega(\rho'') \leq \omega(\rho') - \frac{\epsilon}{2}$, which contradicts the monotonicity of ω. $\qquad \square$

We recall the following result about p-harmonic functions.

Theorem 5. Let $1 < p < +\infty$, and let $w \in W^{1,p}(B_1)$ be a local minimizer of the functional

$$\int_{B_1} |Du|^p \, dy,$$

such that $\|Dw\|_{L^p} = 1$. Then w has a representative with continuous first derivatives and there exists an absolute constant $c_{n,p}$ such that for every $\rho \leq \frac{1}{2}$

$$\max_{y \in B_\rho} |Dw(y)| \leq c_{n,p}.$$

Proof. The result follows immediately by e.g. Theorem 1 in [Lew83] and by the classical Poincaré inequality. $\qquad \square$

The following decay theorem is a key step in the proof of Theorem 1

Theorem 6 (Decay). *For every $n \in \mathbf{N}$, $n \geq 2$, and every $c > 0$, $\alpha \in (0, \frac{1}{2})$ and $\beta \in (0, 1)$ such that $\alpha^{-\beta} > \omega_n c_{n,p}{}^p$, there exist $\epsilon = \epsilon(n, c, \alpha, \beta) > 0$ and $\theta = \theta(n, c, \alpha, \beta) > 0$ such that if $\Omega \subset \mathbf{R}^n$ is open, $\rho > 0$, $\overline{B}_\rho(x) \subset \Omega$ and $v \in SBV(\Omega)$ with*

$$\mathcal{H}^m(S_v \cap \overline{B}_\rho(x)) \leq \epsilon^p \rho^m,$$

$$\Psi(v, c, \overline{B}_\rho(x)) \leq \theta F(v, c, \overline{B}_\rho(x)),$$

then

$$F(v, c, \overline{B}_{\alpha\rho}(x)) \leq \alpha^{n-\beta} F(v, c, \overline{B}_\rho(x)).$$

Proof. The decay result can be proved by contradiction. Were the theorem false, it would be possible to find c, α, β as above, and sequences $\epsilon_h \downarrow 0$, $\theta_h \downarrow 0$, (v_h) in $SBV(\Omega)$, and $\overline{B}_{\rho_h}(x_h) \subset \Omega$, such that

$$\mathcal{H}^m\left(S_{v_h} \cap \overline{B}_{\rho_h}(x_h)\right) = \epsilon_h^p \rho_h^m,$$

$$\Psi(v_h, c, \overline{B}_{\rho_h}(x_h)) \leq \theta_h F(v_h, c, \overline{B}_{\rho_h}(x_h))$$

and

$$F(v_h, c, \overline{B}_{\alpha\rho_h}(x_h)) > \alpha^{n-\beta} F(v_h, c, \overline{B}_{\rho_h}(x_h)).$$

For each h, translating x_h into the origin and blowing up, *i.e.* setting

$$w_h(x) = \sigma_h^{-\frac{1}{p}} \rho_h^{\frac{1-p}{p}} v_h(x_h + \rho_h x) \quad \text{for } x \in B_1$$

where $\sigma_h = \rho_h^{-m} F(v_h, c, \overline{B}_{\rho_h}(x_h))$, we obtain a sequence of functions $w_h \in SBV(B_1)$ such that, by Lemma 5,

$$\mathcal{H}^m\left(S_{w_h} \cap \overline{B}_1\right) = \epsilon_h^p,$$

$$F(w_h, \frac{c}{\sigma_h}, \overline{B}_1) = 1$$

$$\Psi\left(w_h, \frac{c}{\sigma_h}, \overline{B}_1\right) \leq \theta_h$$

and

$$F\left(w_h, \frac{c}{\sigma_h}, \overline{B}_\alpha\right) > \alpha^{n-\beta}.$$

Since $\lim_h \epsilon_h = 0$ and $\lim_h \Psi\left(w_h, \frac{c}{\sigma_h}, \overline{B}_1\right) = 0$, by Theorem 4, there exist $w \in W^{1,p}(B_1)$ and a subsequence, that we still denote by (w_h), such that $\lim_h \left(w_h - \text{med}(w_h, B_1)\right) = w$ a.e. on B_1, w is p-harmonic in B_1 and for every $\rho < 1$

$$\lim_h F(w_h, \frac{c}{\sigma_h}, B_\rho) = \int_{B_\rho} |Dw|^p \, dy.$$

By the properties of p-harmonic functions (see Theorem 5), we have

$$\int_{B_\alpha} |Dw|^p \, dy \le \omega_n \alpha^n c_{n,p}{}^p,$$

so that

$$\alpha^{n-\beta} \le \limsup_h F\left(w_h, \frac{c}{\sigma_h}, \overline{B}_\alpha\right) = \int_{B_\alpha} |Dw|^p \, dy$$
$$\le \omega_n \alpha^n c_{n,p}{}^p,$$

and this contradicts the assumption on α and β. □

Theorem 7. *Assume that v_0 is a minimizer in $SBV_{loc}(\Omega)$ of the functional $\mathcal{G}$ with $g \in L^\infty(\Omega) \cap L^q(\Omega)$ and define*

$$\Omega_0 = \left\{ x \in \Omega \; : \; \lim_{\rho \to 0} \rho^{-m} F(v_0, \lambda, B_\rho) = 0 \right\}.$$

Then Ω_0 is open.

Proof. By Lemma 3, there exist two positive constant c_0 and c_1 such that for every $B_\rho(x) \subset \Omega$ we have

$$F(v_0, \lambda, B_\rho(x)) \le c_0 \rho^m, \qquad \Psi(v_0, \lambda, B_\rho(x)) \le c_1 \rho^n. \tag{4.6}$$

Let α, β, ϵ and θ be as in Theorem 6. Define $\rho_0 = \epsilon^p \theta \lambda \alpha^{n-\beta}/c_1$. Assume that $x_0 \in \Omega_0$ and fix $\rho \le \rho_0/2$ such that $B_{2\rho}(x_0) \subset \Omega$ and

$$F(v_0, \lambda, B_{2\rho}(x_0)) \le \epsilon^p \lambda \rho^m.$$

Then for every $y \in B_\rho(x_0)$ we have

$$F(v_0, \lambda, B_\rho(y)) \le \epsilon^p \lambda \rho^m. \tag{4.7}$$

If

$$\Psi(v_0, \lambda, B_\rho(y)) \le \theta F(v_0, \lambda, B_\rho(y)), \tag{4.8}$$

then by Theorem 6 we get

$$F(v_0, \lambda, B_{\alpha\rho}(y)) \le \epsilon^p \lambda \alpha^{1-\beta}(\alpha\rho)^m.$$

If inequality (4.8) is false, then we have

$$F(v_0, \lambda, B_\rho(y)) < \frac{1}{\theta} \Psi(v_0, \lambda, B_\rho(y)) \le \frac{c_1 \rho}{\theta} \rho^m \le \epsilon^p \lambda \alpha^{1-\beta}(\alpha\rho)^m.$$

Hence we get the same estimate as before, and by induction we have, for every $h \in \mathbf{N}$,

$$F(v_0, \lambda, B_{\alpha^h \rho}(y)) \le \epsilon^p \lambda \alpha^{h(1-\beta)}(\alpha^h \rho)^m.$$

Since for every $0 < t < \rho$ there exists an integer h such that $\alpha^{h+1}\rho \le t < \alpha^h \rho$ then

$$t^{-m} F(v_0, \lambda, B_t(y)) \le (\alpha^{h+1}\rho)^{-m} F(v_0, \lambda, B_{\alpha^h \rho}(y)) \le \epsilon^p \lambda \alpha^{h(1-\beta)-m}$$

hence, passing to the limit as $t \to 0$, so that $h \to +\infty$, we obtain $B_\rho(x_0) \subset \Omega_0$ and the proof is achieved. □

Now we can prove the existence of a minimizing pair for the functional G.

Proof of Theorem 1. Let v_0 be a minimizer of $\mathcal{G}$ and let Ω_0 be the open set given by Theorem 7. By Lemma 2 we know that $\mathcal{H}^m(S_{v_0} \cap \Omega_0) = 0$, hence $v_0 \in W^{1,p}(\Omega_0)$ and it is a local minimizer in $W^{1,p}(\Omega_0)$ of the functional

$$\int_{\Omega_0} (|Du|^p + \mu|u - g|^q) \, dy,$$

hence by standard regularity results for elliptic problems we get $\tilde{v}_0 \in C^1(\Omega_0)$ and $S_{v_0} \cap \Omega_0 = \emptyset$. Define $K_0 = \overline{\Omega} \setminus \Omega_0$. By Lemma 2 we have $\mathcal{H}^m(\overline{S_{v_0}} \cap \Omega) = \mathcal{H}^m(S_{v_0})$ so that $(K_0, \tilde{v}_0) \in \mathcal{A}$ and

$$G(K_0, \tilde{v}_0) = \mathcal{G}(v_0).$$

By Lemma 1, $(K_0, \tilde{v}_0)$ is a minimizing pair for G and the proof is complete. $\square$

Remark 3. By Lemma 1 we have conversely that if (K, u) is a minimizing pair for the functional G then $u \in SBV_{loc}(\Omega)$, u is a minimizer for $\mathcal{G}$, $\overline{S_u} \subset K$ is $(\mathcal{H}^m, m)$ rectifiable and $\tilde{u} \in C^1(\Omega \setminus \overline{S_u})$.

Finally we prove a lower density estimate for the singular set.

Theorem 8. *Assume that v_0 is a minimizer in $SBV(\Omega)$ of the functional $\mathcal{G}$ with $g \in L^\infty(\Omega) \cap L^q(\Omega)$. There exists $\rho_0 > 0$ and $\epsilon_1 > 0$ such that*

$$\mathcal{H}^m(S_{v_0} \cap \overline{B}_\rho(x)) \geq \epsilon_1 \rho^m$$

for every $x \in \overline{S_{v_0}}$ and for every $\rho \leq \rho_0$ with $\overline{B}_\rho(x) \subset \Omega$.

Proof. Let α, β, ϵ and θ be as in Theorem 6 and let Ω_0 be as in Theorem 7. Define $\rho_0 = \epsilon^p \theta \lambda \alpha^{n-\beta} / c_1$ and fix $k \in \mathbf{N}$ such that $c_0 \alpha^{k(1-\beta)} \leq \epsilon^p \lambda$, where c_0 is given in (4.6). We define $\epsilon_1 = \epsilon^p \alpha^{km}$, and we prove the thesis by contradiction. Assume that $x \in \overline{S_{v_0}}$, $\rho \leq \rho_0$, $\overline{B}_\rho(x) \subset \Omega$ and

$$\mathcal{H}^m(S_{v_0} \cap \overline{B}_\rho(x)) < \epsilon_1 \rho^m. \tag{4.9}$$

If

$$\Psi(v_0, \lambda, B_\rho(x)) \leq \theta F(v_0, \lambda, B_\rho(x)), \tag{4.10}$$

then by Theorem 6 and by (4.6) we get

$$F(v_0, \lambda, B_{\alpha\rho}(x)) \leq \alpha^{n-\beta} F(v_0, \lambda, B_\rho(x)) \leq c_0 \alpha^{1-\beta}(\alpha\rho)^m.$$

If inequality (4.10) is false, then by (4.6) and $\rho \leq \rho_0$ we have

$$F(v_0, \lambda, B_\rho(x)) < \frac{1}{\theta}\Psi(v_0, \lambda, B_\rho(x)) \leq \epsilon^p \lambda \alpha^{1-\beta}(\alpha\rho)^m.$$

Hence, by assumption (4.9), we can apply k times Theorem 6 to achieve

$$F(v_0, \lambda, B_{\alpha^k\rho}(x)) \leq (c_0 \vee \epsilon^p \lambda)\alpha^{k(1-\beta)}(\alpha^k\rho)^m \leq \epsilon^p \lambda(\alpha^h\rho)^m.$$

As in the proof of Theorem 7, after inequality (4.7), we deduce $x \in \Omega_0$, contradicting the assumption $x \in \overline{S_{v_0}}$. $\square$

Remark 4. Many regularity properties for the minimizers of G in the case $p = 2$ have been proved in [AFP97] (see also [DS96] and [MS94]). The existence of a minimizing pair for the functional G with $p = 2$ and g possibly unbounded has been proved in [Lea94].

Free discontinuity problems and their non-local approximation

A. Braides

1 Free discontinuity problems

Following a notation introduced by De Giorgi, we denote by "free discontinuity problems" all the problems in the calculus of variations where the unknown is a pair (u, K) with K varying in a class of closed subsets of a fixed open set $\Omega \subset \mathbf{R}^n$, and $u : \Omega \setminus K \to \mathbf{R}^m$ is a function in some function space (e.g., $u \in C^1(\Omega \setminus K)$ or $u \in W^{1,p}(\Omega \setminus K)$). Such problems are usually of the form

$$\min\{E_v(u, K) + E_s(u, K) + \text{ lower order terms }\}, \tag{1.1}$$

with E_v, E_s being interpreted as *volume* and *surface* energies, respectively.

Example 1. (i) *Fractured hyperelastic media* . In this case $\Omega \subset \mathbf{R}^3$ is the reference configuration of an elastic body, K is the crack surface, and u represents the elastic deformation in the unfractured part of the body. Following Griffith's theory of fracture, we can introduce a surface energy which accounts for fracture initiation. In the homogeneous case, this energy is simply proportional to the surface area[1] of K if the body is isotropic: $E_s(u, K) = c\,\mathcal{H}^2(K)$, or more in general is an integral on K depending on the orientation ν of the crack surface in the non-isotropic case: $E_s(u, K) = \int_K \varphi(\nu)\,d\mathcal{H}^2$. The volume energy takes the form $E_v(u, K) = \int_{\Omega \setminus K} W(\nabla u)\,dx$, where W is an elastic bulk energy density. (For applications of free discontinuity problems in this framework see for example [AB97], [BC93], [BDV96], [FF95])

(ii) *Signal reconstruction.* A source signal is usually a piecewise smooth function u (which we may think as parameterized on some interval (a, b)). The problem of reconstructing u from a disturbed input g deriving from a distorted transmission, can be modelled as finding the minimum of

$$\min\left\{ \int_{(a,b)\setminus S(u)} |u'|^2\,dt + c_1 \int_a^b |u - g|^2\,dt + c_2\,\#(S(u)) \right\}. \tag{1.2}$$

where $S(u)$ denotes the set of discontinuity points of u. Here c_1 and c_2 are tuning parameter. In this case $\Omega = (a, b)$, $K = S(u)$, $E_v(u, K) = \int_{\Omega \setminus K} |u'|\,dt$, $E_s(u, K) = c_2\,\#(K)$.

---

[1] We denote by $\mathcal{H}^k$ the k-dimensional Hausdorff measure.

(iii) *Image reconstruction.* The formulation above can be extended to model some problem in computer vision introducing the functional (see [MS89b], [MS94], [Gio91b], [MMS92])

$$\int_{\Omega\setminus K} |\nabla u|^2 \, dx + c_1 \int_{\Omega\setminus K} |u - g|^2 \, dx + c_2 \mathcal{H}^1(K) \,. \tag{1.3}$$

In this case g is interpreted as the input picture taken from a camera, u is the "cleaned" image, and K is the relevant contour of the objects in the picture. Again, c_1 and c_2 are contrast parameters. Note that the problem is meaningful also adding the constraint $\nabla u = 0$ outside K, in which case we have a minimal partitioning problem (see e.g. [CT91], [AB90], [BP96b]).

(iv) *Drops of liquid crystals.* (Oseen-Frank energy with surface interaction). In this model $D \subset \Omega$ represents the region occupied by a liquid crystal, whose energy is

$$\int_D W(n, \nabla n) \, dx + \int_{\partial D \cap \Omega} f(n, \nu_D) \, d\mathcal{H}^2 \,. \tag{1.4}$$

In this case $K = \partial D$, n is the orientation of the crystal, $u = (n, \chi_D)$, and ν_D is the normal to ∂D (see [Vir89]).

(v) *Prescribed curvature problems.* As a particular free discontinuity problem we can also recover the problem of finding sets E with boundary of prescribed mean curvature H; i.e., satisfying $H = g\nu$ on ∂E. In this case the energy to be minimized is

$$\int_E g(x) \, dx + \mathcal{H}^{n-1}(\partial E) \,, \tag{1.5}$$

which can be seen as the sum of a volume and a surface energy depending on the unknowns $u = \chi_E$ and $K = \partial E$.

2   A weak formulation: SBV functions

The treatment of free discontinuity problems following the direct methods of the calculus of variations presents many difficulties, due to the dependence of the energies on the surface K. Unless topological constraints are added it is usually not possible to deduce compactness properties from the only information that such kind of energies are bounded. The idea of De Giorgi has been to interpret K as the set of discontinuity points of the function u, and to set the problems in a space of discontinuous functions. The requirements on such a space are of two kinds: (a) *structure properties*: we want to be able to define K as the set of discontinuity points of the function u in such a way that it is (sufficiently) smooth (i.e., a normal ν to K and the traces of u on both sides of K exist, and we can compute surface integrals on K), and u should be "differentiable" almost everywhere so that bulk energy depending on ∇u can be defined; (b) *compactness properties*: it is possible to apply the direct method of the calculus of variations, obtaining compactness of sequences of functions with bounded energy.

The answer to the two requirements above has been De Giorgi and Ambrosio's space of special functions of bounded variation SBV [GA88] We recall that the space $BV(\Omega)$ of functions of bounded variation on a set Ω is the set of all functions $u \in L^1(\Omega)$ whose *distributional derivative* Du is a bounded vector measure.

Definition 1. *Let Ω be an open set in $\mathbf{R}^n$ and let $u \in L^1(\Omega)$. We say that u is a special function of bounded variation on Ω, and we write $u \in SBV(\Omega)$, if there exist $f \in L^1(\Omega; \mathbf{R}^n)$ and $g \in L^1(\Omega, \mathcal{H}^{n-1}; \mathbf{R}^n)$ such that*

$$Du = f\mathcal{L}_n + g\mathcal{H}^{n-1}, \tag{2.1}$$

i.e., we can represent the distributional derivative in a special form which contains only a volume and a surface part.[2]

The structure theorem of $BV(\Omega)$ gives a precise description of the densities f and g in (2.1). We state briefly some well-known definitions and results (see [EG92], [Fed69], [Giu84]).

Definition 2. *Let $u \in BV(\Omega)$. Then we can decompose Du as*

$$Du = D_a u + D_c u + D_j u, \tag{2.2}$$

where $D_a u$ is the Lebesgue part of Du (i.e., the absolutely continuous part of Du with respect to $\mathcal{L}_n$), $D_j u$ is the jump part of Du (concentrated on a σ-finite set with respect to $\mathcal{H}^{n-1}$), and $D_c u$ is the Cantor part of Du (which is characterized as being orthogonal to $\mathcal{L}_n$ and such that $|D_c u|(B) = 0$ if $\mathcal{H}^{n-1}(B) < +\infty$).

The Lebesgue part of Du is characterized as follows.

Proposition 1. *We have $D_a u = \nabla u\, \mathcal{L}_n$, where ∇u is the approximate gradient of u.*

To describe $D_j u$ we need some more definitions.

Definition 3. *Let $u \in BV(\Omega)$. We define the jump set of u, $S(u)$, as* [3]

$$S(u) = \Big\{ x \in \Omega : \exists \nu \in S^{n-1}, \exists u^{\pm} \in \mathbf{R},\ u^+ \neq u^- :$$

$$\lim_{\rho \to 0+} \frac{1}{\rho^n} \int_{B_\rho(x) \cap H_\nu^{\pm}(x)} |u(y) - u^{\pm}| dy = 0 \Big\}, \tag{2.3}$$

where $H^{\pm}(x) = \{ y : \pm\langle y - x, \nu \rangle > 0 \}$. For all $x \in S(u)$ we choose $\nu_u(x)$ and $u^{\pm}(x)$ satisfying (2.3). The function $\nu : S(u) \to S^{n-1}$ is called the normal to $S(u)$, and $u^{\pm}$ are the traces of u on both sides of $S(u)$.

---

[2] $\mathcal{L}_n$ denotes the Lebesgue measure on $\mathbf{R}^n$, and $L^1(\Omega, \mathcal{H}^{n-1}; \mathbf{R}^n)$ denotes the space of $\mathbf{R}^n$-valued functions g with $\int_\Omega |g|\, d\mathcal{H}^{n-1} < +\infty$. Note that this implies that $g = 0$ except on a set of σ-finite $\mathcal{H}^{n-1}$ measure.

[3] $B_\rho(x)$ denotes the open ball of centre x and radius ρ.

Proposition 2. *If $u \in BV(\Omega)$, the jump part of Du can be represented as*[4]

$$D_j u = (u^+ - u^-)\nu_u \, \mathcal{H}^{n-1} \, \llcorner \, S(u).$$

The following propositions provides two alternative definitions of $SBV(\Omega)$.

Proposition 3. *The following statements are equivalent:*

(i) $u \in SBV(\Omega)$;
(ii) $u \in BV(\Omega)$ *and* $D_c u = 0$;
(iii) $u \in BV(\Omega)$ *and* $Du = \nabla u \, \mathcal{L}_n + (u^+ - u^-)\nu_u \, \mathcal{H}^{n-1} \, \llcorner \, S(u)$.

Remark 1. Note that for functions $u \in BV(\Omega)$ we can have a weak formulation of energies as in Example 1, simply substituting K by $S(u)$. Note however that functions $u \in BV(\Omega)$ with $D_a u = D_j u = 0$ are dense in $BV(\Omega)$ in the $L^1(\Omega)$-topology. This fact suggests that the extension of free discontinuity energies as in Example 1 to $BV(\Omega)$ cannot be straightforward, and it motivates the introduction of the space $SBV(\Omega)$.

3   Ambrosio's compactness theorem

In the sequel we take as a model the n-dimensional "weak version" of the so-called Mumford–Shah functional:

$$E(u) = \int_\Omega |\nabla u|^2 \, dx + \mathcal{H}^{n-1}(S(u)), \qquad (3.1)$$

which is obtained from the energy in (1.3) substituting K by $S(u)$. We drop the lower order term $c_1 \int_\Omega |u - g|^2 \, dx$, and we set $c_2 = 1$ in (1.3) for the sake of simplicity. Note that the volume integral can be computed on the whole Ω since ∇u is defined $\mathcal{L}_n$-a.e. In this case $\Omega \subset \mathbf{R}^n$. We will state a compactness theorem for sequences of $SBV(\Omega)$ functions suited to treat minimum problems involving E. More general theorems are available in the literature (see [Amb89], [Amb90], [Amb94], [AM97]).

Theorem 1 (Ambrosio's Compactness Theorem). *Let (u_j) be a sequence of $SBV(\Omega)$ functions. Assume that*

(i) (u_j) *is uniformly bounded in* $BV(\Omega)$;
(ii) (∇u_j) *is bounded in* $L^2(\Omega; \mathbf{R}^n)$;
(iii) $\sup_j \mathcal{H}^{n-1}(S(u_j)) < +\infty$;
    then there exists a sequence (not relabelled) such that
(a) *there exists* $u \in SBV(\Omega)$ *such that* $u_j \to u$ *in* $L^1(\Omega)$;
(b) $\nabla u_j \rightharpoonup \nabla u$ *weakly in* $L^1(\Omega; \mathbf{R}^n)$;
(c) $\mathcal{H}^{n-1}(S(u)) \le \liminf_j \mathcal{H}^{n-1}(S(u_j))$.

---

[4] If μ is a measure and S is a measurable set, we define the measure $\mu \llcorner S$ by setting $\mu \llcorner S(B) = \mu(B \cap S)$.

Example 2. We can apply the theorem above to find weak solutions for problems in computer vision. Let $g \in L^\infty(\Omega)$, and let (v_j) be a minimizing sequence for

$$\inf\left\{ E(u) + c_1 \int_\Omega |u - g|^2\, dx : u \in SBV(\Omega) \right\}. \tag{3.2}$$

Let $u_j = -\|g\|_\infty \vee (\|g\|_\infty \wedge v_j)$; then $E(u_j) \leq E(v_j)$ and $\int_\Omega |u_j - g|^2\, dx \leq \int_\Omega |v_j - g|^2\, dx$, so that (u_j) is still minimizing and $\|u_j\|_\infty \leq \|g\|_\infty$. Taking $u = 0$ in (3.2) we see that we can suppose $\sup_j E(u_j) \leq c_1 \int_\Omega |g|^2\, dx$. In particular $\sup_j \mathcal{H}^{n-1}(S(u_j)) < +\infty$ and $\sup_j \int_\Omega |\nabla u_j|^2\, dx < +\infty$. This shows that (u_j) satisfies conditions (ii) and (iii) of Theorem 3.1. Moreover $\|u_j\|_{BV(\Omega)} = \int_\Omega |u_j|\, dx + \int_\Omega |\nabla u_j|\, dx + \int_{S(u)} |u_j^+ - u_j^-|\, d\mathcal{H}^{n-1} \leq |\Omega|\|g\|_\infty + |\Omega|^{1/2}\|\nabla u_j\|_{L^2(\Omega)} + 2\|g\|_\infty \mathcal{H}^{n-1}(S(u_j))$ is equibounded so that also (i) is verified. Hence, we can suppose that $u_j \to u \in SBV(\Omega)$ in $L^2(\Omega)$. By Theorem 1 and the continuity of $v \mapsto \int_\Omega |v - g|^2$ we see that u is a minimum point for (3.2). "Classical" solutions (u, K) can be obtained setting $K = \overline{S(u)}$ and taking the restriction u to $\Omega \setminus K$ (see [GCL89], [CL91]). Recently regularity results for K have been obtained (see [AFP97], [AP97], [Bon96], [Dib94]).

We outline a proof of Theorem 1 due to Alberti and Mantegazza ([AM97]), whose main idea is a characterization of functions in $SBV(\Omega)$ by means of an integration by parts formula. Similar arguments can be found in [Amb95b] and [ABG98]. Note that for $u \in BV(\Omega)$ a *chain rule* formula holds as follows: if $\phi \in C^1(\mathbf{R})$ is a bounded Lipschitz function then $\phi(u) \in BV(\Omega)$ and

$$D(\phi(u)) = \phi'(\tilde{u})(D_a u + D_c u) + (\phi(u^+) - \phi(u-))\nu_u\, \mathcal{H}^{n-1} \lfloor S(u), \tag{3.3}$$

where $\tilde{u}(x) = \limsup_{\rho \to 0+}(\omega_n \rho^n)^{-1} \int_{B_\rho(x)} u(y)\, dy$ is the *precise representative* of u; in particular

$$|D(\phi(u)) - \phi'(\tilde{u})(D_a u + D_c u)| = \int_{S(u)} |\phi(u^+) - \phi(u^-)|\, d\mathcal{H}^{n-1}. \tag{3.4}$$

If we introduce the auxiliary set of functions

$$X = \{\phi \in C^1(\mathbf{R}) : \|\phi'\|_\infty < +\infty, \|\phi\|_\infty \leq 1/2\}, \tag{3.5}$$

then we have by (3.4)

$$\mathcal{H}^{n-1}(S(u)) = \sup_{\phi \in X} |D(\phi(u)) - \phi'(\tilde{u})(D_a u + D_c u)|. \tag{3.6}$$

From this observation we derive the following useful lemma ([AM97] Proposition 2.3).

Lemma 1. *Let $u \in BV(\Omega)$, and let λ be a $\mathbf{R}^n$-valued measure with $|\lambda|(S(u)) = 0$ and*

$$\sup_{\phi \in X} |D\phi(u) - \phi'(\tilde{u})\lambda| < +\infty. \tag{3.7}$$

Then $\lambda = D_a u + D_c u$

Proof (Sketch). Let $\mu = \lambda - (D_a u + D_c u)$. Plugging (3.3) into (3.7) we obtain $\sup_{\phi \in X} \int_\Omega \phi'(\tilde{u}) \, d|\mu| < +\infty$. As we have no bound on ϕ' if $\phi \in X$, it is easy to see that we must have $|\mu| = 0$.

Proof (Theorem 1). As (u_j) is precompact in $BV(\Omega)$ and (∇u_j) is weakly precompact in $L^2(\Omega)$, without loss of generality we have that there exists $u \in BV(\Omega)$ such that $u_j \to u$ in $L^1(\Omega)$ and $Du_j \rightharpoonup Du$ weakly in the sense of measures. Moreover there exists $g \in L^2(\Omega; \mathbf{R}^n)$ such that $\nabla u_j \rightharpoonup g$ weakly in $L^2(\Omega; \mathbf{R}^n)$. If $\phi \in X$ then by (3.4)

$$|D(\phi(u_j)) - \phi'(u_j)\nabla u_j \mathcal{L}_n| \le \mathcal{H}^{n-1}(S(u_j)). \tag{3.8}$$

We have moreover $D(\phi(u_j)) \rightharpoonup D(\phi(u))$ weakly in the sense of measures and $\phi'(u_j) \to \phi(u)$ in $L^2(\Omega)$, so that

$$(D(\phi(u_j)) - \phi'(u_j)\nabla u_j \mathcal{L}_n) \rightharpoonup (D(\phi(u)) - \phi'(u)g\mathcal{L}_n) \tag{3.9}$$

weakly in the sense of measures. From the lower semicontinuity of the total variation we have by (3.8) and (3.9)

$$|D(\phi(u)) - \phi'(u)g\mathcal{L}_n| \le \liminf_j \mathcal{H}^{n-1}(S(u_j)) < +\infty. \tag{3.10}$$

Hence, we can apply Lemma 1 with $\lambda = g\mathcal{L}_n$, and obtain $g\mathcal{L}_n = D_a u + D_c u$; in particular $D_c u = 0$ (i.e., $u \in SBV(\Omega)$) and $g = \nabla u$. Eventually we obtain (c) from (3.10) recalling (3.6).

4   Non-local approximation of the Mumford-Shah functional

Functionals arising in free discontinuity problems present some serious drawbacks. First of all, the lack of differentiability in any reasonable norm implies the impossibility of flowing those functionals, and dynamic problems can be tackled only in an indirect way. Moreover, numerical problems arise in the detection of the unknown discontinuity surface. Hence, a considerable effort has been spent recently to provide variational approximations of free discontinuity problems with differentiable energies defined on smooth functions. We turn our attention now to the problem of the approximation of the Mumford-Shah functional E defined in (3.1). It is not possible to obtain a variational approximation of $E(u)$, leading to the convergence of minimum points, by means of local integral functionals of the form

$$\int_\Omega f_\varepsilon(\nabla u(x)) \, dx,$$

defined in the Sobolev space $H^1(\Omega)$. In fact, if such an approximation existed, the functional $E(u)$ would be also the variational limit of the sequence of convex functionals

$$\int_\Omega f_\varepsilon^{**}(\nabla u(x)) \, dx,$$

where f_ε^{**} is the convex envelope of f_ε (see, e.g., [Mas93], Proposition 6.1 and Example 3.11), in contrast with the lack of convexity of $E(u)$. A different path can be followed, considering approximations of the form

$$E_\varepsilon(u) = \frac{1}{\varepsilon} \int_\Omega f\left(\varepsilon \fint_{B_\varepsilon(x)\cap\Omega} |\nabla u(y)|^2 \, dy\right) dx, \tag{4.1}$$

defined for $u \in H^1(\Omega)$, where f is a suitable non-decreasing continuous (non-convex) function, and $\fint_B$ denotes the average on B. These functionals are non-local in the sense that their energy density at a point $x \in \Omega$ depends on the behaviour of u in the whole set $B_\varepsilon(x) \cap \Omega$. Note that, even if the term containing the gradient is not convex, the functional E_ε is weakly lower semicontinuous in $H^1(\Omega)$ by Fatou's Lemma. It is possible to prove ([BM97]) the Γ-convergence, as $\varepsilon \to 0$, of the functionals E_ε to the functional E for a suitable choice of the function f. For instance, we can simply take $f(t) = \min\{t, 1/2\}$. We obtain the same result if we replace f by a sequence (f_ε) which converges to f rapidly enough, with the advantage that for a suitable choice of f_ε the functionals E_ε have a minimum point for every $\varepsilon > 0$. By standard arguments this Γ-convergence result implies the convergence, up to a subsequence, of the minimum points of $E_\varepsilon(u) + c_1 \int_\Omega |u - g|^2 \, dx$ to a minimum point of $E(u) + c_1 \int_\Omega |u - g|^2 \, dx$ if $g \in L^2(\Omega)$. We recall that the Γ-convergence (with respect to the $L^1(\Omega)$-topology) of E_ε to E means that:

(i) if $\varepsilon_j \to 0+$, $u_j \to u$ in $L^1(\Omega)$ and $\sup_j E_{\varepsilon_j}(u_j) < +\infty$ then $(-T) \vee (T \wedge u) \in SBV(\Omega)$ for all T (compactness) and $E(u) \le \liminf_j E_{\varepsilon_j}(u_j)$ (lower semicontinuity inequality);

(ii) (existence of a recovery sequence) if $\varepsilon_j \to 0+$ and $u \in SBV(\Omega)$ then there exists a sequence $u_j \to u$ in $L^1(\Omega)$ such that $E(u) \ge \limsup_j E_{\varepsilon_j}(u_j)$.

We prove the convergence result in dimension one. The higher dimensional case is more technical, but it follows the one-dimensional ideas. We consider the functionals

$$E_\varepsilon(u) = \frac{1}{\varepsilon} \int_{\mathbf{R}} f\left(\frac{1}{2}\int_{x-\varepsilon}^{x+\varepsilon} |u'(y)|^2 \, dy\right) dx$$

defined for $u \in H^1_{loc}(\mathbf{R})$ with $u' = 0$ on $\mathbf{R} \setminus (a, b)$ (this is a slight modification of the definition above). Note that $\mathcal{H}^0 = \#$ is the counting measure.

Discretization of the functional E_ε. If $v \in H^1(a, b)$, let

$$g(x) = f\left(\frac{1}{2}\int_{x-\varepsilon}^{x+\varepsilon} |v'(y)|^2 \, dy\right) \qquad x \in \mathbf{R}.$$

The function g is continuous and with compact support, and we can write

$$E_\varepsilon(u) = \sum_{j\in\mathbf{Z}} \frac{1}{\varepsilon} \int_{2\varepsilon j}^{2\varepsilon(j+1)} g(x) \, dx = \sum_{j\in\mathbf{Z}} \frac{1}{\varepsilon} \int_0^{2\varepsilon} g(x + 2\varepsilon j) \, dx =: \int_0^{2\varepsilon} h(x) \, dx.$$

The function h is continuous on $(0, 2\varepsilon)$ (it is a finite sum of continuous functions); hence by the mean value theorem we find $\xi \in (0, 2\varepsilon)$ such that

$$\int_0^{2\varepsilon} h(x)\, dx = 2\varepsilon\, h(\xi).$$

Up to a translation of the interval (a, b) into $(a - \xi, b - \xi)$ we can suppose $\xi = 0$. Hence we have

$$E_\varepsilon(v) = \sum_{j \in \mathbf{Z}} 2f\left(\frac{1}{2} \int_{\varepsilon(2j-1)}^{\varepsilon(2j+1)} |v'|^2\, dy\right).$$

Proof of the compactness and of the lower semicontinuity inequality. Fixed $u \in L^1(a, b)$, let $u_j \to u$ in $L^1(a, b)$ with

$$\sup_j E_{\varepsilon_j}(u_j) < +\infty.$$

For all j we divide $\mathbf{Z}$ into two sets of indices:

$$G_j = \left\{k \in \mathbf{Z} : \int_{\varepsilon_j(2k-1)}^{\varepsilon_j(2k+1)} |u_j'|^2\, dy \le 1\right\},$$

$$B_j = \left\{k \in \mathbf{Z} : \int_{\varepsilon_j(2k-1)}^{\varepsilon_j(2k+1)} |u_j'|^2\, dy > 1\right\}.$$

We can re-write then

$$E_{\varepsilon_k}(u_k) = \sum_{j \in B_k} 2f\left(\frac{1}{2}\int_{\varepsilon_k(2j-1)}^{\varepsilon_k(2j+1)} |u_k'|^2\, dy\right) + \sum_{j \in G_k} 2f\left(\frac{1}{2}\int_{\varepsilon_k(2j-1)}^{\varepsilon_k(2j+1)} |u_k'|^2\, dy\right)$$

$$= \#(B_k) + \sum_{j \in G_k} \int_{\varepsilon_k(2j-1)}^{\varepsilon_k(2j+1)} |u_k'|^2\, dy.$$

Note that $\#(B_k) \le c$ independent of k. We define

$$v_k(x) = \begin{cases} u_k(x) & \text{if } x \in 2\varepsilon_k G_k + (-\varepsilon_k, \varepsilon_k), \\ u_k((2j-1)\varepsilon_k) & \text{if } x \in [(2j-1)\varepsilon_k, (2j+1)\varepsilon_k], \ j \in B_k. \end{cases}$$

Note that

$$\|u_k - v_k\|_{L^1} \le c\|u\|_\infty 2\varepsilon_k \to 0,$$

so that in particular $v_k \to u$ in L^1

$$\mathcal{H}^0(S(v_k)) \le \#(B_k), \qquad \int_{(a,b)} |v_k'|^2\, dx = \sum_{j \in G_k} \int_{\varepsilon_k(2j-1)}^{\varepsilon_k(2j+1)} |u_k'|^2\, dy.$$

Hence,

$$\int_{(a,b)} |v_k'|^2\, dx + \mathcal{H}^0(S(v_k)) \le E_{\varepsilon_k}(u_k) \le c < +\infty.$$

By Ambrosio's compactness theorem we have $u \in SBV(a,b)$, and by the lower semicontinuity of the Mumford-Shah functional

$$\int_{(a,b)} |u'|^2 \, dx + \mathcal{H}^0(S(u)) \leq \liminf_k E_{\varepsilon_k}(u_k).$$

This proves the Γ-liminf inequality.

Construction of the recovery sequence. A recovery sequence for the Γ-limit is easily produced in the following way. Let $u \in SBV(a,b)$ with $\#(S(u)) < +\infty$. It is not a restriction to suppose $(a,b) = (-1,1)$, and $S_u = \{0\}$. Then we define:

$$v_\varepsilon(x) = \begin{cases} u(x) & \text{if } |x| > \varepsilon^2 \\ \text{any } H^1\text{-function with values } u(\pm\varepsilon^2) \text{ at } \pm\varepsilon^2 & \text{otherwise.} \end{cases}$$

It is easy to see that $\lim_{\varepsilon \to 0+} E_\varepsilon(v_\varepsilon) = E(u)$.

Remark 2. (i) If f is any non-decreasing continuous function such that $\alpha, \beta \in \mathbf{R}$ exist such that

$$\lim_{t \to 0+} \frac{f(t)}{t} = \alpha, \qquad \lim_{t \to +\infty} f(t) = \beta, \tag{4.2}$$

then the functionals E_ε defined in (4.1) Γ-converge to

$$\alpha \int_\Omega |\nabla u|^2 \, dx + 2\beta \mathcal{H}^{n-1}(S(u)). \tag{4.3}$$

(ii) A relaxation argument shows that functionals of the form (4.3) are the only possible limits of functionals E_ε also if we drop the hypotheses of continuity and monotonicity. If in place of f we take suitable f_ε in (4.1) we may obtain more general bulk and surface energies (see [Cor98] and [BG98]).

5   Other approximation procedures

Finally, we list some other types of approximation of free discontinuity energies. A comparison between all these different approximations can be found in [Bra98]

5.1   Elliptic approximation with an auxiliary variable

Following an earlier idea developed by Modica and Mortola [MM77], who approximated the perimeter functional by elliptic functionals, Ambrosio and Tortorelli in [AT90] and [AT92] introduced an approximation procedure of $E(u)$ with an auxiliary variable v, which in the limit approaches $1 - \chi_{S(u)}$. A family of functional studied in [AT92] is the following:

$$G_\varepsilon(u,v) = \int_\Omega v^2 |\nabla u|^2 \, dx + \frac{1}{2} \int_\Omega \left(\varepsilon |\nabla v|^2 + \frac{1}{\varepsilon}(1-v)^2 \right) dx, \tag{5.1}$$

defined on functions u, v such that $v \in H^1(\Omega)$, $uv \in H^1(\Omega)$ and $0 \le v \le 1$, which Γ-converges as $\varepsilon \to 0+$ with respect to the $(L^1(\Omega))^2$-topology to the functional

$$G(u, v) = \begin{cases} E(u) & \text{if } v = 1 \text{ a.e. on } \Omega \\ +\infty & \text{otherwise,} \end{cases} \tag{5.2}$$

defined on $(L^1(\Omega))^2$. Clearly, the functional G is equivalent to E as far as minimum problems are concerned. If $u \in SBV(\Omega)$ and $S(u)$ is regular enough, then as a recovery sequence $(u_\varepsilon, v_\varepsilon)$ for $E(u) = G(u, 1)$ we can take $u_\varepsilon = u$ and v_ε with $v_\varepsilon = 0$ on $S(u)$ and $v_\varepsilon = 1$ outside a tubular neighbourhood of $S(u)$. On the transition layer between $v_\varepsilon = 0$ and $v_\varepsilon = 1$ the profile of v_ε is chosen as minimizing the second integral in (5.1).

5.2   Non-local approximation with double integrals

A recent conjecture by De Giorgi, proved by Gobbino [Gob98], provides another type of non-local approximation of the Mumford-Shah functional (in the form (4.3) with suitable α, β), with approximating functionals the family

$$E_\varepsilon(u) = \frac{1}{\varepsilon^{n+1}} \int_{\Omega \times \Omega} \arctan\left(\frac{(u(x) - u(y))^2}{\varepsilon}\right) e^{-|x-y|^2/\varepsilon} \, dx \, dy, \tag{5.3}$$

defined on $L^1(\Omega)$.

5.3   Approximations with higher order perturbations

The "non-local effect" of the average of $|\nabla u|^2$ in (4.1) can be mimicked to some extent by adding a higher order perturbation to a family of local functionals. We consider here only the one-dimensional case $n = 1$. If we take a function f as in Section 4 satisfying (4.2), we can define the functionals

$$E_\varepsilon(u) = \frac{1}{\varepsilon} \int_\Omega f(\varepsilon |u'|^2) \, dx + \varepsilon^3 \int_\Omega |u''|^2 \, dx \tag{5.4}$$

on $H^2(\Omega)$. Note that their Γ-limit would be trivial without the last term. In [ABG98] it has been proven that the family (E_ε) Γ-converges with respect to the $L^1(\Omega)$-topology to the functional defined on $SBV(\Omega)$ by

$$F(u) = \alpha \int_\Omega |u'|^2 \, dx + C \sum_{t \in S(u)} \sqrt{|u^+(t) - u^-(t)|}, \tag{5.5}$$

with C explicitly computable from β. To recover the Mumford-Shah functional with this kind of approximation we must substitute f by a suitable f_ε in (5.4).

Part II

Degree Theory on Convex Sets and Applications to Bifurcation

Introduction to Part II

This part is devoted to a discussion of degree theory, some of its extensions and applications to bifurcation and population problems, and to some related topics such as nonlinear elliptic problems and topological methods in relativistic dynamics.

Classical degree theory due to Brouwer is an important tool in the study of number of solutions of an equation $f(x) = p$ in the interior of the domain (a bounded open set in n-dimensional Euclidean space) of a continuous mapping f with values in the n-dimensional space. An important property of Brouwer's degree is that it is stable under appropriate perturbations. We recall the well-known result of Sard which asserts that the regular values of a smooth mapping are dense in n-dimensional space. Degree has important algebraic and topological properties of excision, product and of homotopy invariance. The degree theory also gives a proof of the famous Brouwer fixed point theorem for continuous maps on compact convex sets and of the theorem of Borsuk that, for continuous odd maps on symmetric bounded open sets, the degree is odd and, in particular, nonzero. These classical results are contained in section 2. Section 3 is concerned with the study of positive solutions of nonlinear elliptic equations and systems, for which we need extensions of Brouwer degree theory to closed cones (more generally to closed convex sets) in infinite-dimensional Banach spaces. Such an extension seems to be a convenient tool to treat problems where there is no variational structure. Degree theory on convex sets is introduced in section 4 and excision, product and homotopy invariance properties are proved for the map $I - A$, where A is a completely continuous mapping from a bounded relatively open set of a closed convex set W in a Banach space into W itself. The notion of index of $I - A$ with respect to an isolated fixed point of A is then introduced as a "limit of degrees" of $I - A$.

The geometric concepts of a wedge and of a cone are introduced in section 5, used in section 6 in the calculation of index, where basic results on index are proved. For completely continuous monotone operators T, a result due to Dancer and Hess states that if K is a cone such that all ordered intervals are bounded and if x_0 and y_0 are two fixed points such that the ordered interval $[x_0, y_0]$ has no other fixed points, then they can be obtained as limits of iterates of T (both positive and negative).

The remaining sections of the lectures by Dancer are concerned with some applications of the theory developed in the above. First of all, section 7 gives an application to the Gelfand problem associated to a semilinear Dirichlet problem associated to the Laplacian, to obtain non-radially symmetric solutions and to get a lower bound for distinct number of non-radially symmetric solutions. Next the index calculations are applied to obtain positive solutions for a coupled semilinear elliptic system of two equations describing the population models. The positive solutions of such a system describe a model of population of two types of competing species. Existence, nonexistence, uniqueness and regularity results for a single semilinear equation $-\Delta u = u(a - u)$ with the Dirichlet boundary condition for the parameter a above or below the first eigenvalue of $-\Delta$ are also studied. The last section 9 discusses the asymptotic behaviour of solutions of competing species system when the interactions are large. Isolated positive solutions with nonzero index of a coupled semilinear system considered above are also considered. Finally some open questions concerning positive solutions when the parameters in the system (the coefficients of the zero order terms) are above the first eigenvalue of $-\Delta$ and problems related to the stability of sign changing solutions are stated.

The chapter by Norman Dancer constitutes the content of the lecture course he delivered and is based on the notes taken by E. Alfarone, A. Groli and A. Pistoia.

The texts by D. Passaseo and G. Cerami following this chapter are concerned with a field that has received great attention in the last few years: existence, nonexistence and multiplicity for semilinear elliptic equations with Dirichlet, Neumann or mixed boundary conditions. The growth of the nonlinear term (subcritical, critical or supercritical) plays an important role in this analysis.

The next complementary text by V. Benci and D. Fortunato deals with the application of topological methods to the theory of solitons and relativistic dynamics.

Part II closes with a presentation by V. Benci of a new, purely algebraic approach to nonstandard analysis.

Degree theory on convex sets and applications to bifurcation

E. N. Dancer

1 Introduction

The aim of these lectures was to give a short introduction to the use of degree theory ideas. In particular, the main emphasis was in the use of degree theory ideas on convex sets. It seems to the author that these ideas are a very convenient tool for a number of problems, especially problems without a variational structure. (For example, many systems do not have a variational structure). Moreover, in many applications, we are only interested in positive solutions (because of the origin of the problem). In these cases, we are looking for solutions in a cone and we naturally find we are looking at problems on a closed convex set. Working directly on a set of positive solutions also has the advantage that we automatically exclude solutions we are not interested in. In many of the applications, we make crucial use of the formula for the index of a non-degenerate solution for a mapping defined on a closed convex set. In the first part of the lectures, §2-3, we discuss rather briefly degree theory for mappings on $\mathbf{R}^n$ and for completely continuous mappings on Banach spaces. This is quite standard material and it is included here for pedagogical reasons. In §4-6, we discuss the basic degree theory of mappings defined on closed convex sets, including the basic index formula, and some abstract applications. In §7-9, we discuss applications to partial differential equations. In §7, we discuss applications to symmetry breaking on two-dimensional annuli. In §8, we discuss the existence of solutions with both components positive of a competing species systems with diffusion. (These are often called coexisting populations). Finally, in §9, we discuss further these models when the two populations interact strongly. We conclude this introduction by recalling our main notations:

- $B_r(x)$ denotes the open ball of radius r centered at x in $\mathbf{R}^n$, $B_r = B_r(0)$;
- $\mathcal{L}^n$ denotes the Lebesgue measure in $\mathbf{R}^n$;
- $\prod_{i=1}^{n}(a_i, b_i)$ is an n-cube in $\mathbf{R}^n$;
- ∇f stands for the gradient of f, J_f stands for the Jacobian determinant of f, while Df stands for the first order distributional derivative;
- $C_c(D)$ stands for the set of continuous functions defined on $D \subset \mathbf{R}^n$ with compact support while $C_0(D)$ stands for the set of continuous functions on $\overline{D}$ which are null on ∂D.

2 Degree theory in $\mathbf{R}^n$

In this section, we define a degree theory for continuous maps $f : \overline{D} \to \mathbf{R}^n$ where D is bounded and open. The degree will provide a "*count*" of the solutions of

$f(x) = p$ in D. We want a "*count*" which is stable to perturbations and relatively easy to count. By considering the map $f(x) = x^2 + t$ on $[-1,1]$ on **R** where the number of solutions changes from zero to 2 as t moves through zero shows that we must count solutions in a more sophisticated way if the "*count*" is to be stable to perturbations.

Definition 1. Let $D \subset \mathbf{R}^n$ be a bounded open set, $f : \overline{D} \to \mathbf{R}^n$ a function in $C^1(D) \cap C^0(\overline{D})$. We say that p is a *regular value* of f if

$$J_f(x) \neq 0, \qquad \forall x \in D \cap f^{-1}(p).$$

Remark 1. If $p \notin f(D)$ then p is a regular value of f.

Assume that p is a regular value of f and $p \notin f(\partial D)$.
If $x \in D \cap f^{-1}(p)$, since $\nabla f(x) \neq 0$, from the inverse function theorem f is locally invertible at x and so in a neighbourhood of x there are no other points in $f^{-1}(p)$ and therefore $D \cap f^{-1}(p)$ is composed of isolated points.
On the other hand $D \cap f^{-1}(p) = \overline{D} \cap f^{-1}(p)$ is a closed subset of $\overline{D}$, then it is a compact set and from the Bolzano-Weierstrass theorem it follows that $D \cap f^{-1}(p)$ is a finite set. Now we can define the degree of the map f with respect to the set D at the point p.

Definition 2. Let D, f be as in definition 1 and $p \in \mathbf{R}^n \setminus f(\partial D)$ a regular value of f.
Then

$$\deg(f, p, D) := \sum_{x \in D \cap f^{-1}(p)} \mathrm{sign}\, J_f(x). \tag{1}$$

Remark 2. The assumption that $p \notin f(\partial D)$ in the definition 2 is made because otherwise under small perturbations of p or f, a solution might move from inside D to out of D and hence any "*count*" of the number of solutions is likely to change. Note also that, by definition, $\deg(f, p, D)$ is integer valued. This is important later.

Now we try to extend the definition of the degree to the points $p \in \mathbf{R}^n \setminus f(\partial D)$ that are not regular.
The idea is to approximate such a point p with a sequence $(p_i)_{i \geq 1}$ of regular points of f and define the degree of f at p as the limit of the degree of f at the points p_i.

There are a number of details to check:

(1) the existence of the sequence $(p_i)_{i \geq 1}$ approaching p;
(2) the existence of $\lim_{i \to \infty} \deg(f, p_i, D)$;
(3) the independence of the limit of the chosen sequence.

The first question is solved by the following result:

Theorem 1 (Sard). *Let $D \subset \mathbf{R}^n$ be an open set, $f : D \to \mathbf{R}^n$ a function in $C^1(D)$. Then the set of the regular values of f is dense in $\mathbf{R}^n$.*

Proof (Sketched). Denoting by

$$S_f := \{x \in D : J_f(x) = 0\}$$

we only need to prove that $\mathcal{L}^n(f(S_f)) = 0$.

First of all we prove that for any n-cube $C \subset D$ such that

(a) $C \cap S_f \neq \emptyset$;
(b) $\exists L > 0$ such that $\forall x, y \in C$ then $|f(x) - f(y)| \leq L|x - y|$;
(c) If $\epsilon > 0$ such that $|f(x) - f(y) - \nabla f(x)(x - y)| \leq \epsilon|x - y|$ for $x, y \in C$,

then $\mathcal{L}^n(f(C)) \leq (2d)^n L^{n-1}\epsilon$ where d is the diameter of C.

Let $x_0 \in C \cap S_f$ and V a proper subspace of $\mathbf{R}^n$ such that $\mathcal{R}(\nabla f(x_0)) \subseteq V$, where $\mathcal{R}$ denotes the range. (We think of $\nabla f(x_0)$ as a linear map from $\mathbf{R}^n$ into $\mathbf{R}^n$ and since $J_f(x_0) = 0$ its range is a subspace of $\mathbf{R}^n$ which is at most $(n-1)$-dimensional). If $x \in C$, by (b),(c) we have

$$|f(x) - f(x_0)| \leq L|x - x_0| \leq Ld$$

$$dist(f(x), f(x_0) + V) \leq |f(x) - f(x_0) - \nabla f(x_0)(x - x_0)| \leq \epsilon|x - x_0| \leq \epsilon d$$

and then

$$\mathcal{L}^n(f(C)) \leq (2Ld)^{n-1}2\epsilon d = (2d)^n L^{n-1}\epsilon.$$

Now we prove that for every closed n-cube $C \subset D$ then $\mathcal{L}^n(f(S_f \cap C)) = 0$.

If $\epsilon > 0$ and $L = \max_{x \in C} |\nabla f(x)|$ then by **Lagrange's** theorem we have that

$$|f(x) - f(y)| \leq L|x - y|, \qquad \forall x, y \in C.$$

Moreover by uniform continuity of ∇f on C

$$\exists \delta_\epsilon > 0 \text{ such that if } x, y \in C \text{ and } |x - y| \leq \delta_\epsilon \text{ then } |\nabla f(x) - \nabla f(y)| \leq \epsilon.$$

Let us choose an integer $k > 0$ such that if we divide C in k^n closed n-cubes C_j we have

$$d_j := \text{diam } C_j = \frac{d}{k} \leq \delta_\epsilon.$$

For $x, y \in C_j$ we have

$$|f(x) - f(y) - \nabla f(x)(x - y)| \leq \epsilon|x - y|.$$

Since C_j satisfies (b),(c) we have

$$\mathcal{L}^n(f(C_j)) \leq (2d_j)^n L^{n-1}\epsilon$$

whenever $C_j \cap S_f \neq \emptyset$.

If $J = \{j : C_j \cap S_f \neq \emptyset\}$ we have

$$
\begin{aligned}
\mathcal{L}^n(f(C \cap S_f)) &\leq \sum_{j \in J} \mathcal{L}^n(f(C_j \cap S_f)) \\
&\leq \sum_{j \in J} \mathcal{L}^n(f(C_j)) \\
&\leq (2kd_j)^n L^{n-1}\epsilon \\
&= (2d)^n L^{n-1}\epsilon.
\end{aligned}
$$

If $\epsilon \to 0$ we conclude that $\mathcal{L}^n(f(C \cap S_f)) = 0$ and since D may be covered by a countable family of closed n-cubes we have that $\mathcal{L}^n(f(S_f)) = 0$.   □

To prove the validity of (2),(3) we proceed indirectly.

Definition 3. Let D, f be as in definition 1, C an n-cube such that $C \cap f(\partial D) = \emptyset$, $\phi \in C_c^\infty(C, \mathbf{R})$ a function such that $\int_C \phi(x)dx = 1$. (Such a ϕ is said to be *admissible*).
We define

$$
d(\phi) := \int_{\overline{D}} \phi(f(x)) J_f(x)dx. \tag{2}
$$

We will show that the value of d is independent of the choice of an admissible ϕ with support in the same cube C and that it coincides with $\deg(f, p, D)$ for every regular value $p \in C$. Note that the definition only depends on C through the requirement ϕ has support in C.

The advantage of using such an integral lies in its greater regularity under perturbations.

Now we recall some results that we will use in the proof. These can be found in Schwartz [Sch69]. Note that Lemma 2 is the crucial lemma and involves a tedious computation. (By using differential geometric ideas as in Nirenberg [Nir74] it is possible to avoid the tedious computation). Lemma 1 is used to prove Lemma 2.

Lemma 1. *If $C \subset \mathbf{R}^n$ is an n-cube and $\psi \in C_c^\infty(C)$ a function such that $\int_C \psi(x)dx$ is null, then there exists $v \in C_c^\infty(C, \mathbf{R}^n)$ such that $\operatorname{div} v = \psi$.*

Lemma 2. *Let $D \subset \mathbf{R}^n$ be an open bounded and $f \in C^0(\overline{D}, \mathbf{R}^n) \cap C^2(D, \mathbf{R}^n)$, C an n-cube such that $C \cap f(\partial D) = \emptyset$, $v \in C_c^\infty(C, \mathbf{R}^n)$, $\int_C v(x)dx = 0$. Then there exists $u \in C_c^1(D, \mathbf{R}^n)$ such that $\forall x \in D$ then*

$$
\operatorname{div} u(x) = \operatorname{div} v(f(x)) J_f(x).
$$

Lemma 3. *If ϕ_1, ϕ_2 are two admissible functions (see def. 3) with support in the same n-cube C, then $d(\phi_1) = d(\phi_2)$.*

Proof (Sketch). Assume first that f is a C^2 function and let $\psi = \phi_1 - \phi_2$. Then, by lemma 2, there exists $u \in C_c^1(D, \mathbf{R}^n)$ such that

$$\operatorname{div} v(f(x))J_f(x) = \operatorname{div} u(x).$$

So we have

$$\int_{\overline{D}} \psi(f(x))J_f(x)dx = \int_{\overline{D}} \operatorname{div} v(f(x))J_f(x)dx = \int_{\overline{D}} \operatorname{div} u(x)dx = \int_{\partial D} u \cdot \nu \, d\sigma = 0.$$

For $f \in C^1$ the proof is made by using a sequence of C^2 functions uniformly approximating f and ∇f. That such an approximation exists follows by Narasimhan [Nar68]. $\square$

Remark 3. By Lemma 3 and our earlier comment $d(\phi)$ is independent of both C and ϕ.

Theorem 2. *Let D, f be as in definition 1, p a regular value for f and ϕ an admissible function with support in C, where C is an n-cube which contains p. Then $d(\phi) = \deg(f, p, D)$.*

Proof. If $f^{-1}(\{p\}) = \{x_1, \ldots, x_k\}$, for $i = 1, \ldots, k$ we choose a neighbourhood D_i of x_i in D such that $D_i \cap D_j = \emptyset$ if $i \neq j$, $f_{|D_i}$ is a diffeomorphism onto a neighbourhood of p and $J_f(x)$ has fixed sign on D_i. (This is possible by shrinking).

Since p does not belong to the compact set $f\left(\overline{D} \setminus \bigcup_{i=1}^{k} D_i\right)$, we can choose the

n-cube C such that $C \subset \bigcap_{i=1}^{k} f(D_i)$, $p \in C$, and $f(x) \notin C$ if $x \in \overline{D} \setminus \bigcup_{i=1}^{k} D_i$. Choose an admissible ϕ for f, with support in C. Then we have

$$d(\phi) = \int_{\overline{D}} \phi(f(x))J_f(x)dx = \sum_{i=1}^{k} \int_{\overline{D}_i} \phi(f(x))J_f(x)dx$$

$$= \sum_{i=1}^{k} \operatorname{sign} J_f(x_i) \int_{\overline{D}_i} \phi(f(x))|J_f(x)|dx = \sum_{i=1}^{k} \operatorname{sign} J_f(x_i) \int_{f(\overline{D}_i)} \phi(y)dy \quad (3)$$

$$= \sum_{i=1}^{k} \operatorname{sign} J_f(x_i) \int_C \phi(x)dx = \deg(f, p, D).$$

$\square$

Choose a sequence p_i of regular values such that $p_i \to p$ as $i \to \infty$ (such a sequence exists by Sard's theorem). Since $d(\phi)$ is locally independent of p, Theorem 2 shows that $\deg(f, p_i, D) = d(\phi)$ for large i and so we can state the following definition.

Definition 4. Let D, f be as in definition 1 and $p \in \mathbf{R}^n \setminus f(\partial D)$, $(p_i)_{i \geq 1}$ a sequence of regular value of f approaching p.
Then

$$\deg(f, p, D) := \lim_{i \to \infty} \deg(f, p_i, D). \tag{4}$$

Remark 4. We have used that $d(\phi)$ is independent of ϕ to choose a special ϕ and then to exploit the close relation between $d(\phi)$ and the change of variable theorem.
We have also proved that both $d(\phi)$ and $\deg(f, p, D)$ are integers (since $\deg(f, p_i, D)$ is an integer).

Now we can give some fundamental properties of the degree.

Theorem 3. *Let D, f be as in definition 1 and $p \in \mathbf{R}^n \setminus f(\partial D)$. Then the following facts hold:*

(i) *If $\deg(f, p, D) \neq 0$ then there exists $x \in D$ such that $f(x) = p$;*
(ii) *(excision) if D_i, $i = 1, \ldots, k$, are disjoint open subsets of D and $f(x) \neq p$ if $x \in \overline{D} \setminus \bigcup_{i=1}^{k} D_i$ then*

$$\deg(f, p, D) = \sum \lim its_{i=1}^{k} \deg(f, p, D_i);$$

(iii) *(product) if $D_1 \subset \mathbf{R}^m$ is a bounded open set, $g : \overline{D}_1 \to \mathbf{R}^m$ is a $C^1(D_1) \cap C^0(\overline{D}_1)$ function and $q \in \mathbf{R}^m \setminus g(\partial D_1)$ then*

$$\deg\left((f, g), (p, q), D \times D_1\right) = \deg(f, p, D) \cdot \deg(g, q, D_1);$$

(iv) *(homotopy invariance) let $F : \overline{D} \times [a, b] \to \mathbf{R}^n$ be a C^1 function. Assume that the function*

$$F_t : \overline{D} \longrightarrow \mathbf{R}^n$$
$$x \longmapsto F(t, x)$$

satisfies $F_t(x) \neq p$ if $x \in \partial D, t \in [a, b]$. Then $\deg(F_t, p, D)$ is independent of t.

Proof. We first prove (i)-(iv) assuming that p (and q for g in (iii)) is a regular value for f.

(i) This is an immediate consequence of the definition;
(ii) since $x \mapsto \|f(x) - p\|$ is continuous and strictly positive on the compact set $\overline{D} \setminus \bigcup_{i=1}^{k} D_i$, then $\exists a > 0$ such that $\|f(x) - p\| \geq a$, $\forall x \in \overline{D} \setminus \bigcup_{i=1}^{k} D_i$.
We point out that $p \notin f(\partial D_i)$ and so $\deg(f, p, D_i)$ is well defined and then

$$\deg(f, p, D) = \sum_{x \in D \cap f^{-1}(p)} \operatorname{sign} J_f(x)$$

$$= \sum_{i=1}^{k} \sum_{x \in D_i \cap f^{-1}(p)} \text{sign } J_f(x)$$

$$= \sum_{i=1}^{k} \deg(f, p, D_i);$$

(iii) if $A = (D \times D_1) \cap (f^{-1}(p) \times g^{-1}(q))$ then

$$\deg\left((f, g), (p, q), D \times D_1\right) = \sum_{(x,y) \in A} \text{sign det} \begin{pmatrix} \nabla f(x) & 0 \\ 0 & \nabla g(y) \end{pmatrix}$$

$$= \sum_{(x,y) \in A} \text{sign } (J_f(x) J_g(y)) = \sum_{x \in D} J_f(x) \cdot \sum_{y \in D_1} J_g(y)$$

$$= \deg(f, p, D) \cdot \deg(g, q, D_1);$$

(iv) we prove that the map

$$[a, b] \ni t \longmapsto \deg(F_t, p, D) \in \mathbf{Z}$$

is continuous and so it is constant (since it is also integer-valued). We do not use here that p is a regular value. By the definition of $\deg(f, p, D)$, $\deg(f, p, D) = d(\phi)$. If ϕ is admissible for F_t, it is easy to see that ϕ is also admissible for F_s for s near t.
For $t \in [a, b]$ and $t_n \to t$, since F and $\nabla_x F$ are uniformly continuous on $\overline{D}$ we may deduce that

$$\lim_{n \to +\infty} \int_{\overline{D}} \phi(F(x, t_n)) J_{F_{t_n}} \, dx = \int_{\overline{D}} \phi(F(x, t)) J_{F_t} \, dx$$

and hence $d(\phi)$ changes continuously with t as required.

Now assume $p \in \mathbf{R}^n \setminus f(\partial D)$ and let $(p_j)_{j \geq 1}$ be a sequence of regular values of f approaching to p.
We prove (i)-(iii) ((iv) has already been proved).

(i) If $\deg(f, p, D) \neq 0$ then $\deg(f, p_j, D) \neq 0$ for $j > j_0$ and then there exists x_j such that $f(x_j) = p$.
But (up to a subsequence) $x_j \to x$ and then $f(x) = p$;

(ii) we know that there exists j_0 such that for $j > j_0$ then $\|p_j - p\| \leq \frac{a}{2}$ and so if $x \in \overline{D} \setminus \bigcup_{i=1}^{k} D_i$ then

$$\|f(x) - p\| \geq \|f(x) - p\| - \|p_j - p\| \geq a - \frac{a}{2} > 0$$

and then $f(x) \neq p_j$ if $x \in \overline{D} \setminus \bigcup_{i=1}^{k} D_i$, so

$$
\deg(f, p, D) = \lim_{j \to \infty} \deg(f, p_j, D) = \lim_{j \to \infty} \sum_{i=1}^{k} \deg(f, p_j, D_i) =
$$
$$
= \sum_{i=1}^{k} \lim_{j \to \infty} \deg(f, p_j, D_i) = \sum_{i=1}^{k} \deg(f, p, D_i).
$$

Point (iii) is immediately solved by passing to the limit in the analogous result already proved for regular values. $\square$

Now we want to extend the notion of degree to continuous maps by approximation with C^1 functions.

Remark 5. Let $f : \overline{D} \to \mathbf{R}^n$ be a continuous function and $p \notin f(\partial D)$.
By the Stone-Weierstrass theorem we may choose a sequence $(f_n)_{n \geq 1} \in C^1(\overline{D})$ uniformly converging to f on $\overline{D}$.
Since by compactness of ∂D there exists $a > 0$ such that $\|f(x) - p\| \geq a$ on ∂D, $p \notin f_n(\partial D)$ for large n and so $\deg(f_n, p, D)$ is defined.

Definition 5. Let $f, p, D, (f_n)_{n \geq 1}$ be as in remark 5.
We define

$$
\deg(f, p, D) := \deg(f_n, p, D)
$$

for any f_n such that $|f_n(x) - f(x)| \leq \frac{1}{2}a$ on $\overline{D}$.

Remark 6. This is a good definition.
In fact if $(g_n)_{n \geq 1}$ is any other C^1 function such that $|g_n(x) - f(x)| \leq \frac{1}{2}a$ on $\overline{D}$, we consider
$$
\begin{aligned}
F^{(n)} : [0, 1] \times \overline{D} &\longrightarrow \mathbf{R}^n \\
(t, x) &\longmapsto tg_n(x) + (1 - t)f_n(x).
\end{aligned}
$$

Let $x \in \partial D, t \in [0, 1]$ then

$$
\|F^{(n)}(t, x) - p\| \geq \|f(x) - p\| - \|t(g_n - f) + (1 - t)(f_n - f)\| \geq
$$
$$
\geq a - (t\|g_n - f\|_\infty + (1 - t)\|f_n - f\|_\infty) \geq \frac{a}{2}
$$

so $p \notin F_t^{(n)}(\partial D)$ and then by homotopy invariance

$$
\deg(f_n, p, D) = \deg(F_0^{(n)}, p, D) = \deg(F_1^{(n)}, p, D) = \deg(g_n, p, D).
$$

Remark 7.
(a) Properties (i)–(iv) of Theorem 3 are still valid for $f \in C^0(\overline{D})$ and $g \in C^0(\overline{D_1})$ in (iii), F only continuous on $\overline{D} \times [a, b]$ in (iv). The proof of these results is easily obtained by using approximation arguments.
(b) Another property deriving from the definitions is that

$$
\deg(f, p, D) = \deg(f(x) - p, 0, D).
$$

(c) If A is an invertible $n \times n$-matrix (thought as linear map from $\mathbf{R}^n$ into $\mathbf{R}^n$) and $0 \in D$, then

$$\deg(A, 0, D) = \text{sign} \det A.$$

The homotopy invariance property of the degree is very important because it often enables to calculate the degree by deforming our map to a simpler map. As an application we prove Brouwer's fixed point theorem.

Theorem 4 (Brouwer). *Let $S \subset \mathbf{R}^n$ be a compact and convex set and $f : S \to S$ be a continuous map.*
Then f has a fixed point.

Proof. Without loss of generality we may assume $\overset{\circ}{S} \neq \emptyset$ because if $\overset{\circ}{S} = \emptyset$, by properties of convex sets, we can find a lower dimensional space in which $\overset{\circ}{S} \neq \emptyset$. We also assume $0 \in \overset{\circ}{S}$. (This is possible by translations).

We prove that if $f(x) \neq x, \forall x \in \partial S$ then $\deg(I - f, 0, \overset{\circ}{S}) \neq 0$ and so the proof is complete by (i) of 3.
We consider the continuous map

$$H : S \times [0, 1] \longrightarrow \mathbf{R}^n$$
$$(x, t) \longmapsto x - tf(x).$$

If there exist $x \in \partial S, t \in (0, 1)$ such that $H(x, t) = 0$ then $x = (1-t)0 + tf(x) \in \overset{\circ}{S}$ and this is a contradiction. We have used here that if C is a convex set, if $a \in \overset{\circ}{C}$, if $b \in C$ and if $0 < t < 1$, then $ta + (1-t)b \in \overset{\circ}{C}$. Note also that we have assumed $H(x, 1) \neq 0$ on ∂C and it is easy to see that $H(x, 0) \neq 0$ on ∂C.
Therefore by homotopy invariance

$$1 = \deg(I, 0, \overset{\circ}{S}) = \deg(H(\cdot, 0), 0, \overset{\circ}{S}) = \deg(H(\cdot, 1), 0, \overset{\circ}{S}) = \deg(I - f, 0, \overset{\circ}{S}).$$

$\square$

Remark 8. This theorem is used in many places. For example it is frequently used to prove the existence of periodic solutions of ordinary differential equations.

There are three other important theorems that may be derived by degree theory which we do not prove. Proofs can be found in the book of Schwartz [Sch69].

Theorem 5 (Borsuk). *Let $D \subset \mathbf{R}^n$ be a bounded, open and symmetric set with $0 \in D$, $f : \overline{D} \to \mathbf{R}^n$ be a continuous and odd function such that $f(x) \neq 0$ for $x \in \partial D$.*
Then $\deg(f, 0, D)$ is odd. (In particular it is not zero).
(A set D is symmetric if $x \in D \Longleftrightarrow -x \in D$).

Remark 9. If f is a smooth function and if 0 is a regular value the proof is immediate because J_f is even and $f(0) = 0$.
The proof in the general case is based on the fact that an odd continuous function may be approximated by a C^1-odd function, which has 0 as regular value.

Two consequences of Borsuk's theorem are the following.

Theorem 6. *Let $m > n$ and $f : \mathbf{R}^m \to \mathbf{R}^n$ be a continuous and odd map .
Then $\forall r > 0, \exists x \in S_r = \partial B_r$ such that $f(x) = 0$.*

Theorem 7. *Let $f : \mathbf{R}^n \to \mathbf{R}^n$ be a $1 - 1$ and continuous map. Then f is an open map.*

3   Degree in infinite dimensional spaces

Before giving the notion of degree in the infinite dimensional case, we extend the theory developed in the previous section to continuous maps defined on open bounded sets of a finite dimensional Banach-space.

Definition 6. Let E be an n-dimensional Banach space, $h : E \to \mathbf{R}^n$ a linear isomorphism, D a bounded open subset of E, $f : \overline{D} \to E$ a continuous map and $p \in E \setminus f(\partial D)$, then we define

$$\deg(f, p, D) = \deg(h \circ f \circ h^{-1}, h(p), h(D)).$$

Remark 10. One can see that the definition is unambigous by reducing to the case that f is C^1 and p is a regular value and then noticing that the sign of the Jacobian determinant does not depend on the choice of the basis in $\mathbf{R}^n$.

For convenience in notation and computation we consider $p = 0$. Moreover since our main interest is for the fixed points of a map A we consider $\deg(I - A, 0, D)$. We now prove a useful reduction theorem that will be used in the definition of the degree in infinite dimension.

Theorem 8. *Let D, E be as in definition 6, M a subspace of E and $A : \overline{D} \to M$ a continuous map such that $x \neq A(x)$ if $x \in \partial D$.
Then*

$$\deg(I - A, 0, D) = \deg((I - A)_{|D \cap M}, 0, D \cap M).$$

Proof. All solutions of $x = A(x)$ are contained in a relatively open subset D_1 of M which can be chosen such that $\overline{D}_1 \subset D$.
We identify E with $M \times M^\perp$ and then choose $\delta > 0$ such that $\overline{D}_1 \times \overline{B}_\delta \subset D$ where for only this proof B_δ denotes the ball in $M^\perp$. With the above identification, A is simply the map S

$$
\begin{aligned}
S : \quad \overline{D} &\longrightarrow M \times M^\perp \\
(x, y) &\longmapsto (A(x, y), 0)
\end{aligned}
$$

We define the homotopy

$$H : \overline{D}_1 \times \overline{B}_\delta \times [0, 1] \longrightarrow M \times M^\perp$$
$$(x, y, t) \longmapsto (x - A(x, ty), y)$$

Note that it is easy to check that $H(x, y, t) \neq 0$ if $(x, y) \in \partial(D_1 \times B_\delta)$ and $t \in [0, 1]$ and that $H(x, y, 0) = (x - A(x, 0), y)$. Then

$$\deg(I - A, 0, D) = \deg(I - S, 0, D_1 \times B_\delta) = \deg(H(\cdot, \cdot, 1), 0, D_1 \times B_\delta) =$$

$$= \deg(H(\cdot, \cdot, 0), 0, D_1 \times B_\delta) = \deg((I - A)_{|M \cap D_1}, 0, D_1 \cap M)$$

$$= \deg((I - A)_{|M \cap D}, 0, D \cap M)$$

where the first and the last equalities derive from the excision property, the third from homotopy invariance and the fourth from the product property.   □

We now try to extend the notion of degree to infinite dimensional Banach spaces. We cannot expect to be able to do this for all maps and the simplest way to see this is to notice that the Brouwer's theorem fails in some Banach spaces.

Example 1. Let $c_0 = \{(x_i)_{i \geq 1}, x_i \in \mathbf{R}, x_i \to 0\}$ which is a Banach space if we endow it with the norm

$$\|(x_i)_{i \geq 1}\| = \sup_{i \geq 1} |x_i|.$$

We consider the continuous function

$$T : c_0 \ni (x_i)_{i \geq 1} \longmapsto (1, x_1, x_2, \ldots) \in c_0$$

which maps the closed unit ball in c_0 into itself, but has no fixed points. (If there were a fixed point then $x_i = 1$, $\forall i$ and so $(x_i)_i \notin (c_0)$).

So we cannot hope that any reasonable degree theory includes all continuous maps. Hence to proceed further we need a restricted class of maps.

Definition 7. *Let W be a closed subset of a Banach space E, we say that a map $A : W \to E$ is completely continuous, if A is continuous and for any bounded subset S of W, then $A(S)$ is relatively compact in E.*

We will construct a degree theory for the map $I - A$ whenever A is completely continuous and the reason that we can do it is that we can approximate completely continuous maps by maps whose range lies in a finite dimensional space.

Lemma 4. *Let K be a compact, convex subset of a Banach space E and $\epsilon > 0$. Then there exists a continuous map $P_\epsilon : K \to K$ such that*

$$\|P_\epsilon(x) - x\|_E \leq \epsilon, \qquad \forall x \in K$$

and $\mathcal{R}(P_\epsilon)$ is contained in a finite dimensional space.

Proof. Since K is a compact then there exist $x_1, \ldots, x_k \in K$ such that $K \subset \bigcup_{i=1}^{k} B_{\frac{\epsilon}{2}}(x_i)$. We define for $x \in K$

$$f_i(x) = (\epsilon - \|x - x_i\|)^+ \qquad i = 1, \ldots, k.$$

Then $f_i(x)$ is a non-negative and continuous function in K and $\forall x \in K$, $\exists i$ such that $f_i(x) > 0$. Let us now define the continuous maps on K

$$g_i(x) = \frac{f_i(x)}{\sum\limits_{i=1}^{k} f_i(x)}, \qquad i = 1, \ldots, k$$

and

$$P_\epsilon(x) = \sum_{i=1}^{k} g_i(x) x_i.$$

Now $P_\epsilon : K \to K$ is continuous and $\mathcal{R}(P_\epsilon) \subset span\{x_1, \ldots, x_k\}$. (Note that $\sum\limits_{i=1}^{k} g_i(x) = 1$ on K and thus $P_\epsilon(x)$ is in the convex hull of $\{x_1, \ldots, x_k\}$).
Then for $x \in K$ we have

$$\|P_\epsilon(x) - x\| = \|\sum_{i=1}^{k} g_i(x)(x - x_i)\| \leq \epsilon$$

since $g_i(x) = 0$ if $\|x - x_i\| > \epsilon$.   □

Proposition 1. *Let E be a Banach space, D an open and bounded subset of E and $A : \overline{D} \to E$ a completely continuous map. Then $\forall \epsilon > 0$ there exists a continuous function $A_\epsilon : \overline{D} \to E$ such that*

$$\|A(x) - A_\epsilon(x)\| \leq \epsilon, \qquad \forall x \in \overline{D}$$

and $\mathcal{R}(A_\epsilon)$ is contained in a finite dimensional subspace of E.

Proof. Since $\overline{A(D)}$ is compact, then $K = \overline{co}(A(D))$ is a convex, compact subset of E. Then by the previous lemma there exists $P_\epsilon : K \to K$ with $\mathcal{R}(P_\epsilon)$ contained in a finite dimensional subspace of E such that

$$\|P_\epsilon(x) - x\| \leq \epsilon.$$

The proof is complete if we define $A_\epsilon = P_\epsilon \circ A$. Now using the approximation found in the previous proposition we define the degree of $I - A$ at 0.   □

Remark 11. Let D be an open, bounded subset of E, $A : \overline{D} \to E$ be a completely continuous function such that $0 \notin (I - A)(\partial D)$.
Then there exists $\alpha > 0$ such that

$$\|x - A(x)\| \geq \alpha, \qquad \forall x \in \partial D. \tag{5}$$

To see this, suppose by way of contradiction, there exists $(x_n)_{n \geq 1} \in \partial D$ such that $x_n - A(x_n) \to 0$. Then we can choose a subsequence $(x_{n_k})_{k \geq 1}$ so that $A(x_{n_k})$ converges to z because A is a completely continuous function.
Since $x_{n_k} - A(x_{n_k}) \to 0$, it follows that $x_{n_k} \to z \in \partial D$.
Passing to the limit, $z = A(z) \in \partial D$ which is impossible by our assumptions.
By Proposition 1 we can choose a sequence $(A_n)_{n \geq 1}$ of continuous maps with $\mathcal{R}(A_n) \subset E_n$, where E_n is a finite dimensional subspace of E and $\|A_n - A\| < \frac{1}{n}$ on $\overline{D}$. We can consider $D_n = D \cap E_n$ as a relatively open subset of E_n whose boundary in E_n is contained in ∂D. Hence by (5) we have that $x \neq A_n(x)$ for $x \in \partial D_n$, if $n > \alpha^{-1}$.
Therefore $\deg((I - A_n)_{|\overline{D}_n}, 0, D_n)$ is defined for $n > \alpha^{-1}$.
Now we prove that if $(B_n)_{n \geq 1}$ is another finite-dimensional continuous mapping with $|B_n(x) - A(x)| < \alpha$ on $\overline{D}$, then degrees of $I - A_n$ and of $I - B_n$ coincide provided $n > \alpha^{-1}$ and the degrees are defined on appropriate spaces.
Let Y_n be a finite dimensional subspace of E which contains $\mathcal{R}(A_n) \cup \mathcal{R}(B_n)$ and $X_n = D \cap Y_n$.
By the reduction theorem we have that

$$\deg((I - A_n)_{|D_n}, 0, D_n) = \deg((I - A_n)_{|X_n}, 0, X_n)$$

and so we only need to verify that

$$\deg((I - A_n)_{|X_n}, 0, X_n) = \deg((I - B_n)_{|X_n}, 0, X_n).$$

We define the homotopy

$$H_n : \overline{X}_n \times [0, 1] \longrightarrow Y_n$$
$$(x, t) \longmapsto t(x - A_n(x)) + (1 - t)(x - B_n(x))$$

We must prove that $H_n(x, t) \neq 0, \forall x \in \partial X_n, t \in [0, 1]$.
We have

$$\|H_n(x, t) - (I - A)(x)\| \leq t\|(I - A_n)(x) - (I - A)(x)\| +$$

$$+ (1 - t)\|(I - B_n)(x) - (I - A)(x)\| \leq t\alpha + (1 - t)\alpha = \alpha$$

and so

$$\|H_n(x, t)\| \geq \|(I - A)(x)\| - \|H_n(x, t) - (I - A)(x)\| > \alpha - \alpha = 0.$$

Then by homotopy invariance we have the desired equality.

This proof also shows that for $n > \alpha^{-1}$ the degree of A_n is constant.

Remark 12. Let D, E, A, A_n, D_n be as in remark 11.
Then we define

$$\deg((I - A), 0, D) = \lim_{n \to \infty} \deg((I - A_n)_{|\overline{D}_n}, 0, D_n)$$

By using finite dimensional approximation and compactness arguments, it is not difficult to check that this degree has analogous properties to those in the finite dimensional case.

Proposition 2. *Let D be a bounded open set in a Banach space E, $A : \overline{D} \to E$ a completely continuous map such that $x \neq A(x)$, for $x \in \partial D$.*
Then the following results hold.

(i) If $\deg(I - A, 0, D) \neq 0$ then there exists $x \in D$ such that $x = A(x)$;
(ii) (excision) if D_i, $i = 1, \ldots, k$ are disjoint open subsets of D and $x \neq A(x)$
if $x \in \overline{D} \setminus \bigcup_{i=1}^{k} D_i$ then

$$\deg(I - A, 0, D) = \sum_{i=1}^{k} \deg(I - A, 0, D_i);$$

(iii) (products) if D_1 is a bounded open set in a Banach space F, $G : \overline{D}_1 \to F$ is a completely continuous function, $0 \in F$ such that $x \neq G(x)$, $\forall x \in \partial D_1$, then

$$\deg(I - (A, G), (0, 0), D \times D_1) = \deg(I - A, 0, D) \cdot \deg(I - G, 0, D_1)$$

(iv) (homotopy invariance) let $H : \overline{D} \times [a, b] \to E$ be a completely continuous function. Assume that $x - H(x, t) \neq 0 \, \forall x \in \partial D, \in [a, b]$, then $\deg(I - H_t, 0, D)$ is independent of t.

Remark 13. Notice that in (iv) it is not sufficient to assume that H is continuous and each H_t is completely continuous. However this is sufficient if we also assume that H_t is uniformly continuous in t.

Now we state a result that may be very easily derived from degree theory in Banach spaces and which is important since it frequently allows to transform an a priori estimate to an existence-theorem. It is widely used to study nonlinear partial differential equations.

Theorem 9. *Let E be a Banach space and $A : E \times [0, 1] \to E$ be a completely continuous map such that $A(x, 0)$ is odd in x and there exists $M > 0$ such that*

$$\forall x \in E, t \in [0, 1] : x = A(x, t) \Rightarrow \|x\| \leq M.$$

Then there exists $x \in E$ such that $x = A(x, 1)$.

Remark 14. If A, D, E are as in Proposition 2, $h : E \to E$ is an homeomorphism such that h, h^{-1} map bounded sets to bounded sets and $h(0) = 0$, then

$$\deg(I - A, 0, D) = \deg(I - h^{-1}Ah, 0, D).$$

The intuitive reason why this holds is that if $x = h^{-1}Ah(x)$ then $Ah(x) = h(x)$ and so the fixed points of $h^{-1}Ah$ are essentially the same as those of A. A formal proof is rather more difficult. It is a special of a rather more general and useful result known as the commutativity theorem for the degree which can be found in the paper of Nussbaum [Nus71].

Remark 15. If x_0 is an isolated zero of $I - A$ then if $\delta > 0$ is small enough, by the excision property, $\deg((I - A, 0, B_\delta(x_0))$ is independent of δ.

Definition 8. Let A, D, E be as in Proposition 2, x_0 be an isolated zero for $I - A$ and $\delta > 0$ be sufficiently small. Then we define

$$\text{index } (I - A, x_0) = \deg(I - A, 0, B_\delta(x_0)).$$

Now we want to find a formula for the degree of a linear map.

Definition 9. If $L : E \to E$ is a linear map, we say that L is *compact* if $\overline{L(B_1)}$ is compact. (If L is linear, this is equivalent to L being completely continuos).

By the theory of linear compact operators (as in Schechter [Sch71] for example), we know that the spectrum of L consists of 0 plus a finite or countable sequence of eigenvalues with 0 as the only possible limit point (and so L has at most a finite number of eigenvalues in $(1, +\infty)$).
If λ_i is a non-zero eigenvalue then, for h large enough

$$E = \ker((\lambda_i I - L)^h) \oplus \mathcal{R}((\lambda_i I - L)^h).$$

We define the algebraic multiplicity of λ_i to be

$$m_i = dim(\ker(\lambda_i I - L)^h)$$

which is a finite number and does not depend on h for h sufficiently large. If 1 is not an eigenvalue of L then

$$\deg(I - L, 0, B_\delta) = (-1)^{\sum_{\{i : \lambda_i \geq 1\}} m_i}. \tag{6}$$

In finite dimension one can see that the previous formula is just equal to sign $\det(I - L)$ (by the formula for $\det(I - L)$ in terms of the eigenvalues of $(I - L)$). To prove (6) we can decompose $E = N \oplus \mathcal{R}$ where N is finite dimensional, $\ker((\lambda_i I - L)^h) \subset N$ if $\lambda_i > 1$, $L(N) \subset N$, $L(\mathcal{R}) \subset \mathcal{R}$, $L_{|\mathcal{R}}$ has no eigenvalues in $(1, +\infty)$.
Then, by the product property of the degree

$$\deg(I - L, 0, B_\delta) = \deg((I - L)_{|N}, 0, B_\delta \cap N) \cdot \deg((I - L)_{|\mathcal{R}}, 0, B_\delta \cap \mathcal{R}).$$

To compute the first factors we notice that $\deg((I - L)_{|N}, 0, B_\delta \cap N)$ is just $(-1)^{\sum_{\{i : \lambda_i \geq 1\}} m_i}$ by the finite dimensional result applied on N. To compute the second factor, note that if $0 < t < 1$, $x \neq 0$ and $(I - tL)(x) = 0$, then $\frac{1}{t}$ is an eigenvalue of $L_{|R}$ and $\frac{1}{t} > 1$. Since this is impossible by construction of $\mathcal{R}$, we see by homotopy invariance that

$$\deg((I - L)_{|R}, 0, B_\delta \cap \mathcal{R}) = \deg((I, 0, B_\delta \cap \mathcal{R}) = 1$$

and the result follows.

4   Degree on convex sets

The notion of degree on convex sets is based on the following extension lemma (See 3.5.8 of [Sch69] for further references):

Lemma 5. *Let E, F be Banach spaces, X a closed subset of E, Y a closed convex subset of F and $A : X \to Y$ a continuous map.*
Then there exists a continuous map $\tilde{A} : E \to Y$ which extends A.

Lemma 6. *Let E be a Banach space, W a closed convex subset of E, U a bounded and relatively open subset of W and $A : \overline{U} \to W$ a completely continuous map such that $\forall x \in \partial_W U$ $x \neq A(x)$ and let $\tilde{U}$ be a bounded open subset of E such that $\tilde{U} \cap W = U$.*
Then the following two facts hold:

(i) $x \neq \tilde{A}(x), \forall x \in \partial_E \tilde{U}$;
(ii) If $\tilde{A}_1 : E \to W$ is another completely continuous extension of A then

$$\deg_W(I - \tilde{A}, 0, \tilde{U}) = \deg_W(I - \tilde{A}_1, 0, \tilde{U}).$$

Proof. (i) Let $x \in Cl_E(\tilde{U})$ be such that $x = \tilde{A}(x)$.
Since $\tilde{A}(x) \in W$ then

$$x \in Cl_E(\tilde{U}) \cap W = Cl_W(U) \implies \tilde{A}(x) = A(x)$$

and then $x \in Cl_W(U)$ and $x = A(x)$, which is impossible.
In particular, if $x \in \partial_E(\tilde{U})$ and $x = \tilde{A}(x)$ then $x \in \partial_W(U)$ and $x = A(x)$ has no solution. (ii) We consider the homotopy $H : [0, 1] \times E \to W$ defined by

$$H(t, x) = t\tilde{A}(x) + (1 - t)\tilde{A}_1(x).$$

Since W is convex and contains the range of $\tilde{A}$ and $\tilde{A}_1$ then $\mathcal{R}(H(t, \cdot)) \subset W$ $\forall t \in [0, 1]$ and then $H(t, \cdot) : E \to W$ is a completely continuous extension of A.
By (i), it follows that

$$x \neq H(t, x), \forall t \in [0, 1], \forall x \in \partial_E \tilde{U}$$

and then by homotopy invariance it holds

$$\deg(I - \tilde{A}, 0, \tilde{U}) = \deg_W(I - \tilde{A}_1, 0, \tilde{U}).$$

$\square$

We can now state the definition of the degree on convex sets.

Definition 10. If $E, W, U, A, \tilde{A}, \tilde{U}$ are as in lemma 6 we define

$$\deg_W(I - A, 0, U) := \deg_W(I - \tilde{A}, 0, \tilde{U}). \tag{7}$$

The usual four properties are easily seen to be still valid.

Proposition 3. *Let E be a Banach space, W a closed convex subset of E, U a bounded and relatively open subset of W and $A : \overline{U} \to W$ a completely continuous map.*
Then the following facts hold.

(i) If $\deg_W(I - A, 0, U) \neq 0$ then there exists $x \in U$ such that $x = A(x)$;
(ii) (excision) if $U_i, i = 1, ..., k$ are disjoint open subsets in W and $x \neq A(x)$ if
$$x \in \overline{U} \setminus \bigcup_{i=1}^{k} U_i \text{ then}$$

$$\deg_W(I - A, 0, U) = \sum_{i=1}^{k} \deg_W(I - A, 0, U_i);$$

(iii) (product) if W_1 is a closed convex subset of a Banach space F, U_1 a bounded and relatively open subset of W_1 and $A_1 : \overline{U_1} \to W_1$ a completely continuous map such that $x \neq A_1(x)$ if $x \in \partial_{W_1} U_1$, then

$$\deg_{W \times W_1}(I - (A, A_1), (0, 0), U \times U_1) = \deg_W(I - A, 0, U) \cdot \deg_{W_1}(I - A_1, 0, U_1);$$

(iv) (homotopy invariance) if $H : \overline{U} \times [0, 1] \to W$ is a completely continuos function such that $x \neq H(t, x), \forall x \in \partial_W U, \forall t \in [0, 1]$, then $\deg(I - H(t, \cdot), 0, U)$ is independent of t.

Definition 11. Let E, W, U, A be as in definition 10 and let x_0 be an isolated fixed point of A in W.
We define

$$\text{index}_W(I - A, x_0) := \lim_{\delta \to 0} \deg_W(I - A, 0, B_\delta(x_0) \cap W) \tag{8}$$

We point out that by the excision property, $\deg_W(I - A, 0, B_\delta(x_0) \cap W)$ is independent of δ for small positive δ.

Remark 16. If $\overset{\circ}{W} \neq \emptyset$ and $x_0 \in \overset{\circ}{W}$ is an isolated fixed point of A in W, then

$$\text{index}_W(I - A, x_0) = \text{index}_E(I - A, x_0)$$

Finally we sketch a proof of Schauder's fixed point theorem.

Proposition 4. *Let W be a closed bounded convex set in a Banach space E. Then every completely continuous map $A : W \to W$ has a fixed point.*

Proof (Sketch). By (i) of proposition 3 it is enough to prove $\deg_W(I-A,0,W) \neq 0$. If $\tilde{A}$ is the usual extension of A and B is a ball containing W then $\tilde{A}: B \to B$ and by using the homotopy $I - t\tilde{A}$ much as in the proof of Brouwer's fixed point theorem, we have that

$$\deg_W(I - A, 0, W) = \deg(I - \tilde{A}, 0, B) = 1.$$

□

Note that is this case we can not directly reduce to the case where W has non-empty interior because W may have empty interior even though it not contained in any proper closed hyperplane in E. Note also that our arguments implies $\deg_W(I - A, 0, W) = 1$ if W is closed, convex and $A(W) \subset W$.

5   Geometry of convex sets

We briefly discuss some conditions on convex sets that we need for the index formula of Theorem 10. Most of this material comes from [Dan83].

Definition 12. Let W be a closed convex set in a Banach space E and $y \in W$. We denote by

$$W_y := \{z \in E : \exists t > 0 \text{ s.t. } y + tz \in W\} \tag{9}$$

$$S_y := \{z \in \overline{W}_y : -z \in \overline{W}_y\} \tag{10}$$

An element $y \in W$ is said to be *demi-interior* to W if $\overline{W}_y = E$.

Note that $\overline{W}_y$ is some form of "linear approximation" to W near y.

Definition 13. A convex subset C of a real vector space is said to be a *wedge* if $0 \in C$ and $\alpha C \subset C, \forall \alpha \geq 0$.
A wedge C is said to be a *cone* if $x \in C \setminus \{0\}$ implies that $-x \notin C$.

We now state some properties of the sets W_y and S_y.

Proposition 5. *Let E, W be as in definition 12. Then the following facts hold:*

(a) $W_y, \overline{W}_y$ *are wedges;*

(b) S_y *is a closed subspace of E;*

(c) *If $y \in \overset{\circ}{W}$ then $\overline{W}_y = S_y = E$;*

(d) *If $\overset{\circ}{W} \neq \emptyset$ and ∂W is a smooth manifold near $y \in \partial W$ then $\overline{W}_y$ is a half space bounded by the tangent hyperplane to C at y (but translated to zero);*

(e) *If W is a cone then $\overline{W}_0 = W$ and $S_0 = \{0\}$;*

(f) *If E is separable and W is not contained in a proper closed subspace of E then the set of demi-interior points of W is dense in W.*

Remark 17. Usually it is easiest to calculate $\overline{W}_y$ directly but it is also not difficult to use the Hahn-Banach theorem to give a dual characterization of $\overline{W}_y$. Define

$$W_y^* = \{f \in E^* : f(x) \geq f(y), \forall x \in W\}.$$

Then

$$\overline{W}_y = \{x \in E : f(x) \geq 0, \forall f \in W_y^*\}$$

and hence

$$\overline{S}_y = \{x \in E : f(x) = 0, \forall f \in W_y^*\}.$$

Example 2. Let $\Omega \subset \mathbf{R}^n$ be a bounded open set and define

$$K = \{u \in C_0(\Omega) : u(x) \geq 0, \forall x \in \Omega\}.$$

Then K is not contained in a proper closed hyperplane of $C_0(\Omega)$. To see this, suppose by way of contradiction that $K \subset X \subset C_0(\Omega)$ with X a closed hyperplane of codimension 1. Then there exists a non-trivial subspace $Y \subset C_0(\Omega)$ such that $C_0(\Omega) = Y \oplus X$.
If $u \in Y \setminus \{0\}$ then $u = u^+ - u^-$ with $u^+, -u^- \in X$ not simultaneously null and then $u \in X$ which is a contradiction.
The following useful results can be easily proved by either characterization:

(1) if $u \in C_0(\Omega)$ is strictly positive in Ω, then

$$\overline{K}_u = C_0(\Omega) \text{ and } \overline{S}_u = C_0(\Omega).$$

(2) if $u \in C_0(\Omega)$ is strictly positive in $\Omega \setminus A$ and $u = 0$ in A, with $A \subset \Omega$ a closed set, then

$$\overline{K}_u = \{v \in C_0(\Omega) : v(x) \geq 0, \forall x \in A\}.$$

(3) if $1 \leq p < \infty$, $u \in L^p(\Omega)$ and $u(x) > 0$ a.e. in Ω, then $\overline{K}_u = L^p(\Omega)$ where K now denotes the set of non-negative functions in $L^p(\Omega)$.

Remark 18. Assume W be a closed convex subset of E, $y \in W$ and L is a bounded linear operator on E such that $L(\overline{W}_y) \subset \overline{W}_y$. It follows easily that $L(S_y) \subseteq S_y$ and hence L induces a bounded linear map $\tilde{L}$ of E/S_y into itself. This will be important for us. In particular if there is a closed complement Y to S_y in E, then it is easy to see that $Y \cong E/\overline{S}_y$ and $\tilde{L} \cong PL_{|Y}$ where P is the projection of E onto Y annihilating S_y. (In our applications, we usually find S_y is complemented).

6 A basic Index calculation

Generally if $A : E \to E$ is completely continuous, x_0 is an isolated fixed point of A and A is differentiable at x_0 (that is there is a bounded linear operator V

such that $|A(x_0 + h) - A(x_0) - V(h)| / |h| \to 0$ as $h \to 0$ and we let $A'(x_0)$ denote V) then

$$\text{index}_E(I - A, x_0) = \pm 1.$$

By generally we mean that this holds when $I - A'(x_0)$ is invertible (and in this case $\text{index}_E(I - A, x_0) = \text{index}_E(I - A'(x_0), 0)$). Krasnosel'skii first realized that in the case of convex sets (in particular in the case of a cone), generally the index may also take the value zero.

Now we give a basic index formula.

Theorem 10. *Assume that:*

(i) $W \subseteq E$ *is a closed convex set and* $y \in W$;
(ii) $A : W \to W$ *is extensible to a map* $A : U \to E$, *where* $W \subseteq U \subseteq E$ *and* U *is an open set;*
(iii) $A(y) = y$ *and* A *is differentiable at* y;
(iv) A *is completely continuous on* $\text{Cl}_E U$;
(v) W *is not contained in a proper closed hyperplane of* E;
(vi) $z \neq A'(y)z, \forall z \in \overline{W}_y \setminus \{0\}$.

We denote by $r(\widetilde{A'(y)})$ *the spectral radius of the reduced map* $\widetilde{A'(y)}$ *as in 18. Then the following results hold.*

(1) y *is an isolated fixed point of* A *in* W;
(2) (a) if $x \neq A'(y)x, \forall x \in E \setminus \{0\}$ *and* $r(\widetilde{A'(y)}) < 1$ *then*

$$\begin{aligned}
\text{index}_W(I - A, y) &= \text{index}_E(I - A, y) \\
&= \text{index}_E(I - A'(y), 0) \\
&= \text{index}_{\overline{S}_y}(I - A'(y), 0);
\end{aligned}$$

(b) if $r(\widetilde{A'(y)}) > 1$ *then*

$$\text{index}_W(I - A, y) = 0;$$

(c) if $\ker(I - A'(y)) \cap (E \setminus \overline{W}_y) \neq \emptyset$ *then*

$$\text{index}_W(I - A, y) = 0.$$

Remark 19. Note that since $A(y) = y$ and $A(W) \subseteq W$ it is easy to prove (as in [Dan83]) that $A'(y)$ maps $\overline{W}_y$ into itself and hence $\overline{S}_y$ into itself. Note that by Krasnosel'skii and Zabrieko [KZ84], $A'(x_0)$ is a compact linear map and hence $\text{index}_E(I - A', x_0)$ is well defined and we can use Theorem 2.4.5 in Lloyd [Llo78] to evaluate it.
If $r(\widetilde{A'(y)}) = 1$, we can argue as in [Dan83] to prove that

$$\ker(I - A'(y)) \cap \overline{W}_y \neq \{0\}$$

(by using the Krein-Rutman theorem).

Hence our other assumptions ensure that $r(\widetilde{A'(y)}) \neq 1$. Thus the above theorem gives a formula for the index in all cases where

$$\ker(I - A'(y)) \cap \overline{W}_y = \{0\}.$$

A detail proof of Theorem 10 can be found in [Dan83] except that one missing case is proved in Proposition 2 of [Dan86]. Lastly it can be shown that $\overline{W}_y / S_y$ is a closed cone in E / S_y which is invariant under $\widetilde{A'(y)}$. Hence the Krein-Rutman theorem as in Amann [Ama76] implies that

$$r(\widetilde{A'(y)}) = \sup\{\lambda > 0 : \lambda \text{ is an eigenvalue of } \widetilde{A'(y)}\}$$

where $r(\widetilde{A'(y)}) = 0$ if there is no such eigenvalue. If y is a demi-interior point of W, it follows that index $_W(I - A, y) = $ index $_E(I - A'(y), 0)$ provided that A is differentiable at Y and $I - A'(y)$ is invertible. We conjecture that this still holds if the last condition is replaced by y is an isolated point in W.

Definition 14.
(a) *Let K be a cone in a Banach space E.*
    We define:

$$x \geq y \Longleftrightarrow x - y \in K.$$

   (We point out that $x \geq y$, $y \geq x \Longrightarrow x = y$.)
(b) *An operator $T : E \to E$ is said to be* monotone *if:*

$$x \geq y \Rightarrow T(x) \geq T(y).$$

(c) *Given x, y such that $x \leq y$ we call* order interval *the set*

$$[x, y] = \{u \in E : x \leq u \leq y\}.$$

As a simple application of these ideas, we prove the following result from Dancer and Hess [DH91].

Theorem 11. *Let $K \subseteq E$ be a cone such that all the order intervals are bounded, $T : E \to E$ is a monotone and completely continuous operator and x_0, y_0 fixed points of T such that $x_0 \leq y_0$ and T has no other fixed points in $I_0 := [x_0, y_0]$. Then the following alternative holds:*

(1) *There exists a sequence $(\tilde{x}_k)_{k \in Z}$ in I_0 such that*

$$\tilde{x}_{k+1} = T(\tilde{x}_k), \qquad \tilde{x}_{k+1} \geq \tilde{x}_k$$

$$\lim_{k \to -\infty} \tilde{x}_k = x_0, \qquad \lim_{k \to +\infty} \tilde{x}_k = y_0;$$

(2) *There exists a sequence $(\tilde{x}_k)_{k \in Z}$ in I_0 such that*

$$\tilde{x}_{k+1} = T(\tilde{x}_k), \qquad \tilde{x}_{k+1} \leq \tilde{x}_k$$

$$\lim_{k \to +\infty} \tilde{x}_k = x_0, \qquad \lim_{k \to -\infty} \tilde{x}_k = y_0.$$

Proof. Since T is monotone and x_0, y_0 are fixed points for it, then $T : I_0 \to I_0$.

Step 1. We prove that the following alternative holds:

(a) index $_{I_0}(I - T, x_0) = 1$ or
(b) $\forall \epsilon > 0 \ \exists x_\epsilon \in \partial B_\epsilon(x_0) \cap I_0$ such that $T(x_\epsilon) \geq x_\epsilon$ (i.e. x_ϵ is a subsolution arbitrarily close but different from x_0).
If (b) does not hold then

$$\exists \epsilon > 0 : \forall x \in \partial B_\epsilon(x_0) \cap I_0 \Rightarrow T(x) - x \notin K.$$

If we consider the homotopy

$$\begin{aligned} H : [0,1] \times (B_\epsilon(x_0) \cap I_0) &\longrightarrow I_0 \\ (t, x) &\longmapsto (1-t)x_0 + tT(x) \end{aligned}$$

then

$$H(t, x) \neq x, \qquad \forall t \in [0,1], x \in \partial B_\epsilon(x_0) \cap I_0$$

because if by contradiction

$$\exists t \in [0,1], x \in \partial B_\epsilon(x_0) \cap I_0 : H(t, x) = x$$

then

$$x - x_0 = t(T(x) - x_0) \leq T(x) - x_0$$

and so $T(x) - x \in K$.
By homotopy invariance we have that

$$\deg_{I_0}(I - H(t, \cdot), 0, B_\epsilon(x_0) \cap I_0)$$

is independent of t and then we can deduce that

$$\deg_{I_0}(I - T, 0, B_\epsilon(x_0) \cap I_0) = \deg_{I_0}(I - x_0, 0, B_\epsilon(x_0) \cap I_0) = 1$$

(For the last equality, we use one of the comments after the proof of Theorem 4, noting that x_0 is the only zero of $I - x_0$). Hence (a) follows.
Step 2. The following alternative holds:
(c) index $_{I_0}(I - T, y_0) = 1$ or
(d) $\forall \epsilon > 0 \ \exists y_\epsilon \in \partial B_\epsilon((y_0) \cap I_0$ such that $T(y_\epsilon) \leq y_\epsilon$
   (y_ϵ is a supersolution arbitrarily close but different from y_0).
The proof is analogous to step (1).
Step 3. We prove that

$$\text{index}_{I_0}(I - T, x_0) \neq 1 \text{ or index}_{I_0}(I - T, y_0) \neq 1.$$

Since I_0 is closed, convex and bounded, by Theorem 4 applied to $T : I_0 \to I_0$ and by the excision property we have

$$\begin{aligned} 1 = \deg_{I_0}(I - T, 0, I_0) &= \deg_{I_0}(I - T, 0, B_\epsilon(x_0) \cap I_0) + \\ &+ \deg_{I_0}(I - T, 0, B_\epsilon(y_0) \cap I_0) = \text{index}_{I_0}(I - T, x_0) + \text{index}_{I_0}(I - T, y_0) \end{aligned}$$

which proves our assertion. (Note that we have used here that T has x_0, y_0 as its only fixed points in I_0).

Step 4. From steps 1,2,3, we deduce that the following alternative holds:
(b) there exists a subsolution arbitrarily close but different from x_0 or
(d) there exists a supersolution arbitrarily close but different from y_0.
We prove that if (b) holds then the condition (1) of the theorem holds. Let
$x \neq x_0$ be a subsolution such that $T(x) \geq x$ and consider the sequence in
I_0, $\tilde{x}_n = T^n(x), n \geq 1$.
Then by monotonicity of T we have

$$\tilde{x}_{n+1} = T^n(T(x)) \geq T^n(x) = \tilde{x}_n.$$

We prove that $\lim\limits_{n \to +\infty} \tilde{x}_n = y_0$.
Since $(\tilde{x}_n)_{n \geq 1}$ is precompact then a subsequence of it has a limit $y \in I_0$ and
being T monotone then all the sequence has limit y and then

$$y = \lim_{n \to +\infty} \tilde{x}_{n+1} = \lim_{n \to +\infty} T(T^n(x)) = T(y).$$

Then we have that $y = y_0$ since x_0, y_0 are the only fixed points of T in I_0
and since

$$\tilde{x}_n = T^{n-1}(T(x)) \geq T^{n-1}(x) \geq x$$

and then

$$y = \lim_{n \to +\infty} \tilde{x}_n \geq x \neq x_0.$$

□

Note that we have not quite proved the full strength of Theorem 11. We have
an increasing sequence starting from near x_0. It is now a tedious diagonalising
argument to obtain the full strength of the theorem. We omit this. It can be
found in [DH91]. When (d) holds, the argument is similar.

7 Applications of Index Formula

In this section we discuss rather briefly one application of the index formula.
The results come from [Dan92].
Assume that D is an annulus in $\mathbf{R}^2$. We consider the solutions of the Gelfand
problem

$$\begin{cases} -\Delta u = \lambda \exp(u) \text{ in } & D \\ u = 0 & \text{on } \partial D \end{cases} \tag{11}$$

for $\lambda \geq 0$.(The case where $\lambda \leq 0$ is simple because it is not difficult to show that
the solution is unique.)
    Moreover if $\lambda > 0$, the maximum principle implies that any solution is posi-
tive in D.
    We first consider the radially symmetric solutions though our main interest
is in non-radially symmetric solutions. It is proved in [Lin89] that there exists
$\alpha > 0$ and continuous maps u_1, u_2 of $(0, \alpha]$ into

$$C_0^r(D) = \{u \in C_0(D) : u \text{ is radially symmetric}\}$$

such that $u_1(t) \to 0$ at $t \to 0^+$, $\|u_2(t)\|_\infty \to \infty$ as $t \to 0^+$, $u_1(\alpha) = u_2(\alpha)$, $u_1(t) < u_2(t)$ for $0 < t < \alpha$ and

$$\{(u_1(t), t) : 0 < t \leq \alpha\} \cup \{(u_2(t), t) : 0 < t \leq \alpha\}$$

are the radially solutions of (11) for $\lambda > 0$.

$u_1(\lambda)$ is known as the minimal solution and general theory for convex mappings (as in [Ama76]) ensures that there are no solutions at all of (11) for $\lambda > \alpha$, no other solutions for $\lambda = \alpha$ and that $u_1(\lambda)$ is a non-degenerate solution (and stable) for $0 < \lambda < \alpha$.

It follows easily from this and the remark after the lemma in Dancer [Dan79] that non-radial solutions cannot bifurcate from $u_1(\lambda)$ for any λ or from $u_2(\lambda)$ for λ near α. It also proved in [Dan92] that $u_2(\lambda)$ is a non-degenerate in the space of radial functions for λ near α.

Fix $n \geq 1$. It is also proved in [Lin89] that there exists $\tau_n \in (0, \alpha)$ such that the smallest eigenvalue $\gamma = \gamma_n(\lambda)$ of the linearized equation

$$\begin{cases} -\Delta h - \lambda \exp(u_2(\lambda))h = \gamma h \text{ in } & D \\ h = 0 & \text{on } \partial D \end{cases} \tag{12}$$

with an eigenfunction of the form $\tilde{h}(r) \cos n\theta$ satisfies $\gamma_n(\lambda) > 0$ if $\tau_n < \lambda < \alpha$ while $\gamma_n(\lambda) < 0$ if $0 < \lambda < \tau_n$.

Moreover τ_n is the only point where (12) has a solution of the form $\tilde{h}(r) \cos n\theta$ and $\tilde{h}$ is non-negative. It is easy to show (by using the equation for the radial part of an eigenfunction) that

$$\tau_{n+1} < \tau_n \text{ for all } n \geq 1 \text{ and } \lim_{n \to +\infty} \tau_n = 0.$$

Note that this can easily be done in this case because there are explicit formulae for $u_1(\lambda), u_2(\lambda)$ and $\gamma_n(\lambda)$.

We now define a suitable narrow cone. Let

$$C_0^n(D) = \left\{ u \in C_0(D) : u\left(r, \theta + \frac{2\pi}{n}\right) = u(r, \theta) \right\},$$

and let K^n be the set of functions $u \in C_0^n(D)$ such that $u \geq 0, u$ is even in $\theta, u(r, \theta)$ is decreasing in θ for $0 \leq \theta \leq \frac{\pi}{n}$, $a \leq r \leq 1$ and $K_0^n = K^n \cap C_0^n(D)$. Here $D = \{x \in \mathbf{R}^2 : a < r < 1\}$ and we are using polar coordinates. It is easy to see that $C_0^n(D)$ is a closed subspace of $C_0(D)$, K_0^n is a cone in $C_0^n(D)$ and $K_0^n - K_0^n$ is dense in $C_0^n(D)$.

Moreover (as in Lemma 1 of [Dan92]) it is not difficult to prove that if $f \in K^n$ the solution of

$$\begin{cases} -\Delta u = f \text{ in } & D \\ u = 0 & \text{on } \partial D \end{cases}$$

belongs to K_0^n.

Moreover it is easy to see that the mapping $u \to \exp(u)$ maps K_0^n into K^n. Hence we see that the map A defined by $A(u) = (-\Delta)^{-1}(\exp u)$ maps K_0^n into itself (and $C_0^n(D)$ into itself). Since $(-\Delta)^{-1}$ maps $C(\overline{D})$ into $C^1(\overline{D})$ (see for example [GT83] or Remark 21) and since the Ascoli-Arzela theorem implies the natural inclusion of $C^1(\overline{D})$ into $C(\overline{D})$ is compact, it is easy to see that A is completely continuous. Note that the inverse of the Laplacian is under Dirichlet boundary conditions.

Theorem 12. *For each $\lambda \in (0, \tau_n)$, there is a non-radially symmetric solution of (11) in K^n and there are at least n distinct non-radially symmetric solutions.(Here distinct means that they can not be obtained from each other by the symmetries.)*

Proof (Sketch). Let $C = K_0^n$. First note that if $\delta > 0$ there is an apriori bound for solutions of (11) with $\lambda \geq \delta$. In particular there is an $M > 0$ such that each such solution (u, λ) satisfies $\|u\|_\infty \leq M$. This can be found in Spruck [Spr88]. We omit the proof. (Note, as ever, the apriori bound is important). By the homotopy invariance of the degree, it follows that $\deg_C(I - \lambda A, B_{M_1} \cap C)$ is independent of λ for $\lambda \geq \delta$. Here $M_1 = M + 1$ and for the moment B_{M_1} denotes the ball of radius M_1 in $C_0^n(D)$. Since there are no solutions for $\lambda > \alpha$, this degree must be zero. Hence, we see that if there are no solutions in C other than $u_1(\lambda)$ and $u_2(\lambda)$ then

$$\text{index}_C(\lambda A, u_2(\lambda)) + \text{index}_C(\lambda A, u_1(\lambda)) = 0. \tag{13}$$

We prove that $\text{index}_C(A, u_2(\lambda))$ is defined for $\lambda \neq \tau_n$ and that $\text{index}_C(A, u_2(\lambda)) = \pm 1$ if $\lambda > \tau_n$ (but is independent of λ if $\lambda > \tau_n$) and is 0 if $\lambda < \tau_n$. In fact the first index is -1 but we do not need this. Since a similar argument shows that the stable solution $u_1(\lambda)$ has index 1 for $\lambda \in (0, \alpha)$, we see that these index calculations contradict (13) if $\lambda \in (0, \tau_n)$ and there must be another solution in C, as required. Now a non-radial solution $u \in K_0^n$ has the property that $\frac{\partial u}{\partial \theta} \leq 0$ for $a < r < 1, 0 < \theta < \frac{\pi}{n}$. By applying the maximum principle to $\frac{\partial u}{\partial \theta}$ (noting that $\frac{\partial u}{\partial \theta}$ solves $-\Delta h = \lambda(\exp u)h$), it follows that $\frac{\partial u}{\partial \theta} < 0$ if $a < r < 1, 0 < \theta < \frac{\pi}{n}$. Hence we see that a non-radial solution cannot be in two different K_0^n. Hence there must be n distinct non-radial solutions if $\lambda \in (0, \tau_n)$. Hence it suffices to establish the index formula. Let $y = u_2(\lambda)$. It is easy to prove that

$$\overline{C_y} = \{\phi \in C_0^n(D) : \phi \text{ is decreasing in } \theta \text{ for } 0 \leq \theta \leq n^{-1}\pi\}$$

and hence that $S_y = \{\phi \in C_0(D) : \phi \text{ is radially symmetric }\}$.
We choose a complement Z to S_y in $C_0^n(D)$ to be those functions in $C_0(D)$ which have a Fourier series expansion $\sum_{j=1}^\infty c_j h_j(r) \cos jn\theta$. It is easy to see that $A'(u_2(\lambda))$ maps S_y and Z into themselves (which is basically due to the symmetries or to self-adjointness). We find that

$$h \neq A'(u_2(\lambda))h \text{ if } h \in \overline{C_y} \setminus \{0\} \text{ and if } \lambda \neq \tau_n$$

and that

$$r(A'(u_2(\lambda))_{|Z}) \begin{cases} < 1 \text{ if } \lambda > \tau_n \\ > 1 \text{ if } \lambda < \tau_n \end{cases}$$

(one needs to use separation of variables and results on the spectrum of ordinary differential operators). Hence we can apply Theorem 10 and obtain our claim. This completes our sketch of the proof. $\qquad\square$

Remark 20. In higher dimensions, it seems difficult to find suitable cones (and the apriori bound fails). One can also obtain related but different results by variational methods. It is possible to obtain unbounded connected nets of solutions in K_0^n by using ideas of Rabinowitz. (See [Rab71])

8   Population models

We discuss the application of our index calculations to prove the existence of positive solutions of population models. Many people have studied these problems. Further references can be found in [Dan91] and [DD95]. We consider the system

$$\begin{cases} -\Delta u = u(a - u - cv) \text{ in } & \Omega \\ -\Delta v = v(d - v - eu) \text{ in } & \Omega \\ u \geq 0, v \geq 0 & \text{in } \Omega \\ u = 0, v = 0 & \text{on } \partial\Omega \end{cases} \tag{14}$$

with $\Omega \subset \mathbf{R}^n$ is connected open set with smooth boundary and a, c, d, e are positive constants.

This is a Lotka-Volterra model for competing species where there is diffusion and we are looking for solutions independent of time. Because u and v represent populations, it is natural to only look at solutions with u, v non-negative on Ω. We point out that 14 has always the *trivial solution* $(0, 0)$. Moreover if a and d are suitable chosen it has also the *semitrivial solutions* $(u, 0)$ and $(0, v)$. (See Proposition 6). We are interested in searching for *coexisting solutions* (u, v), that is with $u(x) > 0$ in Ω, $v(x) > 0$ in Ω. First of all we recall some well known results about regularity and positivity of solutions of linear equations.(See for example [GT83]).

Remark 21. Let $\Omega \subset \mathbf{R}^n$ be a connected and bounded open set with smooth boundary and consider the following problem:

$$\begin{cases} -\Delta u + a(x)u = f(x) \text{ in } & \Omega \\ u = 0 & \text{on } \partial\Omega \end{cases} \tag{15}$$

where $a \in L^\infty(\Omega)$ and $a \geq 0$ a.e. in Ω. Here we mean a weak solution; the condition $u \in W_0^{1,2}(\Omega)$ incorporates the boundary condition and the equation is satisfied in a weak sense. Then if $f \in L^p(\Omega)$ the equation is uniquely solvable and we have

(i) $p < \frac{n}{2} \Rightarrow u \in L^{\frac{np}{n-2p}}(\Omega)$;
(ii) $p > \frac{n}{2} \Rightarrow u \in C_0(\Omega)$;
(iii) $p > n \Rightarrow u \in C^1(\overline{\Omega})$.

We recall that if $a \in C^1(\overline{\Omega})$ and $f \in C^1(\overline{\Omega})$ then $u \in C^2(\overline{\Omega})$ and if $f \geq 0$ a.e. in Ω then $u \geq 0$ a.e. in Ω.
Moreover if $p > n$ and $f \in L^p(\Omega)$ the following maximum principle holds:

$$f \not\equiv 0 \text{ in } \Omega \text{ and } f(x) \geq 0 \Longrightarrow u \in C^1(\overline{\Omega}), \ u > 0 \text{ in } \Omega, \ \frac{\partial u}{\partial \nu} > 0 \text{ on } \partial\Omega.$$

Remark 22. Now let us consider the problem

$$\begin{cases} -\Delta u + a(x)u = 0 \text{ in } & \Omega \\ u \geq 0 & \text{in } \Omega \\ u = 0 & \text{on } \partial\Omega \end{cases} \tag{16}$$

with $a \in L^\infty$. Then, if u does not vanish identically, $u > 0$ in Ω and for every compact $K \subset \Omega$ there exists $c = c(K, \Omega, \|a\|_{L^\infty})$ such that $\sup_K u < c \cdot \inf_K u$. This is a type of Harnack inequality.

Secondly we recall a well known existence result for a semilinear equation. (See for example [Str90]).

Remark 23. Let $g : \mathbf{R} \to \mathbf{R}$ a C^1-function, let us consider the problem

$$\begin{cases} -\Delta u = g(u) \text{ in } & \Omega \\ u = 0 & \text{on } \partial\Omega \end{cases} \tag{17}$$

We say that u is a *regular subsolution* of (17) if

$$u \in C^2(\overline{\Omega}), \ -\Delta u \leq g(x) \text{ in } \Omega \text{ and } u \leq 0 \text{ in } \partial\Omega.$$

We say that u is a *regular supersolution* of (17) if

$$u \in C^2(\overline{\Omega}), \ -\Delta u \geq g(x) \text{ in } \Omega \text{ and } u \geq 0 \text{ on } \partial\Omega.$$

Theorem 13. *If u_1 is a subsolution of (17) and u_2 is a supersolution of (17) such that $u_1 \leq u_2$ in Ω, then there exists a solution u_0 of (17) with $u_1 \leq u_0 \leq u_2$ in Ω.*

Finally we recall some properties of eigenvalues. (See for example [CH62]).

Remark 24. Let us now consider the eigenvalue-problem

$$\begin{cases} -\Delta u + a(x)u = \lambda m(x)u \text{ in } & \Omega \\ u = 0 & \text{on } \partial\Omega \end{cases} \tag{18}$$

with $a \in L^\infty(\Omega)$, $a \geq 0$, $m \in L^\infty(\Omega)$ and $m(x) \geq \mu > 0$ in Ω. It is well known that there exists an increasing sequence of eigenvalues

$$\lambda_1 < \lambda_2 \leq \lambda_3 \leq \ldots \leq \lambda_k \leq \ldots$$

such that $\lim_{k\to\infty} \lambda_k = +\infty$. The first eigenvalue λ_1 is simple and one can choose the associated eigenfunction strictly positive in Ω. We denote $\lambda_i = \lambda_i(a, m)$ and it holds

$$\lambda_1(a, m) = \inf_{u \in H_0^1(\Omega)\backslash\{0\}} \frac{\int |\nabla u|^2 + au^2}{\int mu^2}.$$

It is easy to see from this that $\lambda_1(a, m)$ is strictly decreasing with respect to m and strictly increasing with respect to a. (Here we use crucially that the first eigenfunction can be chosen to be strictly positive on Ω). In particular, we will use λ_1 to denote $\lambda_1(0, 1)$, that is the first eigenvalue of $-\Delta$.

Remark 25. Now we consider the following problem which is closely related to (14)

$$\begin{cases} -\Delta u = u(a - u) \text{ in } & \Omega \\ u \geq 0 & \text{in } \Omega \\ u = 0 & \text{on } \partial\Omega \end{cases} \tag{19}$$

with $\Omega \subset \mathbf{R}^n$ connected and bounded open set with smooth boundary and $a \in \mathbf{R}$. Then the following results hold.

(a) **non-existence**: if $a \leq \lambda_1$ then (19) has no non-trivial solutions. It suffices to multiply the equation for e_1 (where $e_1 > 0$ is the first eigenfunction of $-\Delta$) and to integrate by parts;

(b) **existence**: if $a > \lambda_1$ then (19) has a non-trivial solution (see [Dan84]). For s small and positive and c a large real number, se_1 is a subsolution and c is a supersolution of (19) and so by Theorem 13 there exists a solution u such that

$$0 < se_1(x) \leq u(x) \leq c \text{ in } \Omega;$$

(c) **uniqueness**: if $a > \lambda_1$ then there exists unique non-trivial solution. (See Theorem 15 below);

(d) **regularity**: a weak-solution u in $C_0(\Omega)$ is a classical solution and $0 \leq u(x) \leq a$ in Ω.

Proposition 6. *If $a > \lambda_1$ then (14) has a semitrivial solution $(\overline{u}, 0)$ with $\overline{u} > 0$ in Ω.*
If $d > \lambda_1$ then (14) has a semitrivial solution $(0, \overline{v})$ with $\overline{v} > 0$ in Ω.
If $a, d \leq \lambda_1$ then (14) has no semitrivial solution.

Proof. It follows from Remark 25.   □

Now we are going to apply Theorem 10 to obtain coexisting solutions of (14). We start with two useful lemmas.

Lemma 7. *If $a \leq \lambda_1$ or $d \leq \lambda_1$ then there is no coexisting solution for (14).*

Proof. If $(\tilde{u}, \tilde{v})$ is a coexisting solution of (19), then

$$-\Delta\tilde{u} = \tilde{u}(a - \tilde{u} - c\tilde{v}) \leq \tilde{u}(a - \tilde{u}) \text{ in } \Omega.$$

So $\tilde{u}$ is a subsolution of (19).
Since a constant $s \gg 0$ is a supersolution of (19), then there exists a solution u with $\tilde{u} \leq u \leq s$.
Then u is a nontrivial solution and thus $a > \lambda_1$.   □

Lemma 8. *There exists $M > 0$ such that for any solution (u, v) of (14) then $0 \le u \le M$ and $0 \le v \le M$ in Ω.*

Proof. We choose $M > \max(a, d)$.
If $(\bar{u}, 0)$ and $(0, \bar{v})$ are the semitrivial solutions, then by 25(d) $0 \le \bar{u}(x) \le a$ and $0 \le \bar{v}(x) \le d$ in Ω. Assume that $(\tilde{u}, \tilde{v})$ is a solution, then

$$-\Delta\tilde{u} = \tilde{u}(a - \tilde{u} - c\tilde{v}) \le \tilde{u}(a - \tilde{u}) \text{ in } \Omega.$$

Therefore $\tilde{u}$ is a subsolution of (19). Since a constant $s \gg 0$ is a supersolution, then there exists a solution u with $\tilde{u} \le u \le s$ and in particular $0 \le \tilde{u}(x) \le u(x) \le a$ in Ω. (Here we are using the last part of 19(d)). Likewise one can prove that $0 \le \tilde{v}(x) \le d$ in Ω. $\qquad\square$

Remark 26. Consider the following eigenvalue problem:

$$\begin{cases} -\Delta w + (e\bar{u} - d)w = \lambda w \text{ in } & \Omega \\ w = 0 & \text{on } \partial\Omega \end{cases} \tag{20}$$

where $e \in \mathbf{R}$ and $\bar{u}$ is the positive solution of (19) whose existence has been proved in remark 25.
Let us denote by $\lambda(e)$ the first eigenvalue of (20), which is a strictly increasing function of e by 24.
Then $\lambda(0) = \lambda_1 - d$ and it can be shown (see [Dan85]) that, if $e \gg 0$ then $\lambda(e) > 0$ and then we infer that

$$d > \lambda_1 \Longrightarrow \exists! \, \bar{e} > 0 : \lambda(\bar{e}) = 0. \tag{21}$$

Now we consider the eigenvalue problem

$$\begin{cases} -\Delta v + (c\bar{v} - a)v = \tilde{\lambda} v \text{ in } & \Omega \\ v = 0 & \text{on } \partial\Omega \end{cases} \tag{22}$$

where $c \in \mathbf{R}$ and $\bar{v}$ is the positive solution of (19) when a is replaced by d.
If we denote with $\tilde{\lambda}(c)$ the first eigenvalue of (22) then

$$a > \lambda_1 \Longrightarrow \exists! \, \bar{c} > 0 : \tilde{\lambda}(\bar{c}) = 0. \tag{23}$$

At this point the following existence theorem holds:

Theorem 14. *Assume that $a > \lambda_1$ and $d > \lambda_1$.*
Then (14) has a coexisting solution $(\tilde{u}, \tilde{v})$ if $e < \bar{e}$ and $c < \bar{c}$, or if $e > \bar{e}$ and $c > \bar{c}$. (See (21) and (23))

Proof. **Step 1.** Construction of the operator A.
Let $E = C_0(\Omega) \oplus C_0(\Omega)$, $K = \{u \in C_0(\Omega) : u \ge 0\}$ and $K_1 = K \oplus K$.
By using (2), it is easy to see that K_1 is a closed, convex subset of E with empty interior, not contained in any closed and proper hyperplane of E. We

are interested in the solutions of (14), that is, the fixed points of the operator of K_1 into E which maps $\begin{pmatrix} u \\ v \end{pmatrix}$ to $\begin{pmatrix} (-\Delta)^{-1}(u(a-u-cv)) \\ (-\Delta)^{-1}(u(d-v-eu)) \end{pmatrix}$. Note that this operator does not map K_1 into itself!

Since we want to use Theorem 10 we need to construct an operator that maps an open subset of the cone K_1 into K_1 whose fixed points are the solutions of (14).

We proceed in this way: since the solutions (u, v) of (14) are such that $\|u\|_\infty \leq M$ and $\|v\|_\infty \leq M$ with $M > \max(a, d)$ (see Lemma 8), we can choose as domain of the operator the set

$$U = \{(u, v) \in K_1 : \|u\|_\infty \leq M, \|v\|_\infty \leq M\}$$

Now we note that $\exists \alpha > 0$ such that

$$\forall (u, v) \in U \qquad a + \alpha - u - cv \geq 0, \, d + \alpha - v - eu \geq 0 \text{ in } \Omega$$

and then for suitable positive M and α it holds

$$(u, v) \in U \Rightarrow (u(a + \alpha - u - cv), v(d + \alpha - v - eu)) \in K_1.$$

Now we are able to define $A : U \to K_1$ by

$$A(u, v) = ((-\Delta + \alpha I)^{-1}(u(a+\alpha-u-cv)), (-\Delta+\alpha I)^{-1}(v(d+\alpha-v-eu))). \tag{24}$$

(The inverses are under Dirichlet boundary conditions)

Step 2. The assumptions of Theorem 10 are satisfied.
(1) if $\alpha > 0$ then
$$(-\Delta + \alpha I)^{-1} : C^0(\Omega) \to C^1(\overline{\Omega})$$

is continuous and monotone. This ensures that $\mathcal{R}(A) \subseteq K_1$;

(2) A is easily seen to be completely continuous since (1) and the Ascoli-Arzela theorem ensures that $(-\Delta + \alpha I)^{-1}$ is a completely continuous map of $C_0(\Omega)$ into itself;

(3) it is easy to see that the operator $A : E \to E$ is differentiable and for $(u, v), (\varphi, \psi) \in E$ we have

$$A'(u, v)(\varphi, \psi) = ((-\Delta + \alpha I)^{-1}(\varphi(a + \alpha - 2u - cv) - cu\psi),$$
$$(-\Delta + \alpha I)^{-1}(\psi(d + \alpha - 2v - eu) - ev\varphi));$$

(4) it is immediate to verify that the fixed points of A are all the solutions of (14);

Step 3. Computation of $\text{index}_{K_1}(I - A, (0, 0))$.

In this case the fixed point of A is $(0, 0)$ and $A'(0, 0) : E \to E$ is

$$A'(0, 0)(\varphi, \psi) = ((-\Delta + \alpha I)^{-1}((a + \alpha)\varphi), (-\Delta + \alpha I)^{-1}((d + \alpha)\psi).$$

Moreover $\overline{K_1}_{\{(0,0)\}} = K_1$, and $S_{\{(0,0)\}} = \{(0, 0)\}$. Thus $E/\overline{S}_{\{(0,0)\}} = E$ and then the reduced operator

$$\widetilde{A'(0, 0)} = A'(0, 0).$$

First of all we must verify (vi) of Theorem 10, that is

$$A'(0,0)(\Phi,\psi) \neq (\Phi,\psi) \qquad \forall(\Phi,\psi) \in K_1 \setminus \{(0,0)\}$$

is equivalent to

$$\begin{cases} -\Delta\Phi = a\Phi & \text{in} \quad \Omega \\ \Phi \geq 0 & \text{in} \quad \Omega \\ \Phi = 0 & \text{on } \partial\Omega \end{cases} \text{ and } \begin{cases} -\Delta\psi = d\psi & \text{in} \quad \Omega \\ \psi \geq 0 & \text{in} \quad \Omega \\ \psi = 0 & \text{on } \partial\Omega \end{cases} \tag{25}$$

If $a \neq \lambda_1$ and $d \neq \lambda_1$, this is not possible.
Now we must calculate the spectral radius of $A'(0,0)$.
We note that

λ is eigenvalue of $A'(0,0) \iff \exists (\Phi,\psi) \neq 0$ such that
$A'(0,0)(\Phi,\psi) = \lambda(\Phi,\psi)$

$$\iff \begin{cases} (-\Delta + \alpha I)^{-1}((a+\alpha)\Phi) = \lambda\Phi \\ (-\Delta + \alpha I)^{-1}((d+\alpha)\psi) = \lambda\psi \end{cases} \iff \begin{cases} -\Delta\Phi = (\frac{a+\alpha}{\lambda} - \alpha)\Phi \\ -\Delta\psi = (\frac{d+\alpha}{\lambda} - \alpha)\psi. \end{cases}$$

If $(\lambda_k)_{k\geq 1}$ is the sequence of the eigenvalues of $-\Delta$ in $H_0^1(\Omega)$ then

λ is eigenvalue of $A'(0,0) \iff \exists k \geq 1$ such that $\lambda_k = \frac{a+\alpha}{\lambda} - \alpha$ or
$$\lambda_k = \frac{d+\alpha}{\lambda} - \alpha$$

$$\iff \exists k \geq 1 \text{ such that } \lambda = \frac{a+\alpha}{\lambda_k+\alpha} \text{ or } \lambda = \frac{d+\alpha}{\lambda_k+\alpha}.$$

So

$$r(A'(0,0)) = \max\left(\frac{a+\alpha}{\lambda_1+\alpha}, \frac{d+\alpha}{\lambda_1+\alpha}\right)$$

and then

$$a < \lambda_1 \text{ and } d < \lambda_1 \Rightarrow r(A'(0,0)) < 1$$

$$a > \lambda_1 \text{ or } d > \lambda_1 \Rightarrow r(A'(0,0)) > 1$$

Now we can use Theorem 10 and deduce that

$$\text{index}_{K_1}(I - A, (0,0)) = \begin{cases} 0 \text{ if } a > \lambda_1 \text{ or } d > \lambda_1 \\ 1 \text{ if } a < \lambda_1 \text{ and } d < \lambda_1 \end{cases}. \tag{26}$$

Step 4. Computing of $\deg_{K_1}(I - A, (0,0), B_M \times B_M)$.
We consider the homotopy

$$H : [0,1] \times U \to K_1$$

defined by

$$H(t,(u,v)) = ((-\Delta+\alpha I)^{-1}(u(ta+\alpha-u-cv)), (-\Delta+\alpha I)^{-1}(v(td+\alpha-v-eu))).$$

If we choose $\alpha > \max\{M(1+c), M(1+e)\}$ then $\mathcal{R}(H) \subseteq K_1$, and $(u,v) \in B_M \times B_M$ then

$$(u,v) = H(t,(u,v)) \Longleftrightarrow \begin{cases} -\Delta u = u(ta + u - cv) \text{ in } & \Omega \\ -\Delta v = v(td + v - eu) \text{ in } & \Omega \\ u,v \geq 0 & \text{in } \Omega \\ u,v = 0 & \text{on } \partial\Omega \end{cases}$$

So proceeding as in 8 it holds

$$0 \leq u(x) \leq ta < M \text{ and } 0 \leq v(x) \leq td < M \text{ in } \Omega.$$

It follows that

$$H(t,(u,v)) \neq (u,v) \; \forall t \in [0,1], (u,v) \in K_1 \text{ with } \|u\|_\infty = M \text{ or } \|v\|_\infty = M.$$

So by homotopy invariance

$$\deg_{K_1}(I - A, (0,0), B_M \times B_M) = \deg_{K_1}(I - H(0,\cdot), (0,0), B_M \times B_M).$$

By our remarks above, the fixed points of $H(0,\cdot)$ are the solutions of (8.1) in K_1 for $a = d = 0$. By (6) and (7), the trivial solution is the only solution.

$$\deg_{K_1}(I - H(0,\cdot), (0,0), B_M \times B_M) = \text{index}_{K_1}(I - H(0,\cdot), (0,0), B_M \times B_M) = 1.$$

(Here we have used Step 3 and in this case $H(0,\cdot)$ is the operator A with $a = d = 0$).

Step 5. Computing of index $_{K_1}(I - A, (\overline{u}, 0))$ and index $_{K_1}(I - A, (0, \overline{v}))$.

(1) We may use Theorem 10 and in this case the fixed point of A is the semi-trivial solution $(\overline{u}, 0)$ and

$$A'(\overline{u}, 0)(\Phi, \psi) = ((-\Delta + \alpha I)^{-1}((a + \alpha - 2\overline{u})\Phi - c\overline{u}\psi), \\ (-\Delta + \alpha I)^{-1}((d + \alpha - e\overline{u})\psi)).$$

Moreover

$$\overline{K_1}_{\{(\overline{u},0)\}} = C_0(\Omega) \oplus K$$
$$S_{\{(\overline{u},0)\}} = C_0(\Omega) \oplus \{0\}$$

because by 2

$$(K \oplus K)_{(\overline{u},0)} = K_{\overline{u}} \oplus K_0 = C_0(\Omega) \oplus K$$

We point out that $\overline{u} > 0$ in Ω.
Then $E/S_{\{(\overline{u},0)\}} = \{0\} \oplus C_0(\Omega)$ and the reduced operator

$$\widetilde{A'(\overline{u}, 0)} : C_0(\Omega) \to C_0(\Omega)$$

is defined by

$$\widetilde{A'(\overline{u}, 0)}(\psi) = (-\Delta + \alpha I)^{-1}((d + \alpha - e\overline{u})\psi).$$

(Note that $\{0\} \oplus C_0(\Omega)$ is a suitable complement to $S_{\{(\overline{u},0)\}}$)

(2) First of all we check that, if $e \neq \bar{e}$

$$A'(\bar{u}, 0)(\Phi, \psi) \neq (\Phi, \psi), \qquad \forall (\Phi, \psi) \in (C_0(\Omega) \oplus K) \setminus \{(0, 0)\}. \qquad (27)$$

If there exists $\psi \in K \setminus \{0\}$ such that

$$(-\Delta + \alpha I)^{-1}((d + \alpha - e\bar{u})\psi) = \psi$$

then ψ is a solution of

$$\begin{cases} -\Delta\psi + (e\bar{u} - d)\psi = 0 \text{ in } & \Omega \\ \psi \geq 0 & \text{in } \Omega \\ \psi = 0 & \text{on } \partial\Omega \end{cases}$$

and in particular by 26 it follows that $\lambda(e) = 0$ and hence $e = \bar{e}$. Hence if $e \neq \bar{e}$, if $(\Phi, \psi) \in C_0(\Omega) \oplus K$ and if $(\Phi, \psi) = A'(\bar{u}, 0)(\Phi, \psi)$, then $\psi = 0$, and

$$\begin{cases} -\Delta\Phi = (a - 2\bar{u})\Phi \text{ in } & \Omega \\ \Phi = 0 & \text{on } \partial\Omega \end{cases}$$

We delay to Remark 27 the proof that this implies $\Phi = 0$ and hence inequality (8.14) holds.

(3) Now we compute the spectral radius of the operator $\widehat{A'(\bar{u}, 0)}$.
By our earlier comments on the complement of $\overline{S}_{\{(\bar{u},0)\}}$ this is equivalent to calculating the spectral radius of $(-\Delta + \alpha I)^{-1}((d + \alpha - e\bar{u})I)$ on $C_0(\Omega)$. Now μ is an eigenvalue of this operator if and only if

$$\exists \psi \in C_0(\Omega), \psi \neq 0 : \frac{(d + \alpha - e\bar{u})}{\mu}\psi = -\Delta\psi + \alpha\psi. \qquad (28)$$

Denote with $\mu = \mu(e)$ the eigenvalue of (28) corresponding to a positive eigenfunction. Note that this eigenvalue problem is of the form in (18) provided by replace μ by μ^{-1} and assume that $d + \alpha - e\bar{u}$ has a positive lower bound on Ω. Note also by (18) $\mu(e)$ will be the largest eigenvalue of $\widehat{A'(\bar{u}, 0)}$ and hence will be the spectral radius. Note also that $\mu(e)$ also depends on α but we suppress this dependence. (We see below that whether $\mu(e) < 1$ is independent of α).
Let us now point out a very important fact that links the eigenvalue $\mu(e)$ with the first eigenvalue $\lambda(e)$ of (20):

$$\mu(e) = 1 \Leftrightarrow \exists \psi \in K \setminus \{0\} : -\Delta\psi + (e\bar{u} - d)\psi = 0 \Leftrightarrow \lambda(e) = 0 \Leftrightarrow e = \bar{e}.$$

Thus $\mu(\bar{e}) = 1$.
Since $\mu(e)$ is strictly decreasing in e,

$$\text{if } e > \bar{e} \Rightarrow \mu(e) < 1 \Rightarrow r\left(\widehat{A'(\bar{u}, 0)}\right) < 1$$

$$\text{if } e < \bar{e} \Rightarrow \mu(e) > 1 \Rightarrow r\left(\widehat{A'(\bar{u}, 0)}\right) > 1.$$

By (b) of Theorem 10

$$\text{if } e < \bar{e} \Rightarrow \text{index}_{K_1}(I - A, (\bar{u}, 0)) = 0.$$

Assume $e > \bar{e}$. Let us now use (a) of Theorem 10. First of all we need to prove that

$$A'(\bar{u}, 0)(\Phi, \psi) \neq (\Phi, \psi), \qquad \forall (\Phi, \psi) \in E \setminus \{(0, 0)\}. \qquad (29)$$

By a similar argument to that (2) of Step 5, this reduces to showing that the following problem has only the trivial solution in $C_0(\Omega)$. Assume that $\psi \in C_0(\Omega) \setminus \{0\}$ is a solution of

$$\begin{cases} -\Delta\psi + (e\bar{u} - d)\psi = 0 \text{ in } & \Omega \\ \psi = 0 & \text{on } \partial\Omega \end{cases}$$

Now $\lambda(e) \leq 0$ (since ψ is an eigenfunction corresponding to the eigenvalue zero). But if $e > \bar{e}$ then $\lambda(e) > \lambda(\bar{e}) = 0$, which is a contradiction. This proves 29. Then

$$\text{index}_{K_1}(I - A, (\bar{u}, 0)) = \text{index}_{S_{\{(\bar{u}, 0)\}}}(I - A'(\bar{u}, 0), 0)$$
$$= \deg\left(I - \left((-\Delta + \alpha I)^{-1}((a + \alpha - 2\bar{u})I)\right), 0, B_\delta\right)$$
$$= 1$$

by remark 27. (See later).
So if $a > \lambda_1$

$$\text{index}_{K_1}(I - A, (\bar{u}, 0)) = \begin{cases} 0 & \text{if } e < \bar{e} \\ 1 & \text{if } e > \bar{e} \end{cases}.$$

Step 6. Computation of $\text{index}_{K_1}(I - A, (0, \bar{v}))$.
It holds that (by proceeding as Step 5) if $d > \lambda_1$

$$\text{index}_{K_1}(I - A, (0, \bar{v})) = \begin{cases} 0 & \text{if } c < \bar{c} \\ 1 & \text{if } c > \bar{c} \end{cases}.$$

Step 7. Completion of the proof.
If $a > \lambda_1$ and $d > \lambda_1$ the problem has the trivial solution $(0, 0)$ and the semi-trivial solutions $(\bar{u}, 0)$ and $(0, \bar{v})$.
By our previous computations it follows that if there are no other solutions then

$$1 = \deg_{K_1}(I - A, (0, 0), B_M \times B_M) = \text{index}_{K_1}(I - A, (0, 0)) +$$

$$+\text{index}_{K_1}(I - A, (\bar{u}, 0)) + \text{index}_{K_1}(I - A, (0, \bar{v})) = \begin{cases} 0 & \text{if } e < \bar{e}, c < \bar{c} \\ 2 & \text{if } e > \bar{e}, c > \bar{c} \end{cases}$$

which gives a contradiction. This completes the proof.

$\square$

We point out that if $e > \bar{e}$ and $c < \bar{c}$ or if $e < \bar{e}$ and $c > \bar{c}$, the method does not prove the existence of a positive solution and indeed there may or may not be a positive solution in this case. These cases are discussed more in [Dan91]. The method is quite flexible and can also be used for many other population problems (for example predator-prey problems as in [Dan85]), and other boundary conditions.

By using the same arguments as in the proof of Theorem 14 we can establish a uniqueness theorem for Problem 19. We start with two useful remarks.

Remark 27. If we define $L : C_0(\Omega) \to C_0(\Omega)$ by

$$L(\varphi) = (-\Delta + \alpha I)^{-1}((a + \alpha - 2\bar{u})\varphi)$$

then $I - L$ is invertible and

$$\deg(I - L, 0, B_\delta) = 1.$$

Consider the homotopy $H : [0, 1] \times B_\delta \to C_0(\Omega)$ defined by

$$H(t, \varphi) = (-\Delta + \alpha I)^{-1}(t(a + \alpha - 2\bar{u})\varphi).$$

We want to prove that

$$H(t, \varphi) \neq \varphi, \qquad \forall t \in [0, 1], \, \varphi \in C_0(\Omega) : \|\varphi\|_\infty = \delta.$$

If, by contradiction, there exists $t \in [0, 1], \varphi \in C_0(\Omega), \|\varphi\|_\infty = \delta$ such that

$$\begin{cases} -\Delta\varphi + \alpha\varphi = t(a + \alpha - 2\bar{u})\varphi \text{ in } & \Omega \\ \varphi = 0 & \text{on } \partial\Omega \end{cases}$$

then $\lambda = 1$ is an eigenvalue of the problem

$$\begin{cases} -\Delta\varphi + \alpha\varphi = \lambda t(a + \alpha - 2\bar{u})\varphi \text{ in } & \Omega \\ \varphi = 0 & \text{on } \partial\Omega \end{cases}.$$

Denote with $\lambda_1\,(t(a + \alpha - 2\bar{u}))$ the first eigenvalue of this problem, where we assume $a + \alpha - 2\bar{u}$ has a positive lower bound on Ω. Since $t(a + \alpha - 2\bar{u}) \leq a + \alpha - 2\bar{u} < a + \alpha - \bar{u}$ then

$$\lambda_1\,(t(a + \alpha - 2\bar{u})) > \lambda_1(a + \alpha - \bar{u})$$

because $\bar{u} > 0$ in Ω and $\alpha > 0$ satisfies $a + \alpha - 2\bar{u} > 0$ in Ω. But $\lambda_1(a + \alpha - \bar{u}) = 1$ since $\bar{u}$ is a positive solution of

$$\begin{cases} -\Delta\bar{u} = \bar{u}(a - \bar{u}) \text{ in } & \Omega \\ \bar{u} = 0 & \text{on } \partial\Omega \end{cases}.$$

It follows that $\lambda_1\,(t(a + \alpha - 2\bar{u})) > 1$ which contradicts the existence of the eigenvalue 1.

Finally, by homotopy invariance it holds

$$\deg(I - L, 0, B_\delta) = \deg(I, 0, B_\delta) = 1.$$

(The requirement that $a + \alpha - 2\bar{u} > 0$ can be replace by $\alpha \geq 0$ by a simple homotopy in α).

Remark 28. If $a < \lambda_1$, $d < \lambda_1$ and $L : C_0(\Omega) \times C_0(\Omega) \to C_0(\Omega) \times C_0(\Omega)$ is such that

$$L(\varphi, \psi) = \left((-\Delta + \alpha I)^{-1}((a + \alpha)\varphi), (-\Delta + \alpha I)^{-1}((d + \alpha)\psi)\right)$$

then

$$\deg(I - L, (0, 0), B_\delta) = 1.$$

If we define $L_a : C_0(\Omega) \to C_0(\Omega)$ such that

$$L_a(\varphi) = (-\Delta + \alpha I)^{-1}((a + \alpha)\varphi)$$

it is sufficient to prove that if $a > \lambda_1$

$$\deg(I - L_a, 0, B_\delta) = 1$$

because then we can use the product property since $L = (L_a, L_d)$. Consider the homotopy $H : [0, 1] \times B_\delta \to C_0(\Omega)$ defined by

$$H(t, \varphi) = (-\Delta + \alpha I)^{-1}(t(a + \alpha)\varphi).$$

We want to prove that

$$H(t, \varphi) \neq \varphi, \qquad \forall t \in [0, 1], \ \varphi \in C_0(\Omega) : \|\varphi\|_\infty = \delta.$$

If, by contradiction, there exists $t \in [0, 1]$, $\varphi \neq 0$ such that

$$\begin{cases} -\Delta\varphi + \alpha\varphi = t(a + \alpha)\varphi \ \text{in} \ \ \Omega \\ \varphi = 0 \qquad\qquad\qquad\quad \text{on} \ \partial\Omega \end{cases}$$

then $t(a + \alpha) - \alpha$ is an eigenvalue of $-\Delta$.
But

$$t(a + \alpha) - \alpha = (t - 1)\alpha + ta \leq ta \leq a < \lambda_1$$

which leads to a contradiction.
Then the homotopy is well defined and

$$\deg(I - L_a, 0, B_\delta) = \deg(I, 0, B_\delta) = 1.$$

Now we get the uniqueness theorem about Problem 19. (Note that this result can be proved in many other ways). Most of the arguments we use are simpler versions of arguments we have already used.

Theorem 15. *Assume that $a > \lambda_1$.*
Then the problem

$$\begin{cases} -\Delta u = u(a - u) \ \text{in} \ \ \Omega \\ u \geq 0 \qquad\qquad\quad in \ \ \Omega \\ u = 0 \qquad\qquad\quad on \ \partial\Omega \end{cases} \tag{30}$$

has a unique non-trivial solution.

Proof. **Step 1.** With the same procedure used before we can construct an operator

$$A : \{u \in K : \|u\|_\infty \le M\} \to K$$

defined by

$$A(u) = (-\Delta + \alpha I)^{-1}((a + \alpha - u)u)$$

whose fixed points are the solutions of (30).

(We recall that M is chosen greater than a and if $0 \le u(x) \le M$ in Ω then $a + \alpha - u > 0$ in Ω.)

Step 2. We prove that each non-trivial fixed point u of A satisfies

$$\text{index}_K(I - A, u) = 1.$$

We use Theorem 10. (It follows that u is also an isolated fixed point of A). Let us point out that for the extended operator $A : C_0(\Omega) \to C_0(\Omega)$ it holds

$$A'(u)(\varphi) = (-\Delta + \alpha I)^{-1}((a + \alpha - 2u)\varphi), \qquad \forall \varphi \in C_0(\Omega).$$

If u is a solution of (30), then $u > 0$ in Ω (by the maximum principle) and thus $\overline{K}_u = C_0(\Omega)$ and $S_u = C_0(\Omega)$. Step 2 now follows by combining Theorem 10 (2)(a) and Remark 27.

Step 3. We prove that

$$\deg_K(I - A, 0, B_M) = 1.$$

Let us consider the homotopy $H : [0, 1] \times \{u \in K : \|u\|_\infty \le M\} \to K$ defined by

$$H(t, u) = (-\Delta + \alpha I)^{-1}((ta + \alpha - u)u).$$

Proceeding as before we prove that the values of H are really taken in K and that

$$H(t, u) \ne u, \qquad \forall t \in [0, 1], u \in K : \|u\|_\infty = M$$

and then we can use homotopy invariance and so

$$\deg_K(I - A, 0, B_M) = \deg_K(I - H(0, \cdot), 0, B_M).$$

Since (30) has only the trivial solution if $a = 0$ then

$$\deg_K(I - H(0, \cdot), 0, B_M) = \text{index}_K(I - H(0, \cdot), 0) = 1$$

by step 4 that we are going to prove. (Remember that $a = 0$ in this case).

Step 4. We prove that

$$\text{index}_K(I - A, 0) = \begin{cases} 1 & \text{if } a < \lambda_1 \\ 0 & \text{if } a > \lambda_1. \end{cases}$$

We want to use Theorem 10.

In this case $A'(0) : C_0(\Omega) \to C_0(\Omega)$ is defined by

$$A'(0)(\Phi) = (-\Delta + \alpha I)^{-1}((a + \alpha)\Phi), \qquad \forall \Phi \in C_0(\Omega).$$

Moreover $\overline{K}_0 = K$ and $S_0 = \{0\}$ and $\widetilde{C_0(\varOmega)/S_0} = C_0(\varOmega)$ and the reduced operator is $\widetilde{A'(0)} = A'(0)$.

Condition (vi) of 10 is verified because $a \neq \lambda_1$ and then it is not possible that there exists $\varPhi \in K \setminus \{0\}$ such that $\varPhi = \widetilde{A'(0)}(\varPhi)$ because this is equivalent to

$$\begin{cases} -\varDelta\varPhi = a\varPhi & \text{in} \quad \varOmega \\ \varPhi \geq 0 & \text{in} \ \partial\varOmega \\ \varPhi = 0 & \text{on} \ \partial\varOmega. \end{cases}$$

Now it is easy to see that

$$r(A'(0)) = \frac{a + \alpha}{\lambda_1 + \alpha}$$

where λ_1 is the first eigenvalue of $-\varDelta$ and so by 10(b) if $a > \lambda_1$ then

$$\text{index}_K(I - A, 0) = 0.$$

Instead by 10(a) if $a < \lambda_1$ then

$$\text{index}_K(I - A, 0) = \text{index}\,(I - A'(0), 0) = 1.$$

The last equality holds because $r(A'(0)) < 1$.

Step 5. The non-trivial solutions of (30) with $a > \lambda_1$, being isolated fixed points of a completely continuous operator A are finite and then using the excision property it holds

$$\deg_K(I - A, 0, B_M) = \text{index}_K(I - A, 0) + \sum_{i=1}^{k} \text{index}_K(I - A, u_i)$$

where $u_1, \ldots, u_k$ are the non-trivial solutions of (30).

But by the previous steps we have that $1 = 0 + k$ and then (30) has unique non-trivial solution.

$\square$

Remark 29 (Open problems).

(1) If $a \neq d$, $a > \lambda_1$ and $d > \lambda_1$, does there always exist c, e with $c < \bar{c}$ and $e > \bar{e}$ (or $c > \bar{c}$ and $e < \bar{e}$) for which there is a coexisting solution? For "most' pairs (a, d) this is proved in [Dan91].

(2) If $e < \bar{e}$ and $c < \bar{c}$ is it possible to decide when uniqueness holds for the co-existing state? If $n \geq 2$, there are counterexamples to uniqueness in [Dan91]. However, we do not know whether uniqueness holds if $\varOmega$ is convex (in particular, in the one dimensional case).

9 Asymptotic behaviour of the solutions

In this section we describe the asymptotic behaviour of the solutions of the competing species system when the interactions are large. This part follows [DD94a]. We study

$$\begin{cases} -\Delta u = u(a - u - cv) \text{ in } & \Omega \\ -\Delta v = v(d - v - eu) \text{ in } & \Omega \\ u \geq 0, v \geq 0 & \text{in } \Omega \\ u = 0, v = 0 & \text{on } \partial\Omega \end{cases} \tag{31}$$

with $\Omega \subset \mathbf{R}^n$ is a bounded domain with smooth boundary and a, c, d, e are positive constants.

In Theorem 14 we have proved that the problem (31) has a "*coexisting solution*" when c and e are large.
Now we are interested in the behaviour of such a solution near $+\infty$.

Theorem 16. *Assume that*

$$\lim_{n \to +\infty} c_n = +\infty, \qquad \lim_{n \to +\infty} e_n = +\infty, \qquad \lim_{n \to +\infty} \frac{c_n}{e_n} = \alpha \in (0, +\infty)$$

and let (u_n, v_n) be a "coexisting solution" of (31).

Then (up to a subsequence) one of the following cases holds:

(I) $\|u_n\|_\infty \geq \alpha > 0, \|v_n\|_\infty \geq \alpha > 0, \qquad \forall n \in \mathbf{N}.$

$\lim_{n \to +\infty} u_n = \hat{u}, \lim_{n \to +\infty} v_n = \hat{v}$ *in* $L^2(\Omega)$, $\hat{u} \neq 0, \hat{v} \neq 0$ *and* $\hat{u} \cdot \hat{v} = 0.$

If $w = \alpha^{-1}\hat{u} - \hat{v}$ *then* $w^+ = \alpha^{-1}\hat{u}$ *and* $w^- = -\hat{v}$ *and* w *is a changing sign solution of*

$$\begin{cases} -\Delta w = aw^+ + dw^- - \alpha(w^+)^2 + (w^-)^2 \text{ in } & \Omega \\ w = 0 & \text{on } \partial\Omega \end{cases} \tag{32}$$

(II) $\lim_{n \to +\infty} e_n u_n = \tilde{u}, \lim_{n \to +\infty} c_n v_n = \tilde{v}$ *in* $H_0^1(\Omega)$, $\tilde{u} \neq 0, \tilde{v} \neq 0.$

The pair $(\tilde{u}, \tilde{v})$ solves the problem

$$\begin{cases} -\Delta u = u(a - v) \text{ in } & \Omega \\ -\Delta v = v(d - u) \text{ in } & \Omega \\ u \geq 0, v \geq 0 & \text{in } \Omega \\ u = 0, v = 0 & \text{on } \partial\Omega \end{cases} \tag{33}$$

(III) $\lim_{n \to +\infty} (\|u_n\|_\infty + \|v_n\|_\infty) = 0, \lim_{n \to +\infty} e_n \|u_n\|_\infty = +\infty \lim_{n \to +\infty} c_n \|v_n\|_\infty = +\infty$

and then

$$\lim_{n \to +\infty} \frac{u_n}{\|u_n\|_\infty} = \overline{u}, \lim_{n \to +\infty} \frac{v_n}{\|v_n\|_\infty} = \overline{v} \text{ in } L^2(\Omega), \overline{u} \neq 0, \overline{v} \neq 0, \overline{u} \cdot \overline{v} = 0.$$

If $w = \overline{u} - \overline{v}$ then $w^+ = \overline{u}$ and $w^- = -\overline{v}$ and w is a solution of

$$\begin{cases} -\Delta w = aw^+ + dw^- \text{ in } \Omega \\ w = 0 \quad\quad\quad\quad\quad \text{ on } \partial\Omega \end{cases} \tag{34}$$

Proof (Sketch). It is possible to prove that (see Lemma (2.1) of [DD94a])

$$\exists M_1, M_2 > 0: M_1 \leq \frac{c_n \|v_n\|_\infty}{e_n \|u_n\|_\infty} \leq M_2, \quad\quad \forall n \in \mathbf{N}$$

and then only two cases may happen:

(a) the sequence $(c_n \|v_n\|_\infty + e_n \|u_n\|_\infty)_{n \geq 1}$ is bounded or
(b) $\lim_{n \to +\infty} c_n \|v_n\|_\infty = \lim_{n \to +\infty} e_n \|u_n\|_\infty = +\infty$.

In the first case, it is easy to see that $\{e_n u_n\}$ and $\{c_n v_n\}$ are bounded in $H_0^1(\Omega)$ and it is easy to pass to the limit. (The only problem is to ceck that $\tilde{u} \neq 0$ and $\tilde{v} \neq 0$). In case (b), the idea is to multiply the first equation of (31) by u_n to bound u_n in $H_0^1(\Omega)$ and similary for v_n. Thus after taking subsequences, we deduce $u_n \to \hat{u}$ and $v_n \to \hat{v}$ weakly in $H_0^1(\Omega)$. If we multiply the first equation of (31) by $\Phi \in C_0^\infty(\Omega)$ and pass to the weak limit (and use that $c_n \to \infty$) we deduce that $\hat{u}\hat{v} = 0$. Hence if we define $w = \alpha^{-1}\hat{u} - \hat{v}$, $w^+ = \alpha^{-1}\hat{u}$, $w^- = -\hat{v}$. Lastly $\frac{e_n}{c_n} u_n - v_n$ converges weakly to w and

$$-\Delta(\frac{e_n}{c_n} u_n - v_n) = \frac{ae_n}{c_n} u_n - dv_n - \frac{e_n}{c_n} u_n^2 + v_n^2$$

and we can pass to the weak limit to obtain the equation (32) for w.(If $\hat{u}$ or $\hat{v}$ are zero, we need to be more careful, we need to use the estimate at the beginning of the proof and we need to consider $(\|u_n\|_\infty)^{-1}u_n$ and $(\|v_n\|_\infty)^{-1}v_n$). □

Let us now see an inverse result, that is the existence of a *"coexisting solution"* generated by the limit equations. These are proved by degree arguments (rather complicated in the case of Theorem 17).

Theorem 17. *Let w_0 be an isolated changing sign solution, with non-zero index in $C_0(\Omega)$ of the problem*

$$\begin{cases} -\Delta w = aw^+ + dw^- - \alpha(w^+)^2 + (w^-)^2 \text{ in } \Omega \\ w = 0 \quad\quad\quad\quad\quad\quad\quad\quad\quad\quad\quad \text{ on } \partial\Omega \end{cases}. \tag{35}$$

Then if c, e are large and $\frac{c}{e}$ is "near" α there exists a "coexisting solution" (u, v) of (31) with u "near" $\alpha^{-1}w_0^+$ and v "near" $-w_0^-$ in $L^2(\Omega)$.

(See Theorem (3.3) of [DD94a]).

Theorem 18. *Let (u_0, v_0) be an isolated positive solution, with non-zero index in $K_1 = K \oplus K$ of the problem*

$$\begin{cases} -\Delta u = u(a - v) \text{ in } \quad \Omega \\ -\Delta v = v(d - u) \text{ in } \quad \Omega \\ u = 0, v = 0 \quad \text{ on } \partial\Omega \end{cases} \tag{36}$$

Then if c, e are large there exists a "coexisting solution" (u, v) of (31) with eu "near" u_0 and cv "near" v_0 in $L^2(\Omega)$.

(See Theorem (3.1) of [DD94a]).

Remark 30 (Open problems).

(1) Has the system (36) a positive solution if $a > \lambda_1$ and $d > \lambda_1$?

We only know (see theorem (3.2) of [DD94a]) that if the problem

$$\begin{cases} -\Delta u = au^+ + du^- \text{ in } \quad \Omega \\ u = 0 \quad \text{ on } \partial\Omega \end{cases}$$

has only the trivial solution and this solution has non-zero index in $C_0(\Omega)$, then (36) has a positive solution. This is by a degree argument. (The corresponding problem for Neumann boundary conditions is trivial. Note that the theory in this section can be extended to Neumann boundary conditions.)

(2) We consider when the problem

$$\begin{cases} -\Delta w = aw^+ + dw^- - \alpha(w^+)^2 + (w^-)^2 \text{ in } \quad \Omega \\ w = 0 \quad \text{ on } \partial\Omega \end{cases} \tag{37}$$

has a sign changing solution. It is easy to see that if $a > \lambda_1$ then there exists a positive solution and if $d > \lambda_1$ then there exists a negative solution. Such solutions are local minima of the functional associated to (37).

We know that in the (a, d)-plane there is a curve γ_2 of the Fučik spectrum) contained in the region $\{(a, d) \in \mathbf{R}^2 : a > \lambda_1, d > \lambda_1\}$, crossing the point (λ_2, λ_2) which has the horizontal line $d = \lambda_1$ as asymptote when $a \to +\infty$ and the vertical line $a = \lambda_1$ as asymptote when $d \to +\infty$.
Now if (a, d) is above γ_2 then (37) has at least a changing sign solution, while if (a, d) is on or under γ_2 then (37) has no changing sign solution. (Theorems (2.5) and (3.2) of [DD94b]). The interesting open problem is when is there a stable sign changing solution. Some partial results appear in [DG95]. The reason for the interest in this problem is that, under natural hypotheses, these generate the stable coexisting solutions of 31. (See [DG94]). By stable, we mean stable for the natural corresponding parabolic system.

Nonlinear elliptic equations involving critical Sobolev exponents

D. Passaseo

1 Introduction and statement of the problems

The purpose of these notes is to present a survey of some recent results dealing with existence, nonexistence and multiplicity of nontrivial solutions for semilinear elliptic equations, whose nonlinear term has critical or supercritical growth. Let us consider, for example, the following Dirichlet problem

$$P(\Omega, p) \begin{cases} \Delta u + |u|^{p-2}u = 0 & \text{in } \Omega \\ u = 0 & \text{on } \partial\Omega \\ u \not\equiv 0 & \text{in } \Omega, \end{cases} \tag{1.1}$$

where Ω is a smooth bounded domain of $\mathbf{R}^n$, $n \geq 3$, and $p \geq \frac{2n}{n-2}$ ($\frac{2n}{n-2} = 2^*$ is the critical Sobolev exponent).

In particular we are interested to find positive solutions of $P(\Omega, p)$ or also sign changing (nodal) solutions with a prescribed number of nodal regions. This equation is a simplified model of some variational problems, coming from Differential Geometry, Mechanics, Mathematical Phisics, Chemistry, whose common feature is the lack of compactness: for example a well known problem in Differential Geometry, the Yamabe's problem (see [Aub76, Sch84, Tru68, Yam60], is related to the solvability of a problem like (1.1) with $p = 2^*$; supercritical nonlinearities arise in some combustion models; lack of compactness also occur in Yang–Mills equations, etc... In our problem the lack of compactness is due to the presence of critical or supercritical exponents: it is well known that $H_0^{1,2}(\Omega)$ is continuously embedded in $L^p(\Omega)$ for $p \leq 2^*$ and that the embedding is compact only for $p < 2^*$. Since the nonlinear term in (1.1) is homogeneous, one can easily verify that solving problem $P_a(\Omega, p)$ is equivalent to finding critical points for the energy functional

$$f(u) = \int_\Omega |Du|^2 dx,$$

constrained on the manifold

$$M_p(\Omega) = \{u \in H_0^{1,2}(\Omega) : \int_\Omega |u|^p dx = 1\}.$$

There is a sharp contrast between the cases $p < 2^*$ and $p \geq 2^*$. If $p < 2^*$, then the infimum $\inf_{M_p(\Omega)} f$ is achieved by a positive function, giving rise to a positive solution of $P(\Omega, p)$, independently of the shape of the domain Ω (indeed, one can find infinitely many solutions exploiting the symmetry properties of f and

$M_p(\Omega)$). On the contrary, if $p \geq 2^*$, the infimum $\inf\limits_{M_p(\Omega)} f$ is not achieved (as we shall see below). Hence the problem cannot be simply solved by minimization arguments and the solutions (when there exist) correspond to higher critical values. But several difficulties also arise when trying to find critical points by means of the usual topological methods of the Calculus of Variations (like Morse Theory, Ljusternik–Schnirelman category, linking methods, etc ...), since the corresponding functional does not satisfy the Palais–Smale compactness condition when $p \geq 2^*$. It is not only a problem of methods: there is a deep reason which explains the impossibility of applying these methods in a standard way. In fact every solution of problem

$$\begin{cases} \Delta u + g(u) = 0 & \text{in } \Omega \\ u = 0 & \text{on } \partial\Omega \end{cases} \tag{1.2}$$

must verify the following Pohozaev's identity (see [Poh65]):

$$(1 - \frac{n}{2}) \int_\Omega g(u)u \, dx + n \int_\Omega G(u) dx = \frac{1}{2} \int_{\partial\Omega} (x \cdot \nu)(\frac{\partial u}{\partial \nu})^2 d\sigma \tag{1.3}$$

where $G(u) = \int_0^u g(t)dt$ and ν denotes the outward normal to $\partial\Omega$. As a consequence (for $g(u) = |u|^{p-2}u$) we have the following nonexistence result:

Theorem 1 (Pohozaev [Poh65]). *If Ω is a star–shaped domain and $p \geq 2^*$, then the problem $P(\Omega, p)$ has no solution.*

After Pohozaev's Theorem, the researches in this topics followed two directions:

(i) exploiting the shape of the domain Ω in order to regain the existence of solutions,

(ii) modifying the equation by lower–order terms .

The first direction of research is supported by the following observation: assume Ω is an annulus (i.e. $\Omega = \{x \in \mathbf{R}^n : 0 < r_1 < |x| < r_2\}$); then, exploiting the radial symmetry of Ω, it is easy to see (as pointed out by Kazdan and Warner [KW75]) that $P(\Omega, p)$ has a positive radial solution and infinitely many nodal radial solutions for all p. This leads to a natural question, pointed out by Nirenberg (see [Bre86]): what happens if Ω has the same topology of an annulus, but not the same radial symmetry properties? is there still a positive solution, at least for $p = 2^*$? and if we assume only that Ω is not contractible, in itself, to a point? This question has been answered by Bahri and Coron in [BC88], where the following theorem is proved (see also [Cor84] and [Rey89b]).

Theorem 2 (Bahri–Coron [BC88]). *Assume Ω is a smooth bounded domain of $\mathbf{R}^n$, having non trivial topology (i.e. there exists an integer $k \geq 1$ such that either $H_{2k-1}(\Omega, \mathbb{Q}) \neq 0$ or $H_k(\Omega, \mathbb{Z}/2\mathbb{Z}) \neq 0$). Then problem $P(\Omega, 2^*)$ has at least one positive solution.*

Remark 1. It is clear that any domain with nontrivial topology is not contractible in itself to a point. When $n = 3$, the converse is also true. On the contrary, when $n \geq 4$ the converse fails (i.e. there exist noncontractible domains with trivial homology groups). Thus, if $n \geq 4$, it is still an open problem whether the conclusion of Theorem 2 holds under the sole assumption that Ω is not contractible. However, note that the assumption "Ω has nontrivial topology" covers a large variety of domains.

After the results of Pohozaev and Bahri–Coron, the following two natural questions arise (see [Bre86]):

Question 1 (Brezis). Assume $p = 2^*$. Can one replace in Pohozaev's Theorem the assumption "Ω is star–shaped" by "Ω has trivial topology"? In other words, are there domains Ω with trivial topology on which $P(\Omega, 2^*)$ has a positive solution?

Question 2 (Rabinowitz). What happens when $p > \frac{2n}{n-2}$? Pohozaev's Theorem still holds. On the other hand, if Ω is an annulus, it is easy to see that $P(\Omega, p)$ has radial solutions for all p. So the question is: assuming Ω is a domain with nontrivial topology, is there still a solution of $P(\Omega, p)$ for all p?

Now let us consider the effect of lower–order terms: we deal with the problem

$$P_a(\Omega, 2^*) \quad \begin{cases} \Delta u - a(x)u + |u|^{2^*-2}u = 0 & \text{in } \Omega \\ u = 0 & \text{on } \partial\Omega \\ u \not\equiv 0 & \text{in } \Omega, \end{cases} \qquad (1.4)$$

where $a(x) \in L^{n/2}(\Omega)$. Motivations for the study of this problem come, for example, from this simple observation: assume Ω is any bounded domain and denote by $\lambda_1 < \lambda_2 \leq \lambda_3 \ldots$ the eigenvalues of $-\Delta$ with zero Dirichlet boundary condition; then general bifurcation results (see [Böh72], [Mar73], [Rab73]) guarantee that $P_a(\Omega, 2^*)$ has solution if $a(x) \equiv -\lambda$, where λ is a constant sufficiently close to the eigenvalues λ_i of $-\Delta$ in $H_0^{1,2}(\Omega)$. In particular, if $\lambda < \lambda_1$ and $|\lambda - \lambda_1|$ is small enough, then there exists a positive solution. On the other hand this equation is related to the solution of Yamabe's problem (see [Aub76], [Sch84] [Tru68], [Yam60]), where the coefficient $a(x)$ represent a scalar curvature.

The first results in this direction have been stated by Brezis and Nirenberg (see [Bre86], [BN83]). The energy functional related to problem $P_a(\Omega, 2^*)$ is

$$f_a(u) = \int_\Omega [|Du|^2 + a(x)u^2]dx; \qquad (1.5)$$

the Pohozaev's identity, satisfied by the solutions of $P_a(\Omega, 2^*)$, becomes

$$\int_\Omega [a + \frac{1}{2}(x \cdot Da)]u^2 dx + \frac{1}{2}\int_\Omega (x \cdot \nu)(\frac{\partial u}{\partial \nu})^2 d\sigma = 0. \qquad (1.6)$$

If $a(x) < 0$ somewhere in Ω and $n \geq 4$, then the infimum $\inf_{M_{2^*}(\Omega)} f_a$ is achieved (the situation is more complicated in the case $n = 3$). On the contrary, if $a(x) \geq 0$

everywhere in Ω, then the infimum is not achieved, but problem $P_a(\Omega, 2^*)$ may still have positive solutions. In fact, as showed in [Bre86], it is easy to construct such an example, where $a(x) > 0$ in Ω and $P_a(\Omega, 2^*)$ has a positive solution (see section 3). This leads to the following natural question:

Question 3 (Brezis [Bre86]). Find general conditions on the nonnegative function $a(x)$ which guarantee the existence of solutions for $P_a(\Omega, 2^*)$, independently of the domain's shape (even in star–shaped domains). Note that, if Ω is star–shaped and $P_a(\Omega, 2^*)$ has solution, then $a(x)$ cannot be a positive constant, because of Pohozaev's identity (1.6).

2   Effect of the domain's shape

In this section we are concerned with the case $a(x) = 0$ and $p \geq 2^*$; Questions 1 and 2 are answered.

Notice that $\inf_{M_p(\Omega)} f = 0$ for all $p > 2^*$ (so this infimum cannot be achieved). For $p = 2^*$ we have

$$\inf_{M_{2^*}(\Omega)} f = S, \qquad (2.1)$$

where S is the best constant for the Sobolev embedding $H_0^{1,2}(\Omega \hookrightarrow L^{2^*}(\Omega))$. It is well known that S is independent of Ω and depends only on the dimension n: this property is an easy consequence of the fact that the ratio $\|Du\|_2 / \|u\|_{2^*}$ is invariant under dilations and translations. Moreover S cannot be achieved in any bounded domain Ω, otherwise (extending a minimizing function by zero outside Ω) it should be achieved even in any star–shaped domain containing Ω, in contradiction with Pohozaev's nonexistence result. S is attained only when $\Omega = \mathbf{R}^n$ and the minimizing function is unique, modulo translations and dilations (see [BL83b], [Lio85], [Tal76]).

Answer to Question 1

The first attempts to answer Question 1 are some results (by Carpio Rodriguez, Comte, Lewandowski, Schaaf) extending Pohozaev's nonexistence theorem to some contractible but non star–shaped domains: in [RCL92], for example, it is proved that $P(\Omega, 2^*)$ has no solution if the domain Ω is obtained removing from a sphere a frustum of cone having vertex outside the sphere (in such a way that the obtained domain is not star–shaped); also, in the case $p > 2^*$, nonexistence results hold in some dumb–bell shaped domains.

However the answer to Question 1 is negative since it is possible to prove existence results in some bounded contractible domains Ω: for example, if Ω is an annulus pierced by removing a cylinder thin enough, then $P(\Omega, 2^*)$ has positive solutions (see [Dan88], [Din89], [Pas89]). Indeed it is possible to find bounded contractible domains Ω where the number of positive solutions of $P(\Omega, 2^*)$ is arbitrarily large:

Theorem 3 (see [Pas89]). *For all positive integer h, there exists a bounded contractible domain Ω_h, such that $P(\Omega_h, 2^*)$ has at least h distinct positive solutions.*

Proof (Sketch). In order to obtain such a domain Ω_h, it suffices to argue as follows. For every positive integer h, let us consider the domain

$$T^h = \{x = (x_1, \ldots, x_n) \in \mathbf{R}^n : \sum_{i=1}^{n-1} x_i^2 < 1, \ 0 < x_n < h+1\}. \quad (2.2)$$

For all $j \in \{1, \ldots, h\}$, put $c_j = (0, \ldots, 0, j) \in \mathbf{R}^n$; fixed $\sigma_1, \ldots, \sigma_h$ such that $0 < \sigma_j < \frac{1}{2}$ for all $j \in \{1, \ldots, h\}$, set

$$D^h = T^h \setminus \bigcup_{j=1}^{h} \overline{B(c_j, \sigma_j)} \quad (2.3)$$

and, for all $\epsilon_1, \ldots, \epsilon_h$ in $]0, 1[$, define

$$\chi_{\epsilon_j}^j = \{x \in \mathbf{R}^n : \sum_{i=1}^{n-1} x_i^2 \leq \epsilon_j^2, \ j \leq x_n \leq j+1\} \quad (2.4)$$

$$\Omega_{\epsilon_1, \ldots, \epsilon_h} = D^h \setminus \bigcup_{j=1}^{h} \chi_{\epsilon_j}^j. \quad (2.5)$$

The assertion of Theorem 3 holds with $\Omega = \Omega_{\epsilon_1, \ldots, \epsilon_h}$ when $\epsilon_1, \ldots, \epsilon_h$ are small enough. In fact $\Omega_{\epsilon_1, \ldots, \epsilon_h}$ is a bounded contractible domain and $P(\Omega_{\epsilon_1, \ldots, \epsilon_h}, 2^*)$ has at least h solutions $u_{\epsilon_1}, \ldots, u_{\epsilon_h}$. These solutions are obtained as local minimum points of the energy functional f on $M_{2^*}(\Omega)$, constrained on the subspace of the functions having radial symmetry with respect to the x_n–axis (notice that the infimum $\inf_{M_{2^*}(\Omega)} f$ is not achieved, not even in the subspace of the radial functions). Moreover, for all $j = 1, \ldots, h$, the method used in the proof shows that, as $\epsilon_j \to 0$, $u_{\epsilon_j} \to 0$ weakly in $H_0^{1,2}(\Omega)$, $f\left(\frac{u_{\epsilon_j}}{\|u_{\epsilon_j}\|_{2^*}}\right) \to S$ and the energy $|Du_{\epsilon_j}|^2$ concentrates like a Dirac mass near a point of the x_n–axis. $\quad \square$

Remark 2. In [Dan88] and [Din89] Dancer and Ding prove that the positive solution one can find in an annulus, persists if the annulus is perturbed removing a subset of small capacity; moreover the solution in the perturbed domain converges to the solution in the annulus, as the capacity of the perturbation tends to zero. Therefore in a pierced annulus, or equivalently in a domain Ω_{ϵ_1} like in the proof of Theorem 3, Dancer and Ding prove the existence of a solution $\tilde{u}_{\epsilon_1}$ (see [Dan88], [Din89]). However, let us point out that the solution $\tilde{u}_{\epsilon_1}$ obtained by Dancer and Ding is distinct from the solution u_{ϵ_1} given by Theorem 3, because $u_{\epsilon_1} \to 0$ weakly in $H_0^{1,2}(\Omega)$, while $\tilde{u}_{\epsilon_1}$ converges strongly to a solution in the limit domain (which is nontrivial in the sense of Bahri–Coron).

Thus the existence of the solutions $u_{\epsilon_1}, \ldots, u_{\epsilon_h}$ of $P(\Omega_{\epsilon_1, \ldots, \epsilon_h}, 2^*)$, which does not depend on the solvability of the limit problem, seems to be related to other new phenomena and suggests that every perturbation of a given domain, which modifies its topological properties and is obtained removing a subset having small capacity , gives rise to solutions vanishing as the capacity of the perturbation tends to zero. Indeed it is also possible to evaluate the number of positive solutions by the topological properties of the perturbation. These results can be summarized as follows (see [Pas90], [Pas98b], [Pas94], for more details).

Definition 1 (see [Fad85], [Pas98b]). *Let X be a topological space and X_1, X_2 two closed subset of X, such that $X_2 \subseteq X_1$. We say that the relative category in X of X_1 with respect to X_2 is m (and write $\mathrm{cat}_X[X_1, X_2] = m$) if m is the smallest positive integer such that*

$$X_1 = \bigcup_{s=0}^{m} F_s , \qquad X_2 \subseteq F_0,$$

where, for all $s = 0, 1, \ldots, m$, F_s is a closed subset and there exists $h_s \in C^0([0,1] \times F_s, X)$ such that

I) $h_s(0, x) = x \quad \forall x \in F_s, \quad \forall s = 0, 1, 2, \ldots, m$
II) $\forall s \geq 1 \ \exists p_s \in X : h_s(1, x) = p_s \quad \forall x \in F_s$
III) $h_0(1, x) \in X_2 \quad \forall x \in F_0;$
   $h_0(t, x) \in X_2 \quad \forall x \in F_0 \cap X_2, \quad \forall t \in [0, 1]$.

Note that $\mathrm{cat}_X[X_1, \emptyset]$ is the well known Ljusternik–Schnirelman category.

Proposition 1 (see [Pas98b]). *Let Ω be a given bounded domain in $\mathbf{R}^n$, $n \geq 3$, and K be a closed subset of Ω. Then, if the capacity of K is small enough, problem $P(\Omega \backslash K, 2^*)$ has at least $\mathrm{cat}_\Omega(\bar{\Omega}, \overline{\Omega \backslash K})$ positive solutions, which converge weakly to zero as the capacity of K tends to zero, and concentrate like a Dirac mass.*

In the previous proposition, as well as in Theorem 3, a basic tool is given by the concentration–compactness principle of Lions (see [Lio85]) or by a global compactness result of Struwe (see [Str84]), which allow us to overcome the difficulties given by the lack of compactness and, in particular, to show that the Palais–Smale condition is satisfied in the energy range $]S, 2^{2/n}S[$ (i.e. every sequence $(u_i)_i$ in $M_{2^*}(\Omega)$, such that $f(u_i) \to c \in]S, 2^{2/n}S[$ and $\mathrm{grad}_{M_{2^*}(\Omega)} f(u_i) \to 0$ in $H_0^{1,2}(\Omega)$, is relatively compact).

Notice that the method of the proof can be iterated in order to show that several independent perturbations produce several distinct positive solutions.

It is clear that this result can be applied in a large variety of geometric situations and allows to obtain an arbitrarily large number of positive solutions. In particular one can obtain multiple positive solutions in domains with several small holes like in [Rey89b] (without requiring, unlike [Rey89b], that the holes are spherical).

On the other hand one can obtain more than one solution even by a unique but topologically complex perturbation:

Example 1. Let $n = 3$ and set

$$C = \{(x_1, x_2, x_3) \in \mathbf{R}^3 : x_3 = 0, \ x_1^2 + x_2^2 = 1\}$$
$$\Omega = \{x \in \mathbf{R}^3 : \operatorname{dist}(x, C) < \frac{1}{2}\}$$

and, for $\epsilon \in \,]0, \frac{1}{2}[$,

$$\Omega_\epsilon = \{x \in \mathbf{R}^3 : \epsilon < \operatorname{dist}(x, C) < \frac{1}{2}\}.$$

Then $\operatorname{cat}_{\bar{\Omega}}(\bar{\Omega}, \bar{\Omega}_\epsilon) = 2$ and so, for $\epsilon > 0$ small enough, $P(\Omega_\epsilon, 2^*)$ has at least two solutions whose energy concentrates like Dirac mass as $\epsilon \to 0$.

Notice that Ω_ϵ has radial symmetry with respect to x_3–axis and so it is easy to find radial solutions; however, let us point out that the solutions given by Proposition 1 cannot have radial symmetry because of their asymptotic behaviour as $\epsilon \to 0$. On the other hand no symmetry assumption is required in Proposition 1, which, for example, can guarantee the existence of two positive solutions, for $\epsilon > 0$ small enough, in a domain of the form

$$\tilde{\Omega}_\epsilon = \{x \in \mathbf{R}^3 : \operatorname{dist}(x, C) < \frac{1}{2}, \ \operatorname{dist}(x, \tilde{C}) > \epsilon\},$$

where

$$\tilde{C} = \{x = (x_1, x_2, x_3) \in \mathbf{R}^3 : x_3 = 0, \ (x_1 - \frac{1}{3})^2 + x_2^2 = 1\},$$

which does not have any symmetry property.

The supercritical case: answer to Question 2

The answer to Question 2 is negative since, as we shall see below, it is possible to find pairs (Ω, p), where Ω is a smooth bounded domain of $\mathbf{R}^n$, nontrivial in the sense of Bahri–Coron, $p > \frac{2n}{n-2}$ and problem $P(\Omega, p)$ has no solution.

In the supercritical case a crucial role seems to be played by the critical exponents

$$2^*(n - k) = \frac{2(n - k)}{(n - k) - 2} \qquad k = 1, \ldots, (n - 3), \tag{2.6}$$

corresponding to lower dimensions. In fact, the possibility of finding a nontrivial domain Ω, such that $P(\Omega, p)$ has no solution, is strictly related to the position of p with respect to the critical exponents (2.6), as showed by the following result:

Theorem 4 (see [Pas93b], [Pas95]). *For every positive integer k there exists a smooth bounded domain Ω in $\mathbf{R}^n$, with $n \geq k + 3$, homotopically equivalent to the k–dimensional sphere S_k (hence nontrivial if $k \geq 1$), such that problem $P(\Omega, p)$ has no solution for $p \geq \frac{2(n-k)}{(n-k)-2}$, while, for $2 < p < \frac{2(n-k)}{(n-k)-2}$, it has infinitely many solutions and at least one of them is positive*

(see [Pas93b] and [Pas95] for detailed statements concerning more general nonlinear terms).

Proof (Sketch). For all $x = (x_1, \ldots, x_n) \in \mathbf{R}^n$, let us set

$$P_1^k(x) = (x_1, \ldots, x_k, x_{k+1}, 0, \ldots, 0) \in \mathbf{R}^n, \qquad P_2^k(x) = x - P_1^k(x)$$

and define

$$S_k = \{x \in \mathbf{R}^n : |x| = 1, \ P_2^k(x) = 0\}$$
$$T_k(\rho) = \{x \in \mathbf{R}^n : \text{dist}(x, S_k) < \rho\}.$$

If $0 < \rho < 1$, then the domain $T_k(\rho)$ is homotopically equivalent to S_k and the conclusion of the theorem holds for $\Omega = T_k(\rho)$. The proof is based on the following generalized Pohozaev's identity : for all $v \in C^1(\bar{\Omega}, \mathbf{R}^n)$, the solutions of problem $P(\Omega, p)$ must satisfy

$$\frac{1}{2} \int_{\partial \Omega} (v \cdot \nu) \left(\frac{\partial u}{\partial \nu} \right)^2 d\sigma = \int_\Omega (dv[Du] \cdot Du) dx + \int_\Omega \frac{|u|^p}{p} \, \text{div} \, v dx - \frac{1}{2} \int_\Omega |Du|^2 \, \text{div} \, v dx.$$

A suitable choice of the function v when $\Omega = T_k(\rho)$ implies that $u \equiv 0$ if $p \geq \frac{2(n-k)}{(n-k)-2}$, which is a contradiction.

On the contrary, if $2 < p < \frac{2(n-k)}{(n-k)-2}$, then one can exploit the radial symmetry of the domain $T_k(\rho)$ with respect to the co-ordinates $x_1, \ldots, x_{k+1}$ in order to find nontrivial solutions in the subspace of the radial symmetric functions (which is compactly embedded in $L^p(T_k(\rho))$ if $0 < \rho < 1$ and $p < \frac{2(n-k)}{(n-k)-2}$).

Note that no symmetry assumption is required for the nonexistence result in the case $p \geq \frac{2(n-k)}{(n-k)-2}$. □

Remark 3. The previous proposition allows us to find a nontrivial domain only when $k \geq 1$ (in fact only in this case $T_k(\rho)$ has nontrivial topology in the sense of Bahri-Coron). Therefore, in the case $n \geq 4$ and $\frac{2n}{n-2} < p < 2^*(n-1) = \frac{2(n-1)}{(n-1)-2}$, or also in the case $n = 3$ and $p > 2^*(3) = 6$, Theorem 4 does not gives counterexamples and Question 2 could have a positive answer: it is still an open problem. Moreover, let us mention the following question, which seems suggested by the proof of Theorem 4: assume Ω has the k-dimensional homology group $H_k(\Omega, \mathbf{Z}/2\mathbf{Z}) \neq 0$; does this assumption guarantee that $P(\Omega, p)$ has (positive) solutions for $2 < p < \frac{2(n-k)}{(n-k)-2}$, if $n \geq k+3$, and for all $p > 2$ if $n < k+3$?

On the other hand there exist examples of bounded contractible domains Ω (the same ones introduced above), such that $P(\Omega, p)$ has positive or nodal solutions for all $p > 2$:

Theorem 5 (see [Pas92b], [Pas96a], [Pas98a]). *Let $p > \frac{2n}{n-2}$ and $\Omega_{\epsilon_1, \ldots, \epsilon_h}$ be the bounded contractible domains above defined (see (2.5)). Then there exists $\bar{\epsilon} > 0$ such that, for all $\epsilon_1, \ldots, \epsilon_h$ in $]0, \bar{\epsilon}[$, problem $P(\Omega_{\epsilon_1, \ldots, \epsilon_h}, p)$ has at least h distinct positive solutions $u_{\epsilon_1}, \ldots, u_{\epsilon_h}$ and at least h^2 nodal solutions $u_{\epsilon_i, \epsilon_j} (i, j = 1, \ldots, h)$, having exactly two nodal regions (i.e. both $u_{\epsilon_i, \epsilon_j}^+$ and $u_{\epsilon_i, \epsilon_j}^-$*

have connected support). Moreover, for all $i = 1, \ldots, h$, $u_{\epsilon_i} \to 0$ strongly in $H_0^{1,2}$ as $\epsilon_i \to 0$ and $\frac{u_{\epsilon_i}}{\|u_{\epsilon_i}\|_p}$ concentrates like a Dirac mass; for all $i, j = 1, \ldots, h$, $u_{\epsilon_i}^+ \to 0$ in $H_0^{1,2}$ as $\epsilon_i \to 0$, $u_{\epsilon_j}^- \to 0$ in $H_0^{1,2}$ as $\epsilon_j \to 0$, and $\frac{u_{\epsilon_i}^+}{\|u_{\epsilon_i}^+\|_p}$, $\frac{u_{\epsilon_j}^-}{\|u_{\epsilon_j}^-\|_p}$ concentrate like Dirac mass.

Proof (Sketch). Let us point out that, even if the solutions we find in the supercritical case present qualitative properties and asymptotic behaviour, as $\epsilon_i \to 0$, analogous to the ones obtained in the critical case (see Theorem 3), there is a deep difference from the point of view of the variational framework. In fact in the critical case the solutions in Theorem 3 correspond to local minimum points, in the subspace of the radial functions, for the energy functional constrained on $M_{2^*}(\Omega)$. On the contrary, when $p > \frac{2n}{n-2}$, the functional f constrained on $M_p(\Omega)$ has no local minimum point, not even in the subspace of the functions having radial symmetry with respect to the x_n-axis.

The solutions in Theorem 5 are obtained using a special device: we modify the functional introducing some obstacle in order to avoid some concentration phenomena, related to the lack of compactness , then we obtain the solutions as local minimum points for the modified functional and, finally, we prove that these modifications do not change the Euler–Lagrange equation, when $\epsilon_1, \ldots, \epsilon_h$ are sufficiently small.

Moreover, the method used in the proof shows that, for all $i = 1, \ldots, h$, $\frac{u_{\epsilon_i}}{\|u_{\epsilon_i}\|_p}$ concentrates near the cylinder $\chi_{\epsilon_i}^i$ (see (2.4)) as $\epsilon_i \to 0$, while $\frac{u_{\epsilon_i, \epsilon_j}^+}{\|u_{\epsilon_i, \epsilon_j}^+\|_p}$ and $\frac{u_{\epsilon_i, \epsilon_j}^-}{\|u_{\epsilon_i, \epsilon_j}^-\|_p}$ concentrate near $\chi_{\epsilon_i}^i$ and $\chi_{\epsilon_j}^j$ respectively, as the size of these cylinders tends to zero. $\qquad\square$

Theorem 5 shows that it is possible to find an arbitrarily large number of nontrivial solutions of $P(\Omega, p)$ in correspondence to several perturbations of a given domain. On the other hand, it is possible to obtain many nontrivial solutions even in correspondence to a unique perturbation, as showed by the following theorem (but these solutions may have, conceivably, more than two nodal regions) .

Theorem 6 (see [Pas]). *Let $p > \frac{2n}{n-2}$ and, for all $\epsilon > 0$, set*

$$A = \{x \in \mathbf{R}^n : 1 < |x| < 2\}$$

$$C_\epsilon = \{x = (x_1, \ldots, x_n) \in \mathbf{R}^n : \sum_{i=1}^{n-1} x_i^2 \leq \epsilon^2, \ x_n > 0\}$$

$$\Omega_\epsilon = A \backslash C_\epsilon.$$

Then, for all positive integer k there exists $\bar\epsilon_k > 0$ such that problem $P(\Omega_\epsilon, p)$ has at least $2k$ distinct solutions for all $\epsilon \in]0, \bar\epsilon_k[$.

Moreover all the solutions tend strongly to zero in $H_0^{1,2}$ as $\epsilon \to 0$ and their energy concentrates like a Dirac mass near a point of the x_n-axis.

Proof (Sketch). The main tool is a truncation method : we modify the nonlinear term near the region where concentration phenomena (and lack of compactness) could occur; then we find infinitely many solutions of the modified problem; finally we analyse the behaviour of these solutions as $\epsilon \to 0$ and show that many solutions of the modified problem (indeed a number which goes to infinity as $\epsilon \to 0$) are solutions of $P(\Omega_\epsilon, p)$ for ϵ sufficiently close to zero, since they tend to concentrate outside the region where the nonlinear term has been modified.□

In conclusion, we have that in the critical case the nontriviality of the domain Ω is only a sufficient but not necessary condition in order to guarentee the existence of nontrivial solutions for $P\left(\Omega, \frac{2n}{n-2}\right)$ (because of the results of [BC88], [Dan88], [Din89], [Pas89], etc...). On the contrary, when $p > \frac{2n}{n-2}$, this condition is neither sufficient, nor necessary (as showed in [Pas93b], [Pas95], [Pas92a], [Pas96a], [Pas98a], [Pas]).

It is a widely open problem to find what kinds of geometrical properties of the domain Ω are related to the solvability of $P(\Omega, p)$ for $p > \frac{2n}{n-2}$.

3   Effect of lower–order terms

The first results in this direction have been stated by Brezis and Nirenberg (see [Bre86], [BL83b]). First, let us observe that, in order to have positive solutions of problem $P_a(\Omega, 2^*)$, it is essential that the linear operator $-\Delta + a$ is positive, i.e. the first eigenvalue

$$\mu_1 = \min\{f_a(u) : u \in H_0^{1,2}(\Omega), \int_\Omega u^2 dx = 1\}$$

is positive.

In fact, denoted by $\varphi_1 > 0$ the corresponding eigenfunction, if u is a positive solution of $P_a(\Omega, 2^*)$, we obtain (multiplying by φ_1 and integrating by parts)

$$\mu_1 \int_\Omega u\varphi_1 dx = \int_\Omega u^{2^*-1}\varphi_1 dx,$$

which implies $\mu_1 > 0$.

It follows that

$$I_a \overset{\text{def}}{=} \inf_{M_{2^*}(\Omega)} f_a \geq 0. \tag{3.1}$$

When this infimum I_a is achieved, a minimizing function gives rise to a positive solution of $P_a(\Omega, 2^*)$. Notice that we have always

$$I_a \leq S,$$

as one can easily verify testing the functional f_a on a suitable sequence of functions (see [Bre86], [BN83]).

A useful tool, in order to prove that I_a is achieved, is given by the following lemma.

Lemma 1 (see [Bre86], [BN83]). *Let* $a \in L^{n/2}(\Omega)$ *and assume that (see (3.1))*

$$I_a < S. \tag{3.2}$$

Then the infimum I_a *is achieved.*

The proof is based on the fact that every minimizing function is relatively compact if $I_a < S$ (indeed Palais–Smale condition holds in the energy range $]-\infty, S[$: see [Bre86], [BL83b], [BN83], [Lio85], [Str84]).

Thus the problem is to find concrete assumptions which guarantee that (3.2) holds. The cases $n = 3$ and $n \geq 4$ are quite different.

In the case $n \geq 4$, the main result is the following

Theorem 7 (see [Bre86], [BN83]). *Assume* Ω *is any bounded domain in* $\mathbf{R}^n$, *with* $n \geq 4$. *Then, the following properties are equivalent:*

1) $a(x) < 0$ *somewhere on* Ω *(it suffices in a neighbourhood of a point)*
2) $I_a < S$ *(and so* I_a *is achieved).*

In the special case where $a(x)$ is a constant function, we obtain the following corollary (other existence and multiplicity results, concerning the case where $a(x)$ is a negative constant, can be found for example in [CFS84] and [CSS86]).

Corollary 1 (see [Bre86], [BN83]). *Let* Ω *be a bounded domain in* $\mathbf{R}^n$, *with* $n \geq 4$, *and assume* $a(x) = -\lambda$ *(constant).*
  Then problem $P_{(-\lambda)}(\Omega, 2^*)$ *has at least one positive solution for* $0 < \lambda < \lambda_1$ *(λ_1 denotes the first eigenvalue of* $-\Delta$ *in* $H_0^{1,2}(\Omega)$); $P_{(-\lambda)}(\Omega, 2^*)$ *has no positive solution for* $\lambda \geq \lambda_1$; *for* $\lambda \leq 0$ $P_{(-\lambda)}(\Omega, 2^*)$ *has no solution if* Ω *is a star–shaped domain.*
  Moreover the solution u_λ, *for* $0 < \lambda < \lambda_1$, *converges to zero in* $H_0^{1,2}(\Omega)$ *as* $\lambda \to \lambda_1$, *while concentrates like a Dirac mass as* $\lambda \to 0$.

The situation is more complicated in the case $n = 3$.

Theorem 8 (see [Bre86], [BN83]). *Assume* Ω *is any bounded domain in* $\mathbf{R}^3$ *and* $a(x) = -\lambda$ *(constant) in* Ω. *Then there exists a constant* $\lambda^*(\Omega) \in]0, \lambda_1[$, *such that*

a) $I_{(-\lambda)} = S$ *for* $\lambda \leq \lambda^*(\Omega)$ *and* $I_{(-\lambda)}$ *is not achieved for* $\lambda < \lambda^*(\Omega)$
b) $I_{(-\lambda)} < S$ *for* $\lambda > \lambda^*(\Omega)$ *and so there exists a positive solution for* $\lambda^*(\Omega) < \lambda < \lambda_1$ *(no positive solution can exist for* $\lambda > \lambda_1$).

It is not clear what happens for $\lambda \leq \lambda^*(\Omega)$; a complete solution is given only in the case where Ω is a ball.

Theorem 9 (see [Bre86], [BN83]). *Assume* Ω *is a ball in* $\mathbf{R}^3$. *Then* $\lambda^*(\Omega) = \frac{1}{4}\lambda_1$ *(see Theorem 8) and there exists no positive solution of problem* $P_{(-\lambda)}(\Omega, 2^*)$ *for* $\lambda \leq \frac{1}{4}\lambda_1$.

When $a(x) \geq 0$ everywhere on Ω, then $I_a = S$ and so I_a cannot be achieved in any bounded domain Ω (otherwise S should be achieved too). Moreover $P_a(\Omega, 2^*)$ cannot have solution in star–shaped domains, if $a(x)$ is a nonnegative constant (because of Pohozaev's identity (1.6)).

However, let us emphasize that, even in the case $a(x) \geq 0$ on Ω, problem $P_a(\Omega, 2^*)$ may still have solutions; but these solutions cannot be obtained by minimization arguments and correspond to higher critical values (obtained by topological methods of Calculus of Variations). In fact, as showed by Brezis in [Bre86], in any bounded domain Ω it is easy to construct an example of a nonnegative (noncostant) function $a(x)$ such that $P_a(\Omega, 2^*)$ has solution: fix $\psi \in C_0^\infty(\Omega)$ such that $\psi \geq 0$, $\psi \not\equiv 0$, and let v be the solution of the problem

$$\begin{cases} \Delta v + \psi = 0 & \text{in } \Omega \\ v = 0 & \text{on } \partial\Omega \end{cases}$$

so that $v > 0$ on Ω. If we set

$$a = t^{2^*-1} v^{2^*-1} - \frac{\psi}{v},$$

then $a(x)$ is a smooth function. Moreover it is easy to verify that $a(x) > 0$ everywhere on Ω, if t is a sufficiently large positive constant, and that $u = tv$ solves problem $P_a(\Omega, 2^*)$. This example motivates Question 3.

Note that Pohozaev's identity (1.6) shows that an obvious necessary condition for the existence of a solution for $P_a(\Omega, 2^*)$ is that $[a(x) + \frac{1}{2}(x \cdot Da(x))]$ should be negative somewhere on Ω.

An answer to Question 3 is given in [Pas96b].

Theorem 10 (see [Pas96b]). *Let Ω be a smooth bounded domain in $\mathbf{R}^n$, with $n \geq 3$, and x_0 be a fixed point in Ω. Let $\bar{\alpha} \in L^{n/2}(\Omega)$ and $\alpha \in L^{n/2}(\mathbf{R}^n)$ be two nonnegative functions and assume that $\|\alpha\|_{L^{n/2}(\mathbf{R}^n)} \neq 0$. Then there exists $\bar{\mu} > 0$ such that, for all $\mu > \bar{\mu}$, problem $P_a(\Omega, 2^*)$, with*

$$a(x) = \bar{\alpha}(x) + \mu^2 \alpha[\mu(x - x_0)],$$

has at least one solution u_μ. Moreover

$$S < f_a\left(\frac{u_\mu}{\|u_\mu\|_{2^*}}\right) < 2^{2/n} S$$

and

$$\lim_{\mu \to +\infty} f_a\left(\frac{u_\mu}{\|u_\mu\|_{2^*}}\right) = S.$$

If we assume in addition that

$$\|\alpha\|_{L^{n/2}(\mathbf{R}^n)} < S(2^{2/n} - 1), \tag{3.3}$$

then $P_a(\Omega, 2^*)$ has at least another solution $\hat{u}_\mu$. Moreover

$$f_a\left(\frac{u_\mu}{\|u_\mu\|_{2^*}}\right) < f_a\left(\frac{\hat{u}_\mu}{\|\hat{u}_\mu\|_{2^*}}\right) < 2^{2/n} S$$

and

$$\liminf_{\mu \to +\infty} f_a \left(\frac{\hat{u}_\mu}{\|\hat{u}_\mu\|_{2^*}} \right) > S.$$

The proof is obtained using topological methods of Calculus of Variations. An important tool is given by the results stated in [Lio85] and [Str84], which give a description of the behaviour of the Palais–Smale sequences and allow us to prove that Palais–Smale condition is satisfied in the energy range $]S, 2^{2/n}S[$.

Remark 4. Note that the assumption on the nonnegative function $a(x)$ in Theorem 10 seems to be fairly general. In fact, if we assume for example that $x_0 = 0$, $\bar{\alpha} \equiv 0$ and

$$a(x) = \begin{cases} 1 & \text{if } |x| \leq 1 \\ |x|^{-\beta} & \text{if } |x| > 1, \end{cases}$$

then, if $\beta \leq 2$ (i.e. $\alpha \notin L^{n/2}(\mathbf{R}^n)$) and Ω is a bounded domain star–shaped with respect to zero, problem $P_a(\Omega, 2^*)$, with $a(x) = \mu^2\alpha(\mu x)$, has no solution for any $\mu > 0$, because of Pohozaev's identity (1.6); on the contrary, if $\beta > 2$ (i.e. $\alpha \in L^{n/2}(\mathbf{R}^n)$), Theorem 10 guarantees the existence of solutions for μ large enough, without any assumption on the shape of Ω (if μ is small enough and Ω is star–shaped, no solution can exist because $a(x)$ is constant in Ω).

Remark 5. The method used in the proof of Theorem 10 can be iterated in order to obtain more general multiplicity results concerning functions $a(x)$ of the form

$$a(x) = \bar{\alpha}(x) + \sum_{i=1}^{h} \mu_i^2 \alpha_i[\mu_i(x - x_i)],$$

where $x_1, \ldots, x_h$ are points in Ω, $\bar{\alpha}$ in $L^{n/2}(\Omega)$ and $\alpha_1, \ldots, \alpha_h$ in $L^{n/2}(\mathbf{R}^n)$ are nonnegative functions and $\lambda_1, \ldots, \lambda_h$ are positive parameters. Indeed, there exist at least h distinct positive solutions when the concentration parameters $\mu_1, \ldots, \mu_h$ are large enough, and at least $2h$ positive solutions when, in addition, the functions $\alpha_1, \ldots, \alpha_h$ satisfy condition (3.3).

Note that it is not necessary to choose distinct concentration points in order to obtain h or $2h$ distinct positive solutions: it suffices to choose only the concentration parameters $\mu_1, \ldots, \mu_h$ in a suitable way (some possible choices of these parameters are described in [Pas96b]).

Remark 6. In [BC89] and [Pas90] one can find some results concerning the case where $\Omega = \mathbf{R}^n$. Consider the problem

$$\begin{cases} \Delta u - [\epsilon + \alpha(x)]u + u^{2^* - 1} = 0 & \text{in } \mathbf{R}^n \\ u > 0 & \text{in } \mathbf{R}^n \\ \int_{\mathbf{R}^n} |Du|^2 dx < \infty, \end{cases}$$

where α in $L^{n/2}(\mathbf{R}^n)$, with $\alpha \not\equiv 0$, is a given nonnegative function and $\epsilon \geq 0$.

In [BC89] it is proved that there exists at least one solution, under the assumption that $\epsilon = 0$ and α satisfies condition (3.3).

In [Pas90] we show that there exist, for $\epsilon > 0$ small enough, at least two solutions, u_ϵ and $\hat{u}_\epsilon$, such that

$$S < f_{a_\epsilon}\left(\frac{u_\epsilon}{\|u_\epsilon\|_{2^*}}\right) < f_{a_\epsilon}\left(\frac{\hat{u}_\epsilon}{\|\hat{u}_\epsilon\|_{2^*}}\right) < 2^{2/n}S,$$

where $a_\epsilon(x) = [\epsilon + \alpha(x)]$.

Moreover u_ϵ vanishes as $\epsilon \to 0$, while $\hat{u}_\epsilon$ converges to a solution of the limit problem.

Finally, let us mention a result (stated in [Pas93a]), where we exploit the combined effect of both the lower-order terms and the domain shape.

Notice that the positive solutions u_λ given by Corollary 1 concentrate as $\lambda \to 0$ like Dirac mass near points of Ω. Indeed, for every family $(\tilde{u}_\lambda)_{\lambda>0}$ of functions in $M_{2^*}(\Omega)$, such that $f_{(-\lambda)}(\tilde{u}_\lambda) \leq S \; \forall \lambda > 0$, there exists a family of points $(x_\lambda)_{\lambda>0}$ in Ω such that the functions $\tilde{u}_\lambda(x - x_\lambda)$ tend, as $\lambda \to 0$, to δ_0, the Dirac mass in zero.

This fact enable us to relate the topological properties of the sublevels of the energy functional $f_{(-\lambda)}$ to the shape of the domain Ω. Thus, taking also into account the behaviour of the Palais–Smale sequences described in [Bre86], [Lio85], [Str84], it is possible to evaluate the number of positive solutions of $P_{(-\lambda)}(\Omega, 2^*)$, when λ is a positive constant sufficiently close to zero, by the Ljusternik–Schnirelman category of Ω, or by other topological invariants (see [Laz92], [Pas93a], [Rey89a]).

Theorem 11 (see [Pas93a]). *Let Ω be a bounded domain in $\mathbf{R}^n$, with $n \geq 4$, and denote by m its Ljusternik–Schnirelman category.*

Then there exists $\bar{\lambda} \in]0, \lambda_1[$ such that, for all $\lambda \in]0, \bar{\lambda}[$, problem $P_{(-\lambda)}(\Omega, 2^*)$ has at least m distinct positive solutions, $u_{1,\lambda}, \ldots, u_{m,\lambda}$, such that

$$f_{(-\lambda)}\left(\frac{u_{i,\lambda}}{\|u_{i,\lambda}\|_{2^*}}\right) < S \quad \forall i = 1, \ldots, m.$$

If we assume in addition that Ω is not contractible in itself (i.e. $m > 1$), then there exists another solution $u_{m+1,\lambda}$ such that

$$S < f_{(-\lambda)}\left(\frac{u_{m+1,\lambda}}{\|u_{m+1,\lambda}\|_{2^*}}\right) < 2^{2/n}S.$$

Remark 7. The lower energy solutions $u_{1,\lambda}, \ldots, u_{m,\lambda}$ converge weakly to zero in $H_0^{1,2}(\Omega)$ as $\lambda \to 0$ and concentrate like Dirac mass near some points of Ω. Moreover (see [Rey90]) the concentration points are the critical points of the regular part of the Green's function for Laplace operator with zero Dirichlet boundary condition.

The higher energy solution $u_{m+1,\lambda}$ either converges as $\lambda \to 0$ to a solution of the limit problem, or converges weakly to zero and can be decomposed as sum of at most two functions whose energy concentrates like Dirac mass near two points of Ω.

Finally, let us point out that, if Ω has some symmetry properties, then the number of solutions may increase considerably: for example, if Ω is a domain homotopically equivalent to the $(k-1)$–dimensional sphere S_{k-1} and is symmetric with respect to a point $x_0 \notin \Omega$, then problem $P_{(-\lambda)}(\Omega, 2^*)$ has, for $\lambda > 0$ small enough, at least $2k+1$ solutions, even if the Ljusternik–Schnirelman categoryof Ω in itself is only 2.

On the existence and multiplicity of positive solutions for semilinear mixed and Neumann elliptic problems

G. Cerami*

1 Introduction

The aim of this paper is to give the main ideas and to summarize the results of some investigations concerning problems of the following type

$$\left.\begin{array}{rcl} -\Delta u + a(x)u = u^{p-1} & in & \Omega \\ u > 0 & in & \Omega \\ u = 0 & on & \Gamma_0 \\ \frac{\partial u}{\partial \nu} = 0 & on & \Gamma_1 \end{array}\right\} \tag{1}$$

where Ω is an open set in $\mathbb{R}^N$, $N \geq 3$, whose boundary $\partial\Omega$ is the union of two disjoint submanifolds Γ_0 and Γ_1, $\Gamma_1 \neq \emptyset$, ν denotes the outer normal to $\partial\Omega$, $p \in (2, \frac{2N}{N-2}]$ and $a \in L^{\frac{N}{2}}$ is a given nonnegative function.

In the past few years there have been several researches concerning the existence and multiplicity of solutions for problems like (1), most of them regarding the pure Neumann case:

$$\left.\begin{array}{rcl} -\Delta u + a(x)u = u^{p-1} & in & \Omega \\ u > 0 & in & \Omega \\ \frac{\partial u}{\partial \nu} = 0 & on & \partial\Omega \end{array}\right\} \tag{2}$$

Before considering the mathematical characteristics of (1), it is worth remarking that the interest in these questions is due to the fact that problems like (1) arise naturally in various studies of pattern formations in mathematical biology. For instance, a positive nontrivial solution of (2), Ω bounded, gives rise to a steady state of the Keller-Segel's chemotaxis model of a cellular slime molds (amoebae)(see e.g. [KS70], [Sch85]). The problem (2) may also be viewed as the "shadow system" for the non saturated case of an activator-inhibitor type system in morphogenesis, due to Gierer-Meinhart (see e.g. [LNT88] and the references therein).

Problem (1) has a variational structure, so it is easy to see that solutions of it can be found looking for positive functions u belonging to the space

$$W^{1,2}(\Omega, \Gamma_0) = \{u \in W^{1,2}(\Omega) : u = 0 \text{ on } \Gamma_0\}$$

which are either critical points of the free energy functional

* Supported by MURST of Italy.

$$I(u) = \frac{1}{2} \int_\Omega (|\nabla u|^2 + a(x)u^2)dx \; - \frac{1}{p} \int_\Omega |u|^p dx$$

or critical points of the functional

$$E(u) = \int_\Omega (|\nabla u|^2 + a(x)u^2)dx$$

constrained to lie on the manifold

$$V = \left\{ u \in W^{1,2}(\Omega, \Gamma_0) : \int_\Omega |u|^p dx = 1 \right\}.$$

Also, it is immediate that a, natural, necessary condition for the solvability of (1) is that $-\Delta + a$ be a positive operator on $W^{1,2}(\Omega, \Gamma_0)$. So, from now on, we assume that the relation

$$\inf \left\{ \int_\Omega (|\nabla u|^2 + a(x)u^2)dx, \; u \in W^{1,2}(\Omega, \Gamma_0), \int_\Omega u^2 \, dx = 1 \right\} > 0 \qquad (3)$$

holds.

The variational approach gives good results, but it is easy to understand that the difficulties that occur are quite different according to Ω is bounded or not, $p < \frac{2N}{(N-2)}$ or $p = 2^* = \frac{2N}{(N-2)}$, $a(x)$ is constant or not..... For instance it is well known that if either $p = 2^*$ or Ω is unbounded (1) has a lack of compactness, and the variational techniques cannot be applied in a standard way. On the other hand if we consider (2) it is clear that, when $a(x) = \lambda > 0$ constant, the constant function $\lambda^{\frac{1}{(p-2)}}$ is a "trivial" positive solution; thus the interesting question is to know if there exist nonconstant solutions.

For the noncompact problems much progress has been made recently and, in this paper, we shall be mainly concerned with results regarding these cases. Nevertheless in section 2 we shall review some results about the compact case. Sections 3 and 4 will be devoted to the critical case: in section 3 the existence of "low energy" solutions is discussed, while in section 4 some recent development regarding the existence and the multiplicity of "high energy" solutions is exposed. We conclude the article with the study, in section 5, of the problem in unbounded domains.

2   The compact case

Throughout this section we suppose the domain $\Omega \subset \mathbb{R}^N$, $N \geq 3$, bounded, with smooth boundary $\partial \Omega$, $2 < p < 2^*$.

For problem(1) the following result can be stated:

Theorem 1. *Assume* $a \in L^{\frac{N}{2}}(\Omega)$ *satisfying* (3). *Then problem* (1) *possesses a least energy solution.*

The proof of the above theorem can be easily obtained using a standard variational argument. We sketch it here, because the approach is very natural and, also, it will be useful, in the sequel, to understand the difficulties arising in the non compact cases.

Let consider the minimization problem

$$\inf \left\{ \int_\Omega (|\nabla u|^2 + a(x)u^2)\, dx, \ u \in W^{1,2}(\Omega, \Gamma_0), \int_\Omega |u|^p dx = 1 \right\} \quad (4)$$

and call m_a the infimum in (4); m_a is nonnegative because $-\Delta + a$ is positive definite. Consider a minimizing sequence $\{u_m\}$ then

$$
\text{and} \qquad
\left.
\begin{array}{c}
u_m \in W^{1,2}(\Omega, \Gamma_0) \quad |u_m|_{L^p(\Omega)} = 1 \\[2mm]
\int_\Omega (|\nabla u_m|^2 + a(x)u_m^2)\, dx = m_a + o(1).
\end{array}
\right\} \quad (5)
$$

$\{u_m\}$ is bounded in $W^{1,2}$ and, up to a subsequence still denoted by $\{u_m\}$, converges weakly to some $\bar{u} \in W^{1,2}(\Omega, \Gamma_0)$

$$u_m \rightharpoonup \bar{u} \quad \text{weakly in} \ W^{1,2}(\Omega, \Gamma_0)$$

and by the compactness of the embedding $W^{1,2}(\Omega, \Gamma_0) \to L^p(\Omega)$

$$u_m \to \bar{u} \quad \text{strongly in} \ L^p(\Omega).$$

Hence passing to the limit in (5) as $m \to +\infty$ we obtain

$$\int_\Omega (|\nabla \bar{u}|^2 + a(x)\bar{u}^2)\, dx \leq m_a \quad |\bar{u}|_{L^p(\Omega)} = 1.$$

Thus $\bar{u}$ is a minimizer and $m_a > 0$. Moreover we can assume $\bar{u} \geq 0$ (otherwise we replace $\bar{u}$ by $|\bar{u}|$) and since it solves

$$-\Delta u + a(x)u = m_a u^{p-1}$$

$\bar{v} = m_a^{\frac{1}{(p-2)}} \bar{u}$ solves (1) and by the strong Hopf maximum principle $\bar{v} > 0$ (standard elliptic regularity results show that $\bar{v}$ is smooth on $\bar{\Omega}$). To see that $\bar{v}$ is a least energy solution, consider another solution v of (1) then

$$\int_\Omega (|\nabla v|^2 + a(x)v^2)\, dx = \int_\Omega |v|^p\, dx \quad (6)$$

and by definition of m_a

$$\int_\Omega (|\nabla v|^2 + a(x)v^2)\, dx \geq m_a \left(\int_\Omega |v|^p\, dx \right)^{\frac{2}{p}}. \quad (7)$$

Hence combining (6) and (7) we obtain

$$\int_\Omega (|\nabla v|^2 + a(x)v^2)\,dx \geq (m_a)^{\frac{p}{p-2}}$$

so

$$I(v) = \tfrac{1}{2}\int_\Omega(|\nabla v|^2 + a(x)v^2)dx - \tfrac{1}{p}\int_\Omega |v|^p\, dx$$

$$= \left(\tfrac{1}{2} - \tfrac{1}{p}\right)\left(\int_\Omega(|\nabla v|^2 + a(x)v^2)dx\right)$$

$$\geq \left(\tfrac{1}{2} - \tfrac{1}{p}\right)(m_a)^{\frac{p}{p-2}} = I(\bar{v}).$$

Of course this result holds in the particular case $\Gamma_0 = \emptyset$, the pure Neumann case. Nevertheless, when $a(x) = \lambda \geq 0$ it is not clear that the found solution is different from the constant solution $u(x) = \lambda^{\frac{1}{p-2}}$. This question as well as the behaviour and the shape of the solutions of

$$\left.\begin{array}{rl} -\Delta u + \lambda\, u = u^{p-1} & in\ \Omega \\ u > 0 & in\ \Omega \\ \frac{\partial u}{\partial \nu} = 0 & on\ \partial\Omega \end{array}\right\} \tag{8}$$

has been widely investigated in a series of papers ([LNT88], [NT91], [NT92], [LN88]), in which has been established that a "least energy" solution possesses a single spike layer at the boundary, and, moreover, that the location of the spike can be determined. Mathematically, this result can be summarized as follows:

Theorem 2. *For every* $\lambda > 0$, (8) *possesses a least energy solution* u_λ *with the following properties*

i) $u_\lambda = \lambda^{\frac{1}{p-2}}$ *is the unique solution for every* λ *sufficiently small. Moreover, there exists a* $\lambda_0 > 0$ *such that for all* $\lambda > \lambda_0$ *the solution* u_λ *is nonconstant and has a unique local (thus global) maximum point* P_λ *that lies on the boundary* $\partial\Omega$.

ii) P_λ *must be situated at the "most curved" part of* $\partial\Omega$, *when* λ *is sufficiently large, i.e.* $H(P_\lambda) \to \max_{P\in\partial\Omega} H(P)$ *as* $\lambda \to +\infty$, *where* $H(P)$ *denotes the mean curvature of* $\partial\Omega$ *at* P.

For the proof of this theorem, which is lenghty and technical, we refer the interested reader to the quoted papers. We only observe that the idea for proving that, for λ large, u_λ must be nonconstant is to give some careful bounds for the number

$$m_\lambda = \inf\left\{\int_\Omega(|\nabla u|^2 + \lambda\, u^2)dx,\ u \in W^{1,2}(\Omega),\ |u|_{L^p(\Omega)} = 1\right\} \tag{9}$$

and to remark that the energy of the constant solution does not satisfy those constraints if λ is large enough.

Subsequently, the researchers' interest was addressed to the study of sufficient conditions for the existence of multiple solutions. In particular, the possibility of obtaining multiplicity results, by exploiting the topological richness of the domain, was examined.

Denoted by $\mathrm{cat}(\partial\Omega)$ the Ljusternik-Schnirelman category of $\partial\Omega$ in itself, in [Wan92a] and [MM92] the following theorem has been proven:

Theorem 3. *There exists* $\lambda_0 > 0$ *such that for any* $\lambda > \lambda_0$, *problem* (8) *has at least* $\mathrm{cat}\ \partial\Omega$ *nonconstant solutions.*

The proof of the above theorem is based on ideas formerly introduced for the Dirichlet problem in [BC91] and [BCP91]. Namely, following the method devised in [BC91], in [Wan92a] and in [MM92] the authors prove that it is possible to relate a suitable sublevel set of E to $\partial\Omega$ and to estimate the topology of such set in terms of the topology of $\partial\Omega$. Then the multiplicity result is a consequence of a classical theorem of the Ljusternik-Schnirelman theory and of the fact that the solutions, so found, correspond to critical points of E on V having energy bounded from above by $2^{\frac{p-2}{p}} m_\lambda$ (the least energy of a changing sign solution).

We remark, also, that a result analogous to Theorem 3 has been obtained in [CL95] for problem (1) with $a(x) = \lambda$, the statement says that if λ is large enough the number of solutions of (1) is bounded from below by the $\mathrm{cat}(\Gamma_1)$.

We end this section reporting that multiple, single peaked, spike layer, solutions for (8) can be constructed taking advantage of the geometry of $\partial\Omega$. Ni and Oh [NO], in fact, assert that for any given point $P \in \partial\Omega$, with the second fundamental form of $\partial\Omega$ at P being nondegenerate, there is a spike layer solution of (8) with its only peak located near P, for λ sufficiently large.

3 The critical case: existence of low energy solutions

Let us turn, now, to the case $p = 2^* = \frac{2N}{N-2}$, $\Omega \subset \mathbb{R}^N$, $N \geq 3$, bounded. In this case the nonlinearity is said to have a "critical" growth because 2^* is the limiting exponent for the Sobolev embedding $j : W^{1,2}(\Omega, \Gamma_0) \rightarrow L^p(\Omega)$, namely 2^* is the exponent such that the embedding is continuous, but not compact.

The lack of compactness, technically, implies that the manifold V is not closed for the weak $W^{1,2}$ topology and the functionals I and $E_{|V}$ do not verify in all the energy levels the classical

Palais–Smale condition : Let $f \in C^1(B, \mathbb{R})$, B real Banach space. f satisfies the P-S condition in c if any sequence $\{u_m\} \subset B$ such that $f(u_m) \rightarrow c$ and $f'(u_m) \rightarrow 0$ as $m \rightarrow \infty$ possesses a convergent subsequence.

The difficulties one meets with, when wants make use of variational methods, can be easily understood trying to repeat the argument used, in section 2, for the subcritical case. In fact, let us consider the minimization problem (4) and let $\{u_m\}$ be a minimizing sequence. Now, for the weak limit $\bar{u}$ of $\{u_m\}$ in $W^{1,2}(\Omega, \Gamma_0)$, we can only say that

$$\int_\Omega (|\nabla \bar{u}|^2 + a(x)\bar{u}^2) dx \leq m_a , \quad |\bar{u}|_{L^{2^*}(\Omega)} \leq 1 ,$$

because $\{u_m\}$ needs not converge strongly in $L^{2^*}(\Omega)$. Thus the weak limit of a minimizing sequence $\{u_m\}$ may not be a minimizer.

The first general existence result, for least energy solutions of (1) (with $p = 2^*$), was obtained by X.J. Wang [Wan91] and Adimurthi-Mancini [AM91], who, to get around the lack of compactness, followed the method introduced by Brezis-Nirenberg [BN83] for the same equation with zero Dirichlet boundary conditions.

Actually, denoting, as usual, by S the best Sobolev constant

$$S = \inf \left\{ \int_{R^N} |\nabla u|^2 \, dx, \ u \in D^{1,2}(R^N), \ \int_{R^N} |u|^{2^*} dx = 1 \right\} \qquad (10)$$

and by m_a the infimum in (4) (with $p = 2^*$), those authors have pointed out that the minimizing sequences are relatively compact if

$$m_a < 2^{-\frac{2}{N}} S \qquad (11)$$

Then, once proven such strict inequality, the existence of a solution of (1) can be obtained by applying the minimization argument explained in section 2.

The existence results contained in [Wan91] and in [AM91] can be stated as follows:

Theorem 4. *Assume that* $\Omega \subset R^N, N \geq 3,$ *is a bounded set verifying the condition*

$(*)$
$\begin{cases}
\text{there is a point } \bar{x} \text{ in the interior part of } \Gamma_1 \text{ such that, in a} \\
\text{neighbourhood } \omega \text{ of } \bar{x}, \ \Gamma_1 \text{ is regular and } \Omega \text{ lies on one side of} \\
\text{the tangent plane at } \bar{x}, \ a(x) \in L^\infty(\omega) \cap L^{\frac{N}{2}}(\Omega) \text{ and the mean curvature} \\
\text{of } \Gamma_1 \text{ at } \bar{x} \text{ with respect to the unit outward normal is positive}
\end{cases}$

then (1) has at least a least energy solution.

Theorem 5. *Assume* $\Omega \subset R^N, N \geq 3$, *bounded with smooth boundary* $\partial\Omega, a(x) \in L^{\frac{N}{2}}(\Omega) \cap L^\infty(\omega)$, ω *being a neighbourhood of* $\partial\Omega$. *Then (2) has at least one least energy solution. Furthermore if* $a(x) = \lambda > 0$ *constant, the least energy solution is nonconstant for* λ *large enough.*

The crucial role played by the geometric condition $(*)$ in proving the relation (11) and, consequently, the Theorem 4 can be understood by the following considerations.

It is well known that S is attained by the family of positive functions

$$\Psi_{\sigma,y}(x) = \sigma^{-\frac{N}{2}} \Psi \left(\frac{x-y}{\sigma} \right) \qquad \sigma \in R^+ \setminus \{0\}, \ y \in R^N$$

where

$$\Psi(x) = \frac{U(x)}{|U|_{L^{2^*}(\mathbb{R}^N)}} \quad \text{and} \quad U(x) = \frac{1}{[1 + |x|^2]^{\frac{N-2}{2}}}$$

It is easy to verify, moreover, that

$$\Sigma = \inf \left\{ \int_{\mathbb{R}^N_+} |\nabla u|^2 \, dx, \ u \in D^{1,2}(\mathbb{R}^N_+), \ \int_{\mathbb{R}^N_+} |u|^{2^*} dx = 1 \right\} \tag{12}$$

where $\mathbb{R}^N_+ = \{x \in \mathbb{R}^N : x = (x_1, x_2, ..., x_N), \ x_N > 0\}$, is achieved by the positive functions defined for $x \in \mathbb{R}^N_+$ by

$$\Phi_{\sigma,y}(x) = 2^{\frac{1}{2^*}} \Psi_{\sigma,y}(x) \quad \sigma \in \mathbb{R}^+ \setminus \{0\}, \quad y \in \partial \mathbb{R}^N_+ \tag{13}$$

so that

$$\Sigma = 2^{-\frac{2}{N}} S.$$

Now let consider the functions defined in Ω by

$$\chi_{\sigma,\bar{x}}(x) = \frac{\eta(|x - \bar{x}|)\Phi_{\sigma,\bar{x}}(x)}{|\eta(|x - \bar{x}|)\Phi_{\sigma,\bar{x}}(x)|_{L^{2^*}(\Omega)}}, \quad x \in \Omega$$

$\bar{x}$ being the point in which (∗) is satisfied, $\sigma \in \mathbb{R}^+ \setminus \{0\}$ and η a $C^\infty(\mathbb{R})$ "cut-off" function.

When $\sigma \to 0, \chi_{\sigma,\bar{x}}$ "concentrates" near the boundary of Ω, around $\bar{x}$. So it is well founded, not only, the expectation that, for σ small enough, because of the boundary curvature at $\bar{x}$, the term $\int_\Omega |\nabla \chi_{\sigma,\bar{x}}|^2 dx$ becomes less than $\int_{\mathbb{R}^N_+} |\nabla \Phi_{\sigma,\bar{x}}|^2 dx$, but, also, the hope that the energy $E(\chi_{\sigma,\bar{x}})$ becomes smaller than Σ. This is what, actually, happens and has been verified (see [Wan91] and [AM91] for the proof) by the following estimate:

Lemma 1. *There exists a positive constant A_N depending only on N such that*

$$E(\chi_{\sigma,\bar{x}}) = 2^{-\frac{2}{N}} S - \begin{cases} A_N H(\bar{x})\sigma \ln(\frac{1}{\sigma^2}) + O(\sigma) & \text{if } N = 3 \\ \\ A_N H(\bar{x})\sigma + O(\sigma^2 \ln(\frac{1}{\sigma^2})) & \text{if } N \geq 4. \end{cases}$$

About the Theorem 5, we observe that the first part of the statement is a straight consequence of the Theorem 4, that holds true in the particular case $\Gamma_0 = \emptyset$, and of the fact that, in this case, the condition (∗) is naturally verified, because Ω is bounded and $\partial\Omega = \Gamma_1$ is smooth. When, in particular, $a(x) = \lambda > 0$ constant, the above result allows to obtain a nonconstant solution. In fact, for any λ, the least energy solution, before found, has energy less than $2^{-\frac{2}{N}} S$, so it cannot coincide with $\lambda^{\frac{1}{2^*-2}}$, the constant solution, if λ is large enough.

The subsequent step in the study of (1) has been, clearly, looking for multiple solutions.

As in the subcritical case, the attention has been focused on the possibility of obtaining multiplicity of solutions taking advantage of the shape of the boundary of the domain Ω. However, the basic part played by the condition (*), in proving the existence results, convinced the researchers that, not only, the topology richness, but also, the geometry of the boundary of Ω could be the cause of the existence of multiple solutions.

The results in [AM94], [Wana] emphasize, once again, the role of the topology. Nevertheless, in this case, what has been proven relevant is a part of the boundary of Ω geometrically significant. In fact, called a point x of $\partial\Omega$ *strictly convex* if all the principal curvatures, with respect to the unit outward normal at x, are positive and denoted by Γ_1^+ the set of strictly convex points of Γ_1, the following theorem has been stated:

Theorem 6. *Let be $a(x) = \lambda$, $\lambda \in \mathbb{R}^+ \setminus \{0\}$. Then there exists a $\lambda_0 > 0$ such that for $\lambda > \lambda_0$ problem (1) admits at least $\mathrm{cat}(\Gamma_1^+, \Gamma_1)$ distinct solutions (nonconstant if $\Gamma_0 = \emptyset$) corresponding to critical points, of E on V, having energy less than Σ.*

From Theorem 6 we deduce, for instance, the existence of k solutions if Γ_1 has k components on each of which the mean curvature is positive somewhere. Also, when $\Gamma_0 = \emptyset$ and Ω is a convex set, we obtain, when λ is large enough, the existence of at least two solutions that are nonconstant because of the bound on their energy.

The proof of Theorem 6 is carried out combining the method, already mentioned, of estimating the topology of suitable sublevel sets of E on V by means of the topology of some part of Γ_1 and the fact that the Palais-Smale condition holds at any energy level $c < \Sigma$.

While Theorem 6 makes use, mainly, of topological tools, in [APY93] and [Wanc] localized low energy solutions for (8) are obtained, by exploiting the geometry of the boundary:

Theorem 7. *Let be $N \geq 5$ and P_0 a strict local maximum point of $H(x)$, such that $H(P_0) > 0$. Then there exists a $\lambda_0 > 0$ such that for all $\lambda > \lambda_0$, problem (8) has a nonconstant solution u_λ, such that $E\left(\frac{u_\lambda}{|u_\lambda|_{L^{2^*}(\Omega)}}\right) < \Sigma$. Moreover the functions u_λ concentrate at P_0, as $\lambda \to +\infty$, in the sense*

$$\lim_{\lambda \to +\infty} \left\| \frac{u_\lambda}{|u_\lambda|_{L^{2^*}}} - \Phi_{\epsilon_\lambda, P_\lambda} \right\|_{W^{1,2}(\Omega)} = 0$$

for some $\epsilon_\lambda > 0$ and $P_\lambda \in \partial\Omega$ with $\epsilon_\lambda \to 0$, $P_\lambda \to P_0$ as $\lambda \to +\infty$.

The above theorem put into relief the role of the mean curvature of the boundary of Ω. From this result we can infer the existence of k solutions of (8) in any bounded domain Ω with k "peaks", i.e. k points on $\partial\Omega$ of positive strict local maximum for the mean curvature $H(x)$.

The proof of the above theorem is rather technical, so we refer the readers to [APY93] for the case $N > 7$ and to [Wanc] for the extension to the cases $N = 5, 6$.

We close this section by mentioning that, as in the subcritical case, the shape of the found solutions has been investigated. Firstly in [NPT92] the least energy solutions have been considered, subsequently, results for low energy solutions have been obtained in [APY93] and [Wanc]. The main results may be summarized in the following

Theorem 8. Let u_λ be a solution of (8) such that $E(\frac{u_\lambda}{|u_\lambda|_{L^{2*}}}) < \Sigma$. Then there exists $\lambda_0 > 0$ such that for all $\lambda > \lambda_0$

a) u_λ attains its maximum at only one point $P_\lambda \in \partial\Omega$ and

$$\lim_{\lambda\to\infty} ||\nabla \left(\frac{u_\lambda}{|u_\lambda|_{L^{2*}}}\right) - \nabla\Phi_{\frac{\beta_\lambda}{\lambda},P_\lambda}||_{L^2(\Omega)} = 0$$

and $\beta_\lambda \to 0$ as $\lambda \to +\infty$.

b) if $N \geq 5$ and u_λ is a least energy solution, then the limit points of P_λ , as $\lambda \to +\infty$, are contained in the set of the points of maximum mean curvature.

4   The critical case: existence of high energy solutions

This section is mainly devoted to the exposition of some recent progress about the question of finding multiple and, if possible, high energy, solutions for the problem

$$\left.\begin{array}{rcll} -\Delta u + a(x)u &=& u^{2^*-1} & in \ \Omega \\ u &>& 0 & in \ \Omega \\ u &=& 0 & on \ \Gamma_0 \\ \frac{\partial u}{\partial \nu} &=& 0 & on \ \Gamma_1 \end{array}\right\} \tag{14}$$

We first recall that, still in the case $a(x) = \lambda$, $\Gamma_0 = \emptyset$, partial, fragmentary results in this direction have been obtained when Ω is a ball ([Ni83]), when Ω is convex ([Gro95]), when Ω is either antipodal invariant or enjoying of a slighty more general kind of symmetry (see [Wanb] for a survey on this subject).

Combining these results with those of the previous section we can observe some relevant facts. All the quoted multiplicity theorems have been obtained when $a(x) = \lambda$, and, moreover, their proof, as well as any existence proof, are strongly dependent either on the topological and geometrical characteristics of $\partial\Omega$, or on the symmetry properties of Ω . Besides, most solutions have been obtained looking for critical points, of E on V, having energy less than Σ . In any case, the existence of multiple solutions has been shown after knowing that the infimum in (4) (with $p = 2^*$) was attained.

The situation when either $a(x) > 0$ is not constant or there are no conditions on the shape of $\partial\Omega$ is more delicate, as the following examples point out. Actually, if we consider (14) when $\Omega = \mathbb{R}^N_+$, $\Gamma_0 = \emptyset$, $a(x) \geq 0$, $|a|_{L^{\frac{N}{2}}(\Omega)} \neq 0$, it is not difficult to show (see [CP97] and next section) that $m_a = \Sigma$ and the corresponding minimization problem has not solutions. So, there is no hope of finding low energy solutions. Furthermore, when Ω is a bounded domain with a spherical hole, Γ_1 is the boundary of this hole and $a(x) = \lambda$ in [EPT89] it is asserted that, as a consequence of an isoperimetric inequality, $m_\lambda = \Sigma$ and this infimum is not achieved.

Thus the following questions arise naturally:

i) Is it possible to find solutions of (14) when the condition (*) is not satisfied or, worse, when $m_a = \Sigma$?

ii) Is it possible to obtain multiplicity results for (14) when $a(x)$ is not constant, without having information about the topology or the geometry of Γ_1 ?

Recently some answers to these problems have been given in [CP96] and [CP97]. In particular, the research, exposed in these papers, has focused the possibility of giving on $a(x)$ conditions sufficient to guarantee the solvability of (14) without imposing geometrical conditions on $\partial\Omega$.

In [CP96] functions of the form

$$a(x) = \lambda^2 [k + \alpha(\lambda(x - x_0))], \tag{15}$$

have been considered, and, denoted by Π_{x_0} the set

$\Pi_{x_0} = \{x \in \mathbb{R}^N : (x|\nu(x_0)) < 0\}$ $\nu(x_0)$ outer normal to $\partial\Omega$ at x_0 the following theorem has been stated:

Theorem 9. Let $a(x)$ be as in (15), $x_0 \in \overset{\circ}{\Gamma}_1$, $\alpha \in L^{\frac{N}{2}}(\mathbb{R}^N)$, $\alpha \geq 0$, $|\alpha|_{L^{\frac{N}{2}}(\Pi_{x_0})} \neq 0$. Then there exists $\bar{k}$ such that to any $k \in (0, \bar{k}]$ there corresponds a $\bar{\lambda}$ in such a way that for all $\lambda > \bar{\lambda}$, problem (14) has at least one solution. Moreover, if α satisfies the additional condition

$$|\alpha|_{L^{\frac{N}{2}}(\Pi_{x_0})} < S [1 - 2^{-\frac{2}{N}}] \tag{16}$$

(14) has at least two solutions. If, further, condition (*) is fulfilled at some point $\bar{x} \in \overset{\circ}{\Gamma}_1$, then there is at least one more solution of (14), corresponding to a minimizer of E on V.

We emphasize, and the sketch of the proof will make it clear, that the first two solutions, whose existence is stated in Theorem 9, are "high energy" solutions, i.e. they correspond to critical points for which $E(u) > \Sigma = 2^{-\frac{2}{N}} S$. The third solution corresponds to a function that realizes m_a.

We remark, also, that the above results can be generalized by considering functions $a(x)$ having several concentration points. Considering, for example,

$$a(x) = a_0(x) + \sum \lambda_i^2 (k + \alpha_i (\lambda_i (x - x_i)))$$

with $x_i \in \overset{\circ}{\Gamma}_1, a_0 \in L^{\frac{N}{2}}(\Omega), \alpha_i \in L^{\frac{N}{2}}(\mathbb{R}^N), \alpha_i \geq 0, |\alpha_i|_{L^{\frac{N}{2}}(\Pi_{x_i})} \neq 0, \lambda_i > 0, i = 1, 2, ...l$ we could obtain for problem (14) a statement of the following type: there is a $\bar{k} > 0$ such that for any $k \in (0, \bar{k}]$ there exist suitable choices of the l numbers $\lambda_1, \lambda_2, ..., \lambda_l$ for which (14) has at least l distinct positive solutions;

furthermore, if each α_i satisfies the condition (16), then the number of distinct solutions is at least $2l$.

Analogous results and considerations can be made in the case of the pure Neumann problem:

$$\left.\begin{array}{rcl} -\Delta u + a(x)u = u^{2^*-1} & in & \Omega \\ u > 0 & in & \Omega \\ \frac{\partial u}{\partial \nu} = 0 & on & \partial\Omega \end{array}\right\} \qquad (17)$$

about which it has been stated the

Theorem 10. Let be $x_0 \in \partial\Omega$, $a(x)$ as in (15), $\alpha \in L^{\frac{N}{2}}(\mathbb{R}^N)$, $|\alpha|_{L^{\frac{N}{2}}(\Pi_{x_0})} \neq 0$. Then there exists $\bar{k}$ such that to any $k \in (0, \bar{k}]$ there corresponds a $\bar{\lambda}$ in such a way that for all $\lambda > \bar{\lambda}$, (17) has at least one solution. Moreover, if α also satisfies condition (16) then (17) has at least two distinct solutions. If, further, $a \in L^\infty(\omega)$, ω being a neighbourhood of $\partial\Omega$, then there is at least one additional solution of (17) corresponding to a minimizer of E on V.

Let us sketch the proof of Theorem 9 in the special case (the worse one!) in which m_a is equal to Σ and is not attained. To avoid technicalities, we suppose $x_0 = 0$, $\Pi_{x_0} = \mathbb{R}^N_+$. Moreover we assume that Γ_1 is the graph of a regular function in a suitable neighbourhood B_ρ of 0.

The first step is a study of the compactness situation, in order to be able to work above the level Σ. This question is settled by the following

Proposition 1. Let $\{u_n\}$ be a Palais-Smale sequence for the functional E constrained on V, i.e.

$$u_n \in V \qquad E(u_n) \to c$$

and

$$\nabla E_{|V}(u_n) \to 0 \quad in(W^{1,2}(\Omega, \Gamma_0))^*$$

if $c < S$ then

$$c = [(E(u_0))^{\frac{N}{2}} + h \, \Sigma^{\frac{N}{2}}]^{\frac{2}{N}}$$

where h can take the values 0 and 1 and u_0 is either 0 or a critical point of E on V.

The above proposition implies, in particular, that if $\{u_n\}$ is a Palais-Smale sequence of E in V and $E(u_n) \to c \in (\Sigma, S)$, $\{u_n\}$ must be relatively compact.

The second step is to show that S is a lower bound for the energy of a sign changing critical point u of E on V. Therefore, to a critical value of E belonging to the interval of the energy values (Σ, S) there corresponds a positive solution of (14).

Then the crucial point is to find a critical level of E on V in the energy range (Σ, S). The proof of this fact is tricky, technically complex and rather long. Here we try to explain only the underlying idea. We start choosing two points in the x_N-axis : $x^* = (0, 0, ...0, (x^*)_N) \in \mathbb{R}^N \setminus \bar\Omega$, and $x_* = (0, 0, ...0, (x_*)_N) \in \Omega$, in such a way that the segment joining them intersects $\partial\Omega$ only at 0. Then we consider the sets

$$K_\lambda = \left\{ (y,\sigma) \in \Gamma_1 \times I\!\!R : |y| \leq \frac{r}{\lambda}, \sigma \in [\sigma_1,\sigma_2] \right\}$$

$$H = \left\{ (y,\sigma) \in I\!\!R^N \times I\!\!R : y \in [x_*,x^*], \sigma \geq \frac{1}{3} \right\}$$

where r, σ_1, σ_2 are positive numbers such that $r < \rho$ and $\sigma_1 < \frac{1}{3} < \sigma_2$.

By construction, for any $\lambda > 0, K_\lambda$ and H link (i.e.: any continuous homotopic deformation that brings K_λ into a set that does not intersect ∂H must, at some time, make ∂K_λ intersect ∂H, and any continuous deformation of ∂K_λ into a set that does not intersect H must at some time make ∂K_λ intersect ∂H). Then we construct two maps

$$\phi_\lambda : (\Gamma_1 \cap B_\rho) \times I\!\!R^+ \to V, \quad \phi_\lambda(y,\sigma) = \chi_{\frac{\sigma}{\lambda},y}(x)$$

and

$$\Theta_\lambda : V \to I\!\!R^N \times I\!\!R^+, \quad \Theta_\lambda(u) = (\frac{1}{\lambda}\beta_\lambda(u), \gamma_\lambda(u))$$

where $\beta_\lambda(u) \in I\!\!R^N$ is a kind of energy asymmetric barycenter and $\gamma_\lambda \in I\!\!R$ measures the concentration of u around $\beta_\lambda(u)$, and we prove that if λ is large enough, $\Theta_\lambda \circ \phi_\lambda$ is homotopically equivalent to the identity of $(\Gamma_1 \cap B_\rho) \times [\sigma_1, \sigma_2]$

$$\Theta_\lambda \circ \phi_\lambda \sim I_{(\Gamma_1 \cap B_\rho) \times [\sigma_1,\sigma_2]} \tag{18}$$

Define then

$$c_{1,\lambda} = \sup \{ E(\phi_\lambda(y,\sigma)(x)), \ (y,\sigma) \in \partial K_\lambda \},$$

$$c_{2,\lambda} = \sup \{ E(\phi_\lambda(y,\sigma)(x)), \ (y,\sigma) \in K_\lambda \},$$

$$b_{1,\lambda} = \inf \{ E(u), u \in V, \ \Theta_\lambda(u) \in H \},$$

$$b_{2,\lambda} = \inf \{ E(u), u \in V, \ \Theta_\lambda(u) \in \partial H \}.$$

Very delicate estimates allow to obtain, if k is small and λ large enough, upper bounds for $c_{1,\lambda}$ and $c_{2,\lambda}$ and lower bounds for $b_{1,\lambda}$ and $b_{2,\lambda}$. Moreover, (18) and the fact that K_λ and H link imply relations between $b_{1,\lambda}$ and $c_{1,\lambda}$, $b_{2,\lambda}$ and $c_{2,\lambda}$ respectively. Precisely it can be shown that for any $\epsilon > 0$,"small",fixed arbitrarily, it is possible to find $\bar{k}$ and $\bar{\lambda}$ so that

$$\Sigma < b_{1,\lambda} \leq c_{1,\lambda} < \Sigma + \epsilon < b_{2,\lambda} \leq c_{2,\lambda} < \Sigma + |\alpha|_{L^{\frac{N}{2}}} + \epsilon < S.$$

for any $\lambda > \bar{\lambda}$ and $0 < k < \bar{k}$.

The argument is completed proving the existence of two distinct critical values in the intervals $[b_{1,\lambda} - \delta, c_{1,\lambda} + \delta]$ and $[b_{2,\lambda} - \bar{\delta}, c_{2,\lambda} + \bar{\delta}]$, where δ and $\bar{\delta}$

are suitably small positive numbers. This last fact is a consequence of the linking of H and K_λ, of the relation (18) and of a well known deformation theorem, obtained using the gradient flow associated to the functional E.

We remark, also, that Theorem 10 can be obtained from Theorem 9 considering the special case $\Gamma_0 = \emptyset$.

5 The problem in unbounded domains

In this section we are concerned with the problem

$$\left. \begin{array}{rcl} -\Delta u + a(x)u = u^{p-1} & \text{in } \Omega \\ u > 0 & \text{on } \Omega \\ \frac{\partial u}{\partial \nu} = 0 & \text{in } \Omega \end{array} \right\} \qquad (19)$$

when $\Omega \subset \mathbb{R}^N, N \geq 3$, is an *unbounded* domain with smooth boundary and $p \in (2, 2^*]$.

We emphasize that problem (19) has a lack of compactness even if $p < 2^*$. In fact when Ω is unbounded the embedding

$$j : W^{1,2}(\Omega) \;\; \rightarrow \;\; L^p(\Omega)$$

is not compact whatever p is.

Let us consider, at first, (19) when $a(x) = \lambda > 0$ constant and $p \in (2, 2^*)$.

An almost immediate observation is that if $\Omega = \mathbb{R}^N_+$ then (19) admits a unique ground state solution, that corresponds to the unique, up to translations, solution of the minimization problem

$$m_{\lambda, \mathbb{R}^N_+} = \inf \left\{ \int_{\mathbb{R}^N_+} (|\nabla u|^2 + \lambda u^2) dx, u \in W^{1,2}(\mathbb{R}^N_+), \int_{\mathbb{R}^N_+} |u|^p dx = 1 \right\}. \quad (20)$$

In fact, it is well known that the analogous problem in $\mathbb{R}^N$ (see for instance [BL83a])

$$M_\infty = \inf \left\{ \int_{\mathbb{R}^N} (|\nabla u|^2 + \lambda u^2) dx, u \in W^{1,2}(\mathbb{R}^N), \int_{\mathbb{R}^N} |u|^p dx = 1 \right\} \quad (21)$$

has a unique (modulo translations) spherically symmetric solution. Denoted by $V(x)$ the solution centered at the origin, it is easy to verify that the function defined by $W(x) = 2^{\frac{1}{p}} V(x)$ for any $x \in \mathbb{R}^N_+$ solves (20), that $m_{\lambda, \mathbb{R}^N_+} = \int_{\mathbb{R}^N_+} |\nabla W|^2 + \lambda W^2 = 2^{\frac{2}{p}-1} M_\infty$ and that $2^{-\frac{1}{p}} M_\infty^{\frac{1}{p-2}} W(x)$ is a ground state solution of (19).

Let us consider, now, the case Ω exterior domain, i.e. $\Omega = \mathbb{R}^N \setminus D$ where D is bounded and ∂D is smooth.

A general existence result of ground state solutions for (19) is given in [Est91], where the following theorem is proven:

Theorem 11. *Let be* $\Omega = \mathbb{R}^N \setminus D$ *with* D *any bounded domain. For every* $p \in (2, 2^*)$ *there exists a ground state solution of* (19).

The proof is carried out by a minimization argument together with a sufficient condition for the relative compactness of the minimizing sequences. Denoted by

$$m_{\lambda,\Omega} = \inf \left\{ \int_\Omega |\nabla u|^2 + \lambda |u|^2 dx, u \in W^{1,2}(\Omega), \int_\Omega |u|^p = 1 \right\}, \qquad (22)$$

it is proven that the strict inequality

$$m_{\lambda,\Omega} < M_\infty \qquad (23)$$

is equivalent to the relative compactness in $W^{1,2}(\Omega)$ of all minimizing sequences. Then the fact that (23) actually holds is verified using suitable test functions. So, the argument is perfectly analogous to that used for the "critical" case and already seen in section 2.

It is worth remarking that, on the contrary, the same equation with zero Dirichlet boundary conditions cannot be solved by minimization (see [BC87]).

In [Est91] the case $\Omega = \mathbb{R}^N \setminus B_R$, $B_R = \{x \in \mathbb{R}^N : |x| < R\}$ has been investigated in a deeper way and the results can be summarized as follows:

Theorem 12. *Let be* $\Omega = \mathbb{R}^N \setminus B_R, a(x) = \lambda > 0$. *Then any ground state solution* u_R *of* (19) *is not radially symmetric. Moreover*

$$\lim_{R \to +\infty} m_{\lambda,\mathbb{R}^N \setminus B_R} = 2^{\frac{2}{p}-1} M_\infty$$

$$\lim_{R \to 0} m_{\lambda,\mathbb{R}^N \setminus B_R} = M_\infty.$$

Furthermore if u_R *is a solution of* (19), *for any* $R > 0$ *there exist* $C_R > 0$ *and* $x_R \in \partial B_R$ *such that*

$$\lim_{R \to +\infty} \|u_R - C_R W(\cdot + x_R)\|_{W^{1,2}(\mathbb{R}^N \setminus B_R)} = 0$$

and for every $R > 0$ *there exists* $x_R \in \mathbb{R}^N \setminus \bar{B}_R$, *such that the sequence of functions* $u_R(\cdot + x_R)$, *conveniently extended to* $\mathbb{R}^N$ *has a subsequence which converges in* $W^{1,2}(\mathbb{R}^N)$ *to* V, *when* R *converges to 0*.

In the case of unbounded domains, also, it has been proven that the number of low energy solutions is affected by the topology of $\partial\Omega$: precisely in [Wan92b] it is stated that if $a(x) = \lambda$ and $\Omega = \mathbb{R}^N \setminus D, D$ bounded and smooth, the number of the nonconstant solutions of (19), if λ is large enough, is bounded from below by the $cat(\partial\Omega)$.

Let us turn, now, to the case $p = 2^*$.

The situation appears, at a first sight, quite different from the subcritical case.

For instance, if $a(x) = \lambda > 0$ constant and $\Omega = \mathbb{R}^N_+$, it is not difficult to realize that (19) has no solutions. In fact, if u were a solution, its extension to whole $\mathbb{R}^N$, u^* obtained by reflection, would solve $-\Delta u + \lambda u = u^{2^*-1}$ in $\mathbb{R}^N$ and, on the contrary, this problem has no nontrivial solutions, as follows from a generalized version of the Pohozaev identity (see [BL83a]). Moreover, it is still not fully clear how to handle critical problems in exterior domains. Indeed, if $\Omega = \mathbb{R}^N \setminus D, D$ bounded, $\partial\Omega$ is smooth and there is a point $\bar{x} \in \partial\Omega$ such that the mean curvature at $\bar{x}$ of $\partial\Omega$ with respect to the outward normal at $\bar{x}$ (inward the bounded set D !), is positive, it is easy to understand that the same argument of section 2 allows to obtain, by minimization, a solution of (19) when either $a(x) = \lambda > 0$ or $a(x) \in L^{\frac{N}{2}}(\Omega) \cap L^\infty(\omega)$, ω being a neighborood of $\bar{x}$. On the contrary, if Ω is the complement of a convex set as a ball, $a(x) = \lambda > 0$ and we consider the minimization problem (22) it is clear that $m_\lambda(\Omega) \le M_\infty = \Sigma$ but it is not clear how to verify wheter the strict inequality or the equality holds. Moreover, since the second conjecture seems more reasonable, if we assume $m_\lambda(\Omega) = \Sigma$ it appears not easy finding a method to prove either existence or nonexistence of solutions.

When $\Omega = \mathbb{R}^N_+$ and $a(x) \ne \lambda$ constant, in spite of the fact that the problem (19) cannot be solved by minimization, in [CP97], it has been shown that there exist high energy solutions. Let us consider functions $a(x)$ satisfying the assumption

$$\left.\begin{array}{ll}
\text{i)} \ \lim_{|x|\to+\infty} a(x) = a_\infty \ge 0 & a_\infty \in \mathbb{R} \\
\text{ii)} \ a(x) \ge a_\infty & \text{for all } x \in \mathbb{R}^N_+ \\
\text{iii)} \ a(x) - a_\infty \in L^{\frac{N}{2}}(\mathbb{R}^N_+), & |a(x) - a_\infty|_{L^{\frac{N}{2}}(\mathbb{R}^N_+)} \ne 0
\end{array}\right\} \quad (24)$$

and set

$$inf\{E(u) : u \in V\} = \Sigma_a \quad (25)$$

then $\Sigma_a = \Sigma$ and the minimization problem has no solution. The proof of the above equality can be obtained by observing that, of course $\Sigma_a \ge \Sigma$, because $a(x) \ge 0$, and, on the other hand, considering the sequence of functions $\psi_{\frac{1}{n},0}(x) = \eta(|x|)\Phi_{\frac{1}{n},0}(x)$, where η is a "cut-off" function and Φ is defined in (13), the properties of $a(x)$ and well known estimations (see f.i. [BN83]), imply

$$\lim_{n\to+\infty} E\left(\frac{\psi_{\frac{1}{n},0}}{|\psi_{\frac{1}{n},0}|_{L^{\frac{N}{2}}(\mathbb{R}^N_+)}}\right) = \Sigma.$$

Furthermore, the infimum in (25) is not achieved . In fact, in the opposite case, if $u \ge 0$ were a function realizing the infimum, denoting by u^* and a^* the extension, by reflection ,to whole $\mathbb{R}^N$ of u and a respectively, we would have

$$S \le \frac{\int_{\mathbb{R}^N} |\nabla u^*|^2 dx}{|u^*|^2_{L^{2^*}(\mathbb{R}^N)}} \le \frac{\int_{\mathbb{R}^N}(|\nabla u^*|^2 + a^*(x)(u^*(x))^2)dx}{|u^*|^2_{L^{2^*}(\mathbb{R}^N)}} = S$$

and, thus, we would obtain the contradiction

$$0 = \int_{\mathbb{R}^N} a^*(x)(u^*(x))^2 dx = \int_{\mathbb{R}^N} a^*(x)\Psi^2_{\sigma,y}(x)dx > 0.$$

In [CP97]the following theorems are proven

Theorem 13. *Let $a(x)$ satisfy* (24) *and let be $a_\infty > 0$. Then there exists a positive number A such that, if $a_\infty \in (0, A)$,* (19) *admits at least a positive solution $v \in W^{1,2}(\mathbb{R}^N_+)$. Moreover, if the further condition*

$$|a(x) - a_\infty|_{L^{\frac{N}{2}}(\mathbb{R}^N_+)} < (1 - 2^{-\frac{2}{N}})S$$

is satisfied, (19) *has at least another solution $u \in W^{1,2}(\mathbb{R}^N_+)$.*

Theorem 14. *Let $a(x)$ satisfy* (24) *and let be $a_\infty = 0$. Assume that*

$$|a|_{L^{\frac{N}{2}}(\mathbb{R}^N_+)} < (1 - 2^{-\frac{2}{N}})S$$

holds. Then (19) *has at least one solution $u \in D^{1,2}(\mathbb{R}^N_+)$.*

Plainly, the solutions, whose existence is stated in the above theorems, correspond to positive functions for which $E(u) > \Sigma$. The idea of the proof of their existence is similar to that, exposed in the previous section, of theorem 14, so, we do not expatiate upon it here: we refer the interested reader to [CP97].

We close this section by remarking the different nature of the solutions above found. The "first" solution, v, found in case $a_\infty > 0$, has energy very near to Σ and vanish when $a_\infty \to 0$. On the contrary, the solution u has energy bounded from below by a number, independent of a_∞, strictly larger than Σ.

Solitons and Relativistic Dynamics

V. Benci and D. Fortunato [*]

1 Introduction

A soliton is a solution of a field equation whose energy travels as a localized packet and which preserves its form under perturbations. In this respect solitons have a particle-like behavior.

In this paper we study a Lorentz invariant equation in three space dimensions which has soliton like solutions. These solitons have some interesting properties since they behave as relativistic bodies, namely:

- they experience a relativisic contraction in the direction of the motion;
- the rest mass is a scalar and not a tensor;
- the celebrated Einstein equation

$$E = mc^2 \tag{1}$$

holds;
- the mass increases with the velocity by the factor γ.

Moreover a topological invariant, consisting of a k-ple of integer numbers, is associated to these solitons. It allows to define the electric charge.

In the last two sections the interaction between the soliton and the electromagnetic field is analysed.

In this analysis we meet with many mathematical questions: some of them are solved rigorously, others require further investigations and in this paper we give just some heuristic arguments.

2 The equation

In our model a field is a function

$$\psi : \mathbf{R}^{3+1} \to \mathbf{R}^n - \Sigma, \qquad \psi = (\psi^1, ..., \psi^n).$$

where $\mathbf{R}^{3+1}$ is the space-time (the space and time coordinates will be respectively denoted by x and t) and the target space $\mathbf{R}^n - \Sigma$ is the internal parameters space; the set Σ is supposed to be closed and it will be called the singular set; moreover we suppose that $0 \notin \Sigma$. Since we require the Lorentz invariance , we shall consider Lagrangian densities of the form

$$\mathcal{L}(\psi, \sigma) = -\frac{1}{2}\alpha(\sigma) - V(\psi), \tag{2}$$

---

[*] Supported by M.U.R.S.T. (40% and 60% funds); the second author was supported also by E.E.C., Program Human Capital Mobility (Contract ERBCHRXCT 940494)

where

$$\sigma = c^2 |\nabla\psi|^2 - |\psi_t|^2$$

$\alpha : \mathbf{R} \to \mathbf{R}$ and the potential function V is defined in $\mathbf{R}^n - \Sigma$. c is the velocity of the light and $\nabla\psi$, ψ_t denote respectively the Jacobian with respect to x and the derivative with respect to t.

The action functional related to (2) is

$$S(\psi) = \int_{\mathbf{R}^{3+1}} \mathcal{L}(\psi,\sigma)\,dxdt = \int_{\mathbf{R}^{3+1}} \left[-\frac{1}{2}\alpha(\sigma) - V(\psi)\right]dxdt.$$

So the Euler-Lagrange equations are

$$\frac{\partial}{\partial t}\left[\alpha'(\sigma)\frac{\partial\psi}{\partial t}\right] - c^2\nabla\cdot[\alpha'(\sigma)\nabla\psi] + \frac{\partial V}{\partial \xi}(\psi) = 0 , \qquad (3)$$

where

$$\frac{\partial V}{\partial \xi} = \left(\frac{\partial V}{\partial \xi_1}, ..., \frac{\partial V}{\partial \xi_n}\right).$$

and $\nabla\cdot[\alpha'(\sigma)\nabla\psi]$ denotes the vector whose j-th component is given by $div\left[\alpha'(\sigma)\nabla\psi^j\right]$.

When $\alpha(\sigma) = \sigma$ the equations (3) reduce to the semilinear wave equation

$$\frac{\partial^2\psi}{\partial t^2} - c^2\Delta\psi + \frac{\partial V}{\partial \xi}(\psi) = 0, \qquad (4)$$

The static solutions of the (3) solve the equation

$$-c^2\nabla\cdot[\alpha'(\sigma)\nabla\psi] + \frac{\partial V}{\partial \xi}(\psi) = 0, \qquad (5)$$

Let $u = u(x_1, x_2, x_3)$ be a solution of the (5), and consider a vector $\mathbf{v}$ with $|\mathbf{v}| < c$. For simplicity we take $\mathbf{v} = (v, 0, 0)$. Then it is easy to verify that

$$\psi_{\mathbf{v}}(x_1, x_2, x_3, t) = u\left(\frac{x_1 - vt}{\sqrt{1 - \left(\frac{v}{c}\right)^2}}, x_2, x_3\right). \qquad (6)$$

is a solution of the equations (3).

Notice that the function u experiences a contraction of a factor

$$\gamma = \frac{1}{\sqrt{1 - \left(\frac{v}{c}\right)^2}}$$

in the direction of the motion: and this is a consequence of the fact that (3) is Lorentz invariant.

Observe that the (5) are the Euler-Lagrange equation relative to the energy for functions independent of t. In fact we have (cf. e.g. [GF63] pg. 184 or [LL70])

$$T_{00} = \sum_{j=1}^n \frac{\partial\mathcal{L}}{\partial\left(\frac{\partial\psi^j}{\partial t}\right)}\frac{\partial\psi^j}{\partial t} - \mathcal{L}$$

$$= -\frac{1}{2}\sum_{j=1}^{n}\alpha'(\sigma)\frac{\partial\sigma}{\partial\left(\frac{\partial\psi^j}{\partial t}\right)}\frac{\partial\psi^j}{\partial t} + \frac{1}{2}\alpha(\sigma) + V(\psi)$$

$$= \alpha'(\sigma)\sum_{j=1}^{n}\left(\frac{\partial\psi^j}{\partial t}\right)^2 + \frac{1}{2}\alpha(\sigma) + V(\psi)$$

$$= \alpha'(\sigma)\left|\frac{\partial\psi}{\partial t}\right|^2 + \frac{1}{2}\alpha(\sigma) + V(\psi)$$

and so

$$E(\psi) = \int T_{00}\,dx = \int \left[\alpha'(\sigma)\left|\frac{\partial\psi}{\partial t}\right|^2 + \frac{1}{2}\alpha(\sigma) + V(\psi)\right]dx \qquad (7)$$

Thus, if ψ is independent of t, we have

$$E(\psi) = \int\left[\frac{1}{2}\alpha\left(c^2\,|\nabla\psi|^2\right) + V(\psi)\right]dx. \qquad (8)$$

Moreover, if we make the following assumptions on α and V :

$$\alpha'(0) = 1$$

$$V''(0)\,[\xi,\xi] = \omega_0^2|\xi|^2$$

the linearized equation (3) at $\psi = 0$ reduces to the Klein–Gordon equation :

$$\psi_{tt} - c^2\,\Delta\psi + \omega_0^2\psi = 0. \qquad (9)$$

3   The topological invariant

In this section we define a homotopic invariant. This invariant consists of an k-ple of integer numbers. First of all this invariant is useful in the proof of the existence of solitons and to classify them. Moreover, and this seems quite interesting, it provides a discrete invariant of the motion just as the quantum numbers related to the superselection rules (e.g. charge, leptonic number, barionic number, etc.).

We consider an homeomorphism

$$\pi : S^3 \to \mathbf{R}^3 \cup \{\infty\}$$

such that $\pi(N) = \infty$, where S^3 is the 3-dimensional sphere and N is its north pole. Now, set

$$C_0(\mathbf{R}^3, \mathbf{R}^n - \Sigma) = \left\{u \in C(\mathbf{R}^3, \mathbf{R}^n - \Sigma)\mid \lim_{|x|\to\infty} u(x) = 0\right\}.$$

For every function $u \in C_0(\mathbf{R}^3, \mathbf{R}^n - \Sigma)$, we can define the continuous map

$$u \circ \pi : S^3 \to \mathbf{R}^n - \Sigma.$$

We denote with $u^{\#}$ the homotopy class of $u \circ \pi$, that is

$$u^{\#} = [u \circ \pi] \in \pi_3(\mathbf{R}^n - \Sigma).$$

Where $\pi_3(\mathbf{R}^n - \Sigma)$ denote the 3-homotopy group of $\mathbf{R}^n - \Sigma$.

Thus, to every field $u \in C_0(\mathbf{R}^3, \mathbf{R}^n - \Sigma)$ is associated the homotopy class $u^{\#} \in \pi_3(\mathbf{R}^n - \Sigma)$.

Clearly this invariant is stable under uniform convergence.

In many cases (e.g. when the group is finitely generated and torsion free)

$$\pi_3(\mathbf{R}^n - \Sigma) \cong \mathbf{Z}^k$$

with $k \in \mathbf{N}$. For example this situation occurs when

$$\mathbf{R}^n - \Sigma = \mathbf{R}^4 - \{\xi_1, \ldots, \xi_k\}.$$

So, if we fix a set of generators $\{e_1, \ldots, e_k\}$, every configuration $u \in C_0(\mathbf{R}^3, \mathbf{R}^n - \Sigma)$ is characterized by a k-tuple of integer numbers. If $u^{\#} = q_1 e_1 + \ldots + q_k e_k$ $(q_1, \ldots, q_k \in \mathbf{Z})$, then we set

$$q(u) = (q_1, \ldots, q_k).$$

If the field u has the energy concentrated in different regions of the space $C_1, \ldots, C_\ell$, then it is possible to associate a k-ple of numbers $q(u, C_j) = (q_1, \ldots, q_k)$ to any of these components C_j in such a way that

$$q(u) = \sum_{j=1}^{\ell} q(u, C_j).$$

In order make these statements rigorous, it is necessary define a local topological invariant and more work is necessary.

Definition 1. *For every $u \in C_0(\mathbf{R}^3, \mathbf{R}^n - \Sigma)$ we call δ-support of u the compact set defined as follows*

$$K_\delta(u) = \{x : |u(x)| \geq \delta\}. \tag{10}$$

where

$$\delta < \delta_0$$

and

$$\delta_0 = dist(0, \Sigma). \tag{11}$$

Now we decompose $K_\delta(u)$ into its connected components (which might be infinite but this fact does not create trouble):

$$K_\delta(u) = \bigcup_{j=1}^{\ell} C_j \tag{12}$$

and we want to define

$$q(u, C_j) \in \pi_3(\mathbf{R}^n - \Sigma)$$

with suitable properties (finite additivity and so on).

Since $\partial C_j \subset \partial K_\delta(u)$ and $\|u(x)\| = \delta$ for $x \in \partial K_\delta(u)$, by the continuity of u there exists an open neighborhood N of ∂C_j such that $u(N) \subset \mathbf{R}^n - \Sigma$. Then, by the Uryson lemma, there exists a continuous function

$$c_j : \mathbf{R}^3 \to [0, 1]$$

such that

$$c_j(x) = \begin{cases} 1 & \text{on } C_j, \\ 0 \text{ on } \mathbf{R}^3 - (C_j \cup N). \end{cases}$$

By this construction, we have that the function

$$\tilde{\varphi}(x) := c_j(x)u(x)$$

takes its values in $\mathbf{R}^n - \Sigma$, i.e. extending $\tilde{\varphi}$ at the point at infinity, $\tilde{\varphi} \circ \pi \in C\left(S^3, \mathbf{R}^n - \Sigma\right)$. Thus it makes sense to give the following definition:

Definition 2. *We set*

$$q(u, C_j) = \tilde{\varphi}^\#.$$

This definition is well posed, *i.e.* it does not depend on $\tilde{\varphi}$ (see [BFP96]).

The local invariant satisfy the finite additivity property. Let C_1, C_2 be connected components of $K_\delta(u)$ with $C_1 \cap C_1 = \emptyset$, then we have

$$q(u, C_1 \cup C_2) = q(u, C_1) + q(u, C_2)$$

(the sum is meant, of course, in the group $\pi_3(\mathbf{R}^n - \Sigma)$).

4   Non existence results

The simplest equation of type (3) is obtained by choosing

$$\alpha(\sigma) = \sigma \tag{13}$$

Then the equation (5) for the static solutions and its energy functional (8) become respectively

$$-c^2 \Delta \psi + \frac{\partial V}{\partial \xi}(\psi) = 0$$

$$E(u) = \int_{\mathbf{R}^3} \frac{c^2}{2} |\nabla u|^2 + V(u) \, dx. \tag{14}$$

If $V \geq 0$, it is not difficult to show, by a rescaling argument, that any local minimizer u of (14) is necessarily trivial, *i.e.* it takes a constant value which is a minimum point of V (Derrick's Theorem, see [DEGM82]).

We shall deduce this non existence result by a more general lemma which will be used in the following to prove some dynamical properties of the solitons:

Lemma 1. *Let u be a critical point of the energy (14) (and hence a weak solution of the equation (5)); then*

$$\frac{1}{2}\int \alpha(|\nabla u|^2)\,dx + \int V(u)\,dx = \int \alpha'(|\nabla u|^2)\left|\frac{\partial u}{\partial x_i}\right|^2 dx, \quad i = 1,2,3 \qquad (15)$$

and, consequently

$$\frac{1}{2}\int \alpha(|\nabla u|^2)\,dx + \int V(u)\,dx = \frac{1}{3}\int \alpha'(|\nabla u|^2)\,|\nabla u|^2\,dx. \qquad (16)$$

whenever the integrals converge (here, to symplify the notation, we have assumed c = 1).

Proof. Let

$$u_\lambda(x) = u(\lambda x_1, x_2, x_3);$$

then, setting $y = (\lambda x_1, x_2, x_3)$

$$E(u_\lambda) = \frac{1}{2}\int \alpha(|\nabla u_\lambda|^2)\,dx + \int V(u_\lambda)\,dx$$

$$= \frac{\lambda^{-1}}{2}\int \alpha\left(\lambda^2\left(\frac{\partial u}{\partial x_1}\right)^2 + \left(\frac{\partial u}{\partial x_2}\right)^2 + \left(\frac{\partial u}{\partial x_3}\right)^2\right) dy + \lambda^{-1}\int V(u)\,dy$$

Since u is a critical point of E,

$$\frac{d}{d\lambda}E(u_\lambda)\bigg|_{\lambda=1} = 0$$

and since

$$\frac{d}{d\lambda}E(u_\lambda) = -\frac{\lambda^{-2}}{2}\int \alpha\left(\lambda^2\left(\frac{\partial u}{\partial x_1}\right)^2 + \left(\frac{\partial u}{\partial x_2}\right)^2 + \left(\frac{\partial u}{\partial x_3}\right)^2\right) dy$$

$$+ \int \alpha'\left(\lambda^2\left(\frac{\partial u}{\partial x_1}\right)^2 + \left(\frac{\partial u}{\partial x_2}\right)^2 + \left(\frac{\partial u}{\partial x_3}\right)^2\right)\left(\frac{\partial u}{\partial x_1}\right)^2 dy$$

$$-\lambda^{-2}\int V(u)\,dy$$

we get

$$-\frac{1}{2}\int \alpha(|\nabla u|^2)\,dy + \int \alpha'(|\nabla u|^2)\left(\frac{\partial u}{\partial x_1}\right)^2 dy - \int V(u)\,dy = 0$$

which gives the 15. The 16 is obtained adding the 15 with $i = 1,2,3$. $\qquad \square$

By the above lemma we get the Derrick's theorem.

Theorem 1. *Let $V \geq 0$ and let u be a solution of the equation*

$$-\Delta u + \frac{\partial V}{\partial \xi}(u) = 0, \tag{17}$$

with finite energy. Then u is a constant c such that $V(c) = 0$.

Proof. If $\alpha(\sigma) = \sigma$ then the 16 gives

$$\frac{1}{2}\int |\nabla u|^2 \, dx + \int V(u) \, dx = \frac{1}{3}\int |\nabla u|^2 \, dx,$$

and so

$$\frac{1}{6}\int |\nabla u|^2 \, dx + \int V(u) \, dx = 0.$$

$\square$

From the above theorem we have:

Corollary 1. *For any C^1 potential V, the energy (14) does not have a smooth nontrivial local minimizer.*

Proof. A smooth nontrivial local minimizer satisfies the equation (17) and hence, if $V \geq 0$, u is trivial by Th. 1. If $V(\xi) < 0$ for some $\xi \in \mathbf{R}^n$, then, it is easy to realise that

$$\inf E(\psi) = -\infty.$$

$\square$

So, by the corollary 1, people have been forced to consider non positive potentials and to look for saddle points, instead of minima. However for these static solutions we have lack of stability . As an example we recall that, if we take scalar fields u with potentials

$$V(\xi) = \frac{1}{2}\xi^2 - \frac{1}{4}\xi^4, \ \xi \in \mathbf{R},$$

critical points of the energy functional

$$E(u) = \int_{\mathbf{R}^3} \left[\frac{c^2}{2}|\nabla u|^2 + \frac{1}{2}u^2 - \frac{1}{4}u^4 \right] dx.$$

have been found in [Ber72] and [Poh65] and for more general potentials in [BL78], [Str77]; but in [AD70] and [BC81] it has been proved that these static solutions are not stable (see also [Str83] and its references)

The difficulties related to Derrick's Theorem can be avoided in other ways: if solutions with infinite energy (i.e. functions in $W_{loc}^1(\mathbf{R}^n)$) are allowed, then the equation (4) may have solutions. Each of these solutions $\psi(t, x)$ can be written as follows

$$\psi(t, x) = u_0(x) + w(t, x);$$

here $u_0(x) \in W^1_{loc}(\mathbf{R}^n)$ is a solution of the (17) which represents the vacuum and $w(t, x)$ is the physically meaningful part of the field. The energy density (whose integral diverges) is replaced by the normalized energy density which is defined as follows:

$$\mathcal{E}(x) = \left[\frac{1}{2}\psi_t^2 + \frac{c^2}{2}|\nabla\psi|^2 + V(\psi)\right] - \left[\frac{1}{2}u_{0t}^2 + \frac{c^2}{2}|\nabla u_0|^2 + V(u_0)\right]$$

Thus the problem consists in finding solutions such that

$$\int \mathcal{E}(x)\,dx < +\infty.$$

Most people in theoretical physics follow this approach (we refer to [Str] for more details).

5 Existence results

To get the existence of solutions of the equation (5) which minimize the energy functional (8), we follow a different approach and we look for functions α (see (2)) different from the (13). Namely we consider α nonlinear

$$\alpha(\sigma) = \sigma + a_2\sigma^2 + a_3\sigma^3 + \dots$$

and look for the simplest choice of α. A priori the simplest choice is

$$\alpha(\sigma) = \sigma + a_2\sigma^2,$$

however this is a bad choice for the evolution equation (3), in fact $\alpha'(\sigma)$ might be negative for some values of σ and this is a "disaster" for (3). Then it is natural to consider the next simplest choice, namely

$$\alpha(\sigma) = \sigma + \frac{\varepsilon}{3}\sigma^3$$

being $\varepsilon > 0$. Of course other choices are possible if we do not assume that the lagrangian splits as in (2) as it is in the Skyrme model ([Sky61]). Then (3) can be written

$$\frac{\partial}{\partial t}\left[(1 + \varepsilon\sigma^2)\,\psi_t\right] - c^2\nabla \cdot \left[(1 + \varepsilon\sigma^2)\,\nabla\psi\right] + \frac{\partial V}{\partial \xi}(\psi) = 0,$$

or

$$\Box\,\psi + \varepsilon\,\Box_6\psi + \frac{\partial V}{\partial \xi}(\psi) = 0 \qquad (18)$$

where

$$\Box_6 \, \psi = \frac{\partial}{\partial t}\left[\left(c^2\,|\nabla\psi|^2 - (\psi_t)^2\right)^2 \psi_t\right] - c^2\nabla\cdot\left[\left(c^2\,|\nabla\psi|^2 - (\psi_t)^2\right)^2 \nabla\psi\right].$$

So the static solutions u solve the system of equations

$$-c^2 \Delta u - \varepsilon c^6 \Delta_6 u + \frac{\partial V}{\partial \xi}(u) = 0, \tag{19}$$

where

$$\Delta_6 u = \nabla\cdot\left(|\nabla u|^4\,\nabla u\right).$$

Then the energy functional becomes

$$E(u) = \int_{\mathbf{R}^3}\left(\frac{c^2}{2}|\nabla u|^2 + \varepsilon\frac{c^6}{6}|\nabla u|^6 + V(u)\right)\,dx \tag{20}$$

We make the following assumptions on V:

- $V \in C^2(\mathbf{R}^n - \Sigma, \mathbf{R})$;
- $V(\xi) \geq 0$ for every $\xi \in \mathbf{R}^n - \Sigma$ and $V(\xi) = 0 \Rightarrow \xi = 0$;
- there exist $k, r > 0$ such that,

$$\text{dist}(\xi, \Sigma) < r \Rightarrow V(\xi) > \frac{k}{\text{dist}(\xi, \Sigma)^6} \tag{21}$$

- $V''(0)\,[\xi, \xi] = \omega_0^2|\xi|^2$, $\omega_0 > 0$.

A simple function V which satisfies the above assumptions is the following:

$$V(\xi) = \frac{1}{2}\cdot\frac{|\xi|^2}{\prod_j |\xi - \xi_j|^6};$$

in this case we have $\Sigma = \{\xi_1, ..., \xi_k\}$, $\xi_j \neq 0$ and

$$\omega_0 = \frac{1}{\prod_j |\xi_j|^3}.$$

Observe that $\alpha'(0) = 1$, then linearizing (3) at 0, we get the Klein–Gordon equation

$$\Box\psi + \omega_0\psi = 0.$$

Of course we can consider our evolution equation 18 as a dynamical system. The configuration space for this system is given by

$$\Lambda = \left\{u \in W^{1,6}(\mathbf{R}^3, \mathbf{R}^n) \cap W^{1,2}(\mathbf{R}^3, \mathbf{R}^n) \mid u(\mathbf{R}^3) \subset \mathbf{R}^n - \Sigma\right\}$$

equipped with the norm

$$\|u\|_\Lambda = \|u\|_{W^{1,6}} + \|u\|_{W^{1,2}}.$$

where $W^{1,6}(\mathbf{R}^3, \mathbf{R}^n)$ and $W^{1,2}(\mathbf{R}^3, \mathbf{R}^n)$ are Sobolev spaces. Observe that the presence of the potential V, which is singular on Σ, implies that the functions, on which the energy is finite, take values on $\mathbf{R}^n - \Sigma$ (see [BFP98]).

By the Sobolev embedding theorem, the functions in Λ are continuous and it is possible to prove that they decay to 0 at infinity (see e.g. [BFP98] Prop. 2.1). So the topological invariant introduced in section 3 is well defined for the functions in Λ. For this reason, the algebraic structure of $\pi_3(\mathbf{R}^n - \Sigma)$ is reflected into a decomposition of the configuration space:

$$\Lambda = \bigcup_{q \in \pi_3(\mathbf{R}^n - \Sigma)} \Lambda_q \tag{22}$$

into connected components $\Lambda_q = \{u \in \Lambda : u^{\#} = q\}$ each of them will be called sector.

We point out that other models (see [Sky61], [Est86], [EL88]) do not contain the singular potential and the topological classification of the fields follows from the fact that they take values in suitable compact manifolds.

It can be shown that the minimizers of E (see (20)) on Λ_q satisfy suitable stability properties (see [BFP98]). Then we give the following definition of soliton :

Definition 3. *A soliton is a function u which minimizes the energy E in a sector Λ_q; the set $K_\delta(u)$ (with $\delta \ll \delta_0 = dist(0, \Sigma)$ which is assumed fixed once for ever) will be called δ-support of the soliton and its diameter will be called diameter of the soliton.*

The definition of diameter of a soliton might appear unsatisfactory since it depends on the choice of δ; on the other hand, since u decays exponentially, the energy outside $K_\delta(u)$ (with δ small) is irrelevant for many problems and an appropriate choice of δ seems possible.

The proof of the existence of solitons is not achieved in a direct fashion since the problem is translation invariant. In fact, due to this invariance, the sectors Λ_q are not weakly closed and therefore we cannot straightforward minimize the energy on each Λ_q.

To overcome the difficulties arising from this lack of compactness we have used an argument in the spirit of the concentration-compactness principle for unbounded domains (see [Lio85], [BC87]), and the following theorem can be proved (see [BFP98] and [BFP96]):

Theorem 2. *Let V satisfy the assumptions (21) and set*

$$\Xi = \{q \in \pi_3(\mathbf{R}^n - \Sigma) \mid E \text{ has a minimum in } \Lambda_q\}$$

Then the subgroup of $\pi_3(\mathbf{R}^n - \Sigma)$ generated by Ξ coincides with $\pi_3(\mathbf{R}^n - \Sigma)$ itself.

Since every local minimizer of E gives rise to a weak solution of the (19), then, if $\pi_3(\mathbf{R}^n - \Sigma)$ is not trivial, there exists at least a non trivial solution of the (19).

In particular, if $\pi_3(\mathbf{R}^n - \Sigma) \simeq \mathbf{Z}^k$, Theorem 2 implies that there exist at least k homotopically distinct static solutions of 18.

Also notice that, if $u(x) \in \Lambda_q$ solves the (19), also $u(-x) \in \Lambda_{-q}$ solves that equation and so, when $\pi_3(\mathbf{R}^n - \Sigma) \simeq \mathbf{Z}^k$, there are at least $2k$ solitons.

6   Mass and energy

In order to show the relativistic behaviour of this kind of solitons, we need to define their inertial mass. The most natural thing to do is to define it as the ratio between the momentum and the velocity of the soliton.

However this definition is possible only if the momentum and the velocity are parallel; otherwise the mass is a tensor. So the first thing to do is to compute the momentum and then the mass tensor of our solitons.

Let $\psi(x,t) = \psi(x_1, x_2, x_3, t)$ be a solution of the 3 (where for simplicity we set $c = 1$). It is well known (see e.g. [GF63]) that the momentum of ψ is given by

$$p_k = \int T_{0,k} \, dx, \quad k = 1, 2, 3 \tag{23}$$

where $T_{i,k}$ is the energy-momentum tensor defined by

$$T_{ki} = \sum_{j=1}^{n} \frac{\partial \mathcal{L}}{\partial \left(\frac{\partial \psi_j}{\partial x_k} \right)} \frac{\partial \psi_j}{\partial x_i} - \delta_{ik} \mathcal{L}; \quad \text{with } x_0 := t \tag{24}$$

In our case, for $k \neq 0$, we have

$$T_{k,0} = -\alpha'(\sigma) \, \psi_{x_k} \cdot \psi_t \tag{25}$$

so we can easily compute the momentum of the soliton 6:

Lemma 2. Let $\psi_{\mathbf{v}}$ be the solution of 3 (with $c = 1$) given by the 6, then the momentum is given by

$$p_1(\psi_{\mathbf{v}}) = v\gamma \int \alpha'(|\nabla u|^2) \left(\frac{\partial u}{\partial x_1} \right)^2 dx$$

$$p_2(\psi_{\mathbf{v}}) = v \int \alpha' \left(|\nabla u|^2 \right) \frac{\partial u}{\partial x_1} \frac{\partial u}{\partial x_2} dx$$

$$p_3(\psi_{\mathbf{v}}) = v \int \alpha' \left(|\nabla u|^2 \right) \frac{\partial u}{\partial x_1} \frac{\partial u}{\partial x_3} dx$$

where

$$\gamma = \frac{1}{\sqrt{1 - v^2}}.$$

Proof. By the 25 we have

$$p_1(\psi_v) = -\int \alpha' \left(|\nabla\psi_v|^2 - \left(\frac{\partial\psi_v}{\partial t}\right)^2 \right) \frac{\partial\psi_v}{\partial x_1} \frac{\partial\psi_v}{\partial t} \, dx;$$

exploiting the Lorentz invariance of $|\nabla\psi_v|^2 - \left(\frac{\partial\psi_v}{\partial t}\right)^2$, we get

$$\left[|\nabla_x\psi_v|^2 - \left(\frac{\partial\psi_v}{\partial t}\right)^2 \right](x_1, x_2, x_3, t) = |\nabla_\xi u|^2 (\xi_1, \xi_2, \xi_3)$$

where

$$\begin{cases} \xi_1 = \gamma(x_1 - vt) \\ \xi_2 = x_2 \\ \xi_3 = x_3 \end{cases}$$

Thus

$$p_1(\psi_v) = -\int \alpha' \left(|\nabla_\xi u|^2 \right) \frac{\partial\psi_v}{\partial x_1} \frac{\partial\psi_v}{\partial t} \, dx_1 \, dx_2 \, dx_3$$

$$= -\int \alpha' \left(|\nabla_\xi u|^2 \right) \frac{\partial u}{\partial \xi_1} \frac{\partial \xi_1}{\partial x_1} \frac{\partial u}{\partial \xi_1} \frac{\partial \xi_1}{\partial t} \frac{1}{\gamma} \, d\xi_1 \, d\xi_2 \, d\xi_3$$

$$= \gamma v \int \alpha' \left(|\nabla_\xi u|^2 \right) \left(\frac{\partial u}{\partial \xi_1}\right)^2 d\xi_1 \, d\xi_2 \, d\xi_3.$$

Now let us compute p_2 :

$$p_2(\psi_v) = -\int \alpha' \left(|\nabla\psi_v|^2 - \left(\frac{\partial\psi_v}{\partial t}\right)^2 \right) \frac{\partial\psi_v}{\partial x_2} \frac{\partial\psi_v}{\partial t} \, dx$$

$$= -\int \alpha' \left(|\nabla_\xi u|^2 \right) \frac{\partial\psi_v}{\partial x_2} \frac{\partial\psi_v}{\partial t} \, dx_1 \, dx_2 \, dx_3$$

$$= \int \alpha' \left(|\nabla_\xi u|^2 \right) \frac{\partial u}{\partial \xi_2} \frac{\partial u}{\partial \xi_1} \gamma v \frac{1}{\gamma} \, d\xi_1 \, d\xi_2 \, d\xi_3$$

$$= v \int \alpha' \left(|\nabla_\xi u|^2 \right) \frac{\partial u}{\partial \xi_2} \frac{\partial u}{\partial \xi_1} d\xi_1 \, d\xi_2 \, d\xi_3.$$

$\square$

The mass matrix $\{m_{ij}\}$ of ψ_v is defined by the following relation:

$$p_i(\psi_v) = \sum_j m_{ij}(\psi_v)v_j,$$

then, by the previous lemma we get:

$$m_{11}(\psi_v) = \gamma \int \alpha'(|\nabla u|^2) \left(\frac{\partial u}{\partial x_1}\right)^2 dx$$

$$m_{12}(\psi_v) = \int \alpha' \left(|\nabla u|^2 \right) \frac{\partial u}{\partial x_1} \frac{\partial u}{\partial x_2} dx$$

$$m_{13}(\psi_v) = \int \alpha' \left(|\nabla u|^2 \right) \frac{\partial u}{\partial x_1} \frac{\partial u}{\partial x_3} dx.$$

and by simmetry arguments we get:

$$m_{ii}(\psi_v) = \gamma \int \alpha'(|\nabla u|^2) \left(\frac{\partial u}{\partial x_i} \right)^2 dx \qquad (26)$$

$$m_{ij}(\psi_v) = \int \alpha' \left(|\nabla u|^2 \right) \frac{\partial u}{\partial x_i} \frac{\partial u}{\partial x_j} dx, \quad for \; i \neq j \qquad (27)$$

So the rest mass matrix is defined as follows:

$$m_{ij}(u) = \int \alpha' \left(|\nabla u|^2 \right) \frac{\partial u}{\partial x_i} \frac{\partial u}{\partial x_j} dx,$$

Now let us compute the energy of a soliton. By the (24) we have

$$E(\psi_v) = \int T_{0,0} \, dx \qquad (28)$$

Since the direct computation of $T_{0,0}$ is quite involved, we start to compute the energy of a static solution; in this case (see (8)) we have that

$$E(u) = \int \left[\frac{1}{2}\alpha(|\nabla u|^2) + V(u) \right] dx \qquad (29)$$

Theorem 3. *If u is a solution of 5 with $c = 1$; then*

$$m_{ii}(\psi_v) = \gamma E(u)$$

$$m_{ij}(\psi_v) = 0 \; if \; i \neq j$$

Proof. Since the matrix $m_{ij}(\psi_v)$ is symmetric, we can put ourselves in a reference frame

$$x' = Tx$$

such that m'_{ij} is diagonal. In this reference frame (we shall omit the '), by the (26)

$$m_{ij}(\psi_v) = \gamma \delta_{ij} \int \alpha'(|\nabla u|^2) \left(\frac{\partial u}{\partial x_i} \right)^2 dx$$

where δ_{ij} is the Kroneker symbol. Then, by the lemma 1 and the (29)

$$m_{ij}(\psi_v) = \gamma E(u)\delta_{ij}$$

Thus $m_{ij}(\psi_v)$ is the identity marix times $\gamma E(u)$. This implies that $m_{ij}(\psi_v) = \gamma E(u)\delta_{ij}$ in any reference frame. $\qquad \square$

By the above theorem we get the following two unexpected facts:

(i) since the mass matrix is a scalar times the identity, the mass is in fact a scalar;

(ii) the mass is equal to the rest energy times γ.

In particular, by the lemma 1, the mass gets the following form:

$$m(\psi_v) = \frac{\gamma}{3} \int \alpha'(|\nabla u|^2) |\nabla u|^2 \, dx \tag{30}$$

and we have the equations

$$\mathbf{p}(\psi_v) = m(\psi_v)\mathbf{v} \tag{31}$$

and

$$m(\psi_v) = \gamma E(u) \tag{32}$$

Notice that

$$(\mathbf{p}(\psi_v), E(\psi_v)) = \left(\int T_{0,1} \, dx, \ \int T_{0,2} \, dx, \ \int T_{0,3} \, dx, \ \int T_{0,0} \, dx \right)$$

is a 4-vector since the equation 3 is Lorentz invariant; so, its Lorentzian norm

$$|\mathbf{p}(\psi_v)|^2 - E(\psi_v)^2 \tag{33}$$

is independent of $\mathbf{v}$. Using this fact we get:

Theorem 4. *If u is a solution of (5) (with $c = 1$) and ψ_v is the function (6), then*

$$E(\psi_v) = \gamma E(u) \tag{34}$$

Proof. Since the (33) is independent of $\mathbf{v}$, $|\mathbf{p}(\psi_v)|^2 - E(\psi_v)^2 = -E(u)^2$; then by (31) and Th. 3

$$E(\psi_v)^2 = E(u)^2 + |\mathbf{p}(\psi_v)|^2 = E(u)^2 + m(u)^2 \gamma^2 v^2$$

$$= E(u)^2 + E(u)^2 \gamma^2 v^2 = \left(1 + \frac{v^2}{1 - v^2} \right) E(u)^2 = \gamma^2 E(u)^2.$$

□

Comparing the (34) with the (32) we get the equality between mass and energy:

$$E(\psi_v) = m(\psi_v) \tag{35}$$

If $c \neq 1$, the above formula gives the celebrated Einstein equation

$$E(\psi_v) = m(\psi_v)c^2$$

7   The dynamics of free solitons

Given $R \in SO(\mathbf{R}^3)$, $\mathbf{v} \in T\mathbf{R}^3$, $\xi \in \mathbf{R}^3$, we set

$$g_{R,\mathbf{v},\xi}(t)\, x = g(t)\, x = RL_{\mathbf{v}}(x - \mathbf{v}t - \xi)$$

where $L_{\mathbf{v}}$ is the relativistic contraction in the direction of $\mathbf{v}$ of the factor $\gamma = \left(\sqrt{1 - |\mathbf{v}|^2}\right)^{-1}$. Also we denote by $\mathcal{G}$ the set of such $g(t)$'s.

Now let S be the set of those solitons u, i.e. solutions u of the eq. (5), which have the baricenter $\beta(u)$ at the origin, where $\beta(u)$ is defined by

$$\beta(u) = \frac{\int \alpha'(|\nabla u|^2)\, |\nabla u|^2\, x\, dx}{\int \alpha'(|\nabla u|^2)\, |\nabla u|^2\, dx}$$

We suppose that S contains one soliton for each sector Λ_q.

If $u \in S$ and $g \in \mathcal{G}$, the function

$$\psi(t, x) = u(g(t)\, x)$$

is a solution of the equation (3); if we "add" ℓ solitons, we obtain a function

$$\psi(t, x) = \sum_{k=1}^{\ell} u_k(g_k(t)\, x), \quad u_k \in S \text{ and } g_k \in \mathcal{G} \tag{36}$$

which we will call multisoliton (time-dependent) function. We will say that the solitons which "build" the $\psi(t, x)$ are located in $\xi_1(t), ..., \xi_\ell(t) \in \mathbf{R}^n$ if these points are the baricenters of the $u_k \circ g_k$'s.

$\psi(t, x)$ is not a solution of the (3) since this equation is nonlinear. However, since the u_k's have exponential decay, $\psi(t, x)$ represents a good approximation of the solution at least for those values of t such that the solitons are far from each other. For example this happens when $t \to \pm\infty$ if

$$g(t)\, x = R_k L_{\mathbf{v}_k}[x - \mathbf{v}_k t - \xi_k]$$

and

$$\mathbf{v}_k \neq \mathbf{v}_h, \ k \neq h$$

If the solitons get close and even more if the solitons collide, the $\psi(t, x)$ is no longer a good approximation of a solution and an additional term must be added. To do this we first define

$$\Gamma = \left\{ h \in W^{1,6}(\mathbf{R}^n) \cap W^{1,2}(\mathbf{R}^n) \mid h(x) = \sum_{k=1}^{\ell} u_k(g_k\, x), \ u_k \in S \text{ and } g_k \in \mathcal{G}_0 \right\}$$

where

$$\mathcal{G}_0 = \{g(0) \mid g(t) \in \mathcal{G}\}\,.$$

A generic element of Γ is called a multisoliton (time-independent) function. For every soliton

$$u_k(g_k\, x) = u_k\left(R_k L_{\mathbf{v}_k}\, [x - \xi_k]\right)$$

its position ξ_k and its velocity $\mathbf{v}_k$ are well defined (actually the velocity is defined up to a sign).

Also, for $q \in \pi_3(\mathbf{R}^n - \Sigma)$, we set

$$\Gamma_q = \left\{ h \in \Gamma \mid h = \sum_{k=1}^{\ell} u_k \circ g_k \; and \; \sum_{k=1}^{\ell} q\,(u_k) = q \right\}$$

Now, given any function $u \in \Lambda_q$, we want to decompose it a "optimal" way as follows:

$$u = h + w \quad with \; h = \sum_{k=1}^{\ell} u_k \circ g_k \in \Gamma_q. \tag{37}$$

The following theorem holds:

Theorem 5. *For any fixed $u_0 \in \Lambda_q$, there is $\bar h \in \Gamma_q$ which minimizes the functional*

$$F(h) = m(u_0 - h); \; h \in \Gamma_q$$

where

$$m(u) = \frac{1}{3} \int \alpha'(|\nabla u|^2)\, |\nabla u|^2 \; dx$$

Thus, by the above theorem, for a generic $u \in \Lambda$ the decomposition (37) is uniquely determined; thus it makes sense to say that the configuration u contains the ℓ solitons u_k.

Using this decomposition, any field $\psi\,(t, x)$ can be written as follows:

$$\psi\,(t, x) = \sum_{k=1}^{\ell(t)} u_k\left(R_k(t) L_{\mathbf{v}_k(t)}\, [x - \xi_k(t)]\right) + w(t, x) \tag{38}$$

Clearly this decomposition, which by theorem 5 can be always produced, is meaningful when $w(t, x)$ is sufficiently small and the distance of the baricenters of the solitons is bigger than their size in such a way that their supports do not intersect.

Since the (36) is an approximated solution of the (3), we expect to find solutions which have the form (38) with $w(t, x)$ small,

$$\frac{d}{dt} R_k(t) = 0$$

and

$$\mathbf{v}_k(t) \cong \frac{d}{dt} \xi_k(t)$$

When the solitons "collide", the functions $\xi_k(t)$ might loose the continuity and the number $\ell(t)$ of solitons contained in the field ψ might change. However, if $\psi(t, x)$ depends continuously on t, the function

$$t \mapsto q(\psi(t, \cdot))$$

is constant, i.e. $\psi(t, \cdot)$ does not change sector in Λ.

Thus the k-ple of integer numbers $q(\psi(t, \cdot))$ can be considered an integral of the motion. Then the numbers $q(\psi(t, \cdot))$ behave as the quantum numbers related to the superselection rules in the elementary particle theory; for this similarity we will call them with this name.

By the previous discussions, if $\psi(t, x)$ is a sufficiently regular solution of the (3) the following quantities are invariant of the motion:

- the global energy
- the global momentum
- the global angular momentum defined by

$$\int M_{0kl}\, dx, \quad (k < l) \tag{39}$$

where M_{ikl} is the angular momentum tensor (cf. e. g. [GF63] pg. 185)
- the global quantum numbers.

If w in the (38) is small, as far as the solitons are far from each other, they move in a straight line since the field ψ decays exponentially out of their supports (i.e. they have a short range interaction with each other). If two or more solitons "collide" and generate other solitons, by virtue of the conservation laws and the equality between mass and energy, the usual laws of the Relativistic Mechanics (cf. [LL70] pg.47-59) are satisfied, moreover the quantum numbers are preserved.

Thus this kind of solitons behave as relativistic particles as long as the "short range" interactions are involved. In the next sections, we will show that these solitons follow a relativistic dynamic even when they are embedded in long range field like the electromagnetic field.

8   The electric charge

In order to show that our solitons behave as relativistic particles under long range interactions, we let them to interact with the electromagnetic field. To carry out this program, we need first to define the electric charge. We recall that, when $\pi_3(\mathbf{R}^n - \Sigma) \simeq \mathbf{Z}^k$, to any configuration $u \in \Lambda$ is associated a $k-$ple of quantum numbers

$$q(u) = (q_1(u), ..., q_k(u)) = q_1(u)e_1 + ... + q_k(u)e_k$$

where $e_1, ..., e_k$ is a set of generators (see section. 3). The first quantum number $q_1(u)$ will be called topological charge. Now we want to define the electric charge

$Q(u)$ of a configuration $u \in \Lambda$ in such a way that $Q(u) = Q_0 q_1(u)$ where Q_0.is a fixed constant.

e_1 is a nontrivial homotopy class in $\pi_3 (\mathbf{R}^n - \Sigma)$; we assume that to this class corresponds a nontrivial cohomology class $[\eta]$ in the De Rham cohomology group $H^3_{DeRham} (\mathbf{R}^n - \Sigma)$; this means that there is a 3-differential form η (closed but non exact) defined in $\mathbf{R}^n - \Sigma$.

In order to fix the ideas and to make this point understandable to those who are not too familiar with topology, we consider the simplest case and assume that,

$$\mathbf{R}^n - \Sigma = \mathbf{R}^4 - \{\bar\xi_1, ..., \bar\xi_k\}.$$

Also, in order to simplify the notation, we set

$$\zeta = \bar\xi_1.$$

In this case, η takes the following aspect

$$\eta = \sum_{1 \leq i < j < k \leq 4} \eta_{ijk}(\xi) d\xi^i \wedge d\xi^j \wedge d\xi^k$$

where

$$\eta_{ijk}(\xi) = \frac{(-1)^{l+1}\xi^l}{|\xi - \zeta|^4}$$

being l the unique index in $\{1, 2, 3, 4\}$ different from i, j, k. If we consider any map $u \in \Lambda$, the differential form

$$u^*(\eta) = \sum_{1 \leq i < j < k \leq 4} \eta_{ijk}(u(x)) \det \frac{\partial(u^i, u^j, u^k)}{\partial(x_1, x_2, x_3)} dx^1 \wedge dx^2 \wedge dx^3,$$

is called pull-back of η by u and it can be shown that (see e.g. [DNF85])

$$q_1(u) = \frac{1}{|S^3|} \int_{\mathbf{R}^3} u^*(\eta) = \frac{1}{|S^3|} \int_{u(\mathbf{R}^3)} \eta.$$

where $|S^3|$ denotes the measure of S^3. Thus we define the electric charge $Q(u)$ as

$$Q(u) = \frac{Q_0}{|S^3|} \int_{\mathbf{R}^3} u^*(\eta); \tag{40}$$

More in general we can define the electric charge contained in a region $\Omega \subset \mathbf{R}^3$ as follows:

$$Q(u, \Omega) = \frac{Q_0}{|S^3|} \int_{\Omega^+} u^*(\eta)$$

where Ω^+ is Ω equipped with the standard orientation of $\mathbf{R}^3$.

9 Solitons and electromagnetic field

In this section we derive the equations describing a soliton interacting with the elettromagnetic field.

We consider the electromagnetic 4-vector potential

$$(\mathbf{A}, \phi) = (A_1, A_2, A_3, \phi)$$

and the associated 1-form in the space-time

$$\omega = A_1 dx_1 + A_2 dx_2 + A_3 dx_3 - \phi dt.$$

The Lagrangian density of the system "soliton-electromagnetic field" is

$$\mathcal{L} = \mathcal{L}_1 + \mathcal{L}_2 + \mathcal{L}_3$$

where

- $\mathcal{L}_1$ is the Lagrangian density of the free field ψ :

$$\mathcal{L}_1 = \mathcal{L}_1(\psi, \sigma) = -\frac{1}{2}\alpha(\sigma) - V(\psi)$$

  with

$$\alpha(\sigma) = \sigma + \frac{\varepsilon}{3}\sigma^3$$

- $\mathcal{L}_2$ is the Lagrangian density of the electromagnetic field (E, H), namely

$$\mathcal{L}_2 = \frac{1}{8\pi}(|E|^2 - |H|^2)$$

  where

$$E = A_t + \nabla\phi \tag{41}$$
$$H = \nabla \times A \tag{42}$$

- $\mathcal{L}_3$ is the Lagrangian density describing the interaction between ψ and the electromagnetic field, and it is defined by

$$\mathcal{L}_3 \, dx^1 \wedge dx^2 \wedge dx^3 \wedge dt = \frac{Q_0}{|S^3|}\psi^*(\eta) \wedge \omega$$

where $\psi^*(\eta)$ is the pull back of η by the function $\psi(.,t)$, namely

$$\psi^*(\eta) = \sum_{\substack{1 \leq i < j < k \leq 4 \\ 1 \leq l < m < n \leq 4}} \eta_{ijk}(\psi(x,t)) \det \frac{\partial(\psi^i(x,t), \psi^j(x,t), \psi^k(x,t))}{\partial(x_l, x_m, x_n)} dx^l \wedge dx^m \wedge dx^n$$

Now we want to write the interaction $\mathcal{L}_3$ in a more physically meaningful way, namely

$$\mathcal{L}_3 = (\mathbf{J}(\psi, \nabla\psi, \psi_t) \mid \mathbf{A}) - \rho(\psi, \nabla\psi)\phi \qquad (43)$$
$$= J_1 A_1 + J_2 A_2 + J_3 A_3 - \phi\rho$$

where ρ will be interpreted as charge density and $\mathbf{J}$ as current density.

To do this we put

$$\frac{Q_0}{|S^3|}\psi^*(\eta) = J_1 dx^2 \wedge dx^3 \wedge dt - J_2 dx^1 \wedge dx^3 \wedge dt + J_3 dx^1 \wedge dx^2 \wedge dt - \rho dx^1 \wedge dx^2 \wedge dx^3$$

and so we get

$$J_i(\psi, \nabla\psi, \psi_t) = (-1)^i \frac{Q_0}{|S^3|} \sum_{\substack{1 \le a < b < c \le 4 \\ 1 \le l < m < n \le 4}} \eta_{abc}(\psi(x,t)) \det \frac{\partial(\psi^a, \psi^b, \psi^c)}{\partial(x_l, x_m, x_n)} \qquad (44)$$

(being i different from l, m, n) and

$$\rho(\psi, \nabla\psi) = \frac{Q_0}{|S^3|} \sum_{1 \le i < j < k \le 4} \eta_{ijk}(\psi(x,t)) \det \frac{\partial(\psi^i, \psi^j, \psi^k)}{\partial(x_1, x_2, x_3)}$$

Notice that, since the form $\psi^*(\eta)$ is closed, $d\psi^*(\eta) = 0$; this gives the continuity equation for $\mathbf{J}$ and ρ :

$$\nabla \cdot \mathbf{J} + \frac{\partial\rho}{\partial t} = 0 \qquad (45)$$

Moreover, by the (44), it is easy to realise that

$$\frac{\partial\psi}{\partial t} = 0 \Rightarrow \mathbf{J} = 0 \qquad (46)$$

namely the current density vanishes whenever the charge carried by the soliton does not move.

This justifies the interpretation of $\mathbf{J}$ as the current relative to ρ, which is indeed a density charge. In fact, by the (40)

$$\int \rho\, dx = \frac{Q_0}{|S^3|} \int \psi^*(\eta) = Q(\psi(.,t)) \qquad (47)$$

is the electric charge. Observe that, if ψ depends continuously on t, the electric charge (like the other quantum numbers) is constant in t, since it is a topological invariant.

Notice that the action describing the interaction between ψ and the electromagnetic field

$$S_3(\psi, \mathbf{A}, \phi) = \int \int \mathcal{L}_3 dx dt$$

is invariant under the gauge transformation

$$\mathbf{A}' = \mathbf{A} + \nabla h, \quad \phi' = \phi - \frac{\partial h}{\partial t}$$

where $h = h(t, x)$ is an arbitrary smooth real function.

In fact, by using (43) and (45), we have

$$S_3(\psi, \mathbf{A}', \phi') = \int \int [(\mathbf{J} \mid \mathbf{A}') - \rho \phi'] \, dx dt$$

$$= \int \int \left[(\mathbf{J} \mid \mathbf{A} + \nabla h) - \rho(\phi - \frac{\partial h}{\partial t}) \right] dx dt$$

$$= S_3(\psi, \mathbf{A}, \phi) + \int \int \left[(\mathbf{J} \mid \nabla h) + \rho \frac{\partial h}{\partial t} \right] dx dt$$

$$= S_3(\psi, \mathbf{A}, \phi) + \int \int (\nabla \cdot \mathbf{J} + \frac{\partial \rho}{\partial t}) h dx dt = S_3(\psi, \mathbf{A}, \phi)$$

Then we conclude that the total action of the system "soliton-electromagnetic field" is

$$S = S(\psi, A, \phi) = S_1(\psi) + S_2(\mathbf{A}, \phi) + S_3(\psi, \mathbf{A}, \phi) \tag{48}$$

$$S = S(\psi, A, \phi) = S_1(\psi) + S_2(\mathbf{A}, \phi) + S_3(\psi, \mathbf{A}, \phi)$$

where

$$S_i = \int \int \mathcal{L}_i dx dt, \quad i = 1, 2, 3$$

The Euler-Lagrange equations

$$dS = 0 \tag{49}$$

give rise to a system whose unknown are ψ, A, ϕ.

Taking the variation of $S(\psi, A, \phi)$ with respect to A and carring out the usual computations, we get

$$dS[\delta A] = 0$$

if and only if

$$\nabla \times (\nabla \times A) = 4\pi J(\psi, \nabla\psi, \psi_t) + \frac{\partial}{\partial t} (A_t + \nabla\phi) \tag{50}$$

Taking the variation of $S(\psi, A, \phi)$ with respect to ϕ, we get

$$dS[\delta\phi] = 0$$

if and only if

$$\nabla \cdot (A_t + \nabla\phi) = 4\pi \rho(\psi, \nabla\psi) \tag{51}$$

Observe that, using (41) and (42), (50) and (51) give respectively

$$\nabla \times H = 4\pi J(\psi, \nabla\psi, \psi_t) + \frac{\partial E}{\partial t} \tag{52}$$

$$\nabla \cdot E = 4\pi \rho(\psi, \nabla\psi) \tag{53}$$

which complete the Maxwell equations (41) and (42).

Finally, if we take the variation with respect to ψ, and set

$$dS[\delta\psi] = 0$$

we get

$$\frac{\partial}{\partial t}\left[\alpha'(\sigma)\frac{\partial\psi}{\partial t}\right] - c^2\nabla\cdot[\alpha'(\sigma)\nabla\psi] + \frac{\partial V}{\partial\xi}(\psi) = F \tag{54}$$

where F depends on A, ϕ, ψ and their derivatives. Notice that when A, ϕ are trivial F is zero and in this case (54) coincides with the equation of the free soliton (see (3)).

The existence of static solutions ψ, A, ϕ with $Q(\psi) \neq 0$ (i.e. with nontrivial charge) of the Euler-lagrange equation (49) has been proved in ([BFMP99]).

10   The dynamics of solitons in the electromagnetic field

Finally we can examine the behavior of solitons in an electromagnetic field. We suppose that they are sufficiently far from each other in such a way that, in the absence of electromagnetic field, the mutisoliton function (36) describes accurately the motion of the solitons. Now let us add a electromagnetic field whose energy is small compared to the mass of the solitons; under this assumption it is reasonable to assume that the shape of the solitons is not deformed in a sensitive way. Also we assume that they do not rotate, namely that

$$\frac{dR_k(t)}{dt} = 0$$

In this case the evolution of the field ψ is described by the following function:

$$\psi(t, x) = \sum_{k=1}^{\ell} u_k\left(R_k L_{\mathbf{v}_k(t)}[x - z_k(t)]\right) \tag{55}$$

with

$$\frac{dz_k(t)}{dt} = \mathbf{v}_k(t);$$

we recall that $z_k(t)$ is the baricenter of the soliton u_k.

Under these assumptions, we want to derive their dynamics, namely the equations which describe the evolution of the $z_k(t)$'s.

To carry out this program, we compute the action of the configuration (55). We recall that the action is given by the (48). Let us begin to compute $\int\int \mathcal{L}_1 dx dt$; we need the following lemma.

Lemma 3. *If*

$$\psi(t, x) = u\left(RL_{\dot{z}(t)}[x - z(t)]\right)$$

then

$$\int\int\left[-\frac{1}{2}\alpha\left(|\nabla\psi|^2 - \psi_t^2\right) - V(\psi)\right]dx dt = -m(u)\int\sqrt{1 - \dot{z}(t)^2}\, dt$$

Proof. First by the lorentz invariance we have that

$$|\nabla\psi(t,\xi)|^2 - \psi_t^2(t,\xi) = |\nabla u(\xi)|^2$$

where

$$\xi = RL_{\dot{z}(t)}\left[x - z(t)\right];$$

then, recalling the (29)

$$\int\int\left[-\frac{1}{2}\alpha\left(|\nabla\psi|^2 - \psi_t^2\right) - V(\psi)\right]dxdt$$

$$= \int\int\left[-\frac{1}{2}\alpha\left(|\nabla u(\xi)|^2\right) - V(u(\xi))\right]\frac{1}{\gamma}d\xi dt$$

$$= -\int\left[\frac{1}{2}\alpha\left(|\nabla u(\xi)|^2\right) + V(u(\xi))\right]d\xi \cdot \int\frac{1}{\gamma}dt$$

$$= -E(u)\int\frac{1}{\gamma}dt$$

where

$$\gamma = \frac{1}{\sqrt{1 - \dot{z}(t)^2}}.$$

The conclusion follows by Einstein formula (35).   □

Consider now the multisoliton field

$$\psi(t, x) = \sum_{k=1}^{\ell} u_k\left(R_k L_{\dot{z}_k}\left[x - z_k\right]\right) \qquad (56)$$

The action of this field is given by

$$\int\int\mathcal{L}_1 dx\,dt = \int\int\left[-\frac{1}{2}\alpha\left(|\nabla\psi|^2 - \left|\frac{\partial\psi}{\partial t}\right|^2\right) - V(\psi)\right]dx\,dt$$

$$= \int\int\left[-\frac{1}{2}\alpha\left(\left|\nabla\left[\sum_{k=1}^{\ell}u_k\right]\right|^2 - \left|\sum_{k=1}^{\ell}\frac{\partial u_k}{\partial t}\right|^2\right) - V\left(\sum_{k=1}^{\ell}u_k\right)\right]dx\,dt.$$

If we assume that the solitons u_k are far from each other, since they decay to zero exponentially at infinity, we can write

$$\int\int\mathcal{L}_1 dx\,dt \cong \sum_{k=1}^{\ell}\int\int\left[-\frac{1}{2}\alpha\left(|\nabla u_k|^2 - \left|\frac{\partial u_k}{\partial t}\right|^2\right) - V(u_k)\right]dx\,dt;$$

finally we can use the above lemma and we get

$$\int\int\mathcal{L}_1 dx\,dt = -\sum_{k=1}^{\ell}m(u_k)\int\sqrt{1 - \dot{z}_k^2}\,dt = -\sum_{k=1}^{\ell}m_k\int\sqrt{1 - \dot{z}_k^2}\,dt$$

where we have used the shortened notation $m(u_k) = m_k$.

Now let us compute $\int \int \mathcal{L}_3 dxdt$; here we shall carry out this computation under the assumption that the potentials $\mathbf{A}$ and ϕ are almost constant in balls $B_r(z(t))$ containing $K(\psi(t, \cdot))$. Under this assumption it is possible to see that,

$$\int \int [\rho(\psi, \nabla\psi)\phi - (\mathbf{J}(\psi, \nabla\psi, \psi_t) \mid \mathbf{A})] \, dxdt \cong Q(u) \int [\phi - (\mathbf{A} \mid \dot{z}(t))] \, dt; \quad (57)$$

In fact, let u be a solution of the (5). Since it decays exponentially with all its derivatives, outside of $B_r(z(t))$, we have:

$$\rho(\psi, \nabla\psi)\phi - (\mathbf{J}(\psi, \nabla\psi, \psi_t) \mid \mathbf{A}) \cong 0;$$

then

$$\int \int [\rho\phi - (\mathbf{J} \mid \mathbf{A})] \, dxdt \cong \int \left[\int_{B_r(z(t))} \rho\phi \, dx - \int_{B_r(z(t))} (\mathbf{J} \mid \mathbf{A}) \, dx \right] dt$$

Since A and ϕ are "almost" constant in $B_r(z(t))$,

$$\int \int [\rho\phi - (\mathbf{J} \mid \mathbf{A})] \, dxdt$$

$$\cong \int \left[\phi(z(t)) \int_{B_r(z(t))} \rho \, dx - \left(\mathbf{A}(z(t)) \mid \int_{B_r(z(t))} \mathbf{J} \, dx \right) \right] dt \quad (58)$$

Now we set

$$\tilde{\rho}(x) = \rho \circ u \circ R \circ L_v$$

so that

$$\rho(\psi, \nabla\psi) = \tilde{\rho}(x - z(t)).$$

Then, by the continuity equation,

$$\nabla \cdot \mathbf{J} = -\frac{\partial\rho}{\partial t} = -\frac{\partial}{\partial t}\tilde{\rho}(x - z(t)) = (\nabla\tilde{\rho} \mid \mathbf{v}) = \nabla \cdot (\tilde{\rho}\mathbf{v})$$

which gives

$$\mathbf{J} = \tilde{\rho}\mathbf{v} + \nabla\times\mathbf{k} = \rho\mathbf{v} + \nabla \times \mathbf{k}$$

where $\mathbf{k}$ is an arbitrary vector field. Since in the frame where $\mathbf{v} = 0$, by the (46) we have $\mathbf{J} = 0$, it follows that $\nabla \times \mathbf{k} = 0$ and so

$$\mathbf{J} = \rho\mathbf{v}$$

Then

$$\int \int [\rho\phi - (\mathbf{J} \mid \mathbf{A})] \, dxdt \cong \int \left[\phi(z(t)) \int_{B_r(z(t))} \rho \, dx - \left(\mathbf{A}(z(t)) \mid \int_{B_r(z(t))} \rho\mathbf{v} \, dx \right) \right] dt$$

$$= \left(\int_{B_r(z(t))} \rho \, dx \right) \int [\phi(z(t)) - (\mathbf{A}(z(t)) \mid \mathbf{v})] \, dt$$

$$= Q(u) \int [\phi - (\mathbf{A} \mid \dot{z}(t))] \, dt.$$

Now let us compute $\int \int \mathcal{L}_3 dx dt$ for a multisoliton field. Let $\psi(t, x)$ be the multisoliton field (56), then, arguing as before, and using the (57), we can write

$$\int \int \mathcal{L}_3 dx \, dt = \int \int [\rho(\psi, \nabla\psi)\phi - (\mathbf{J}(\psi, \nabla\psi, \psi_t) \mid \mathbf{A})] \ dx \, dt$$

$$\cong \sum_{k=1}^{\ell} Q_k \int [\phi(z_k(t)) - (\mathbf{A}(z_k(t)) \mid \dot{z}_k(t))] \, dt$$

where we have used the shortened notation $Q(u_k) = Q_k$.

So we conclude that the total action can be approximated as follows:

$$S(\psi, A, \phi) \cong S(z_1, ..., z_\ell, A, \phi)$$

where

$$S(z_1, ..., z_\ell, A, \phi) = \sum_{k=1}^{\ell} \int \left[-m_k \sqrt{1 - \dot{z}_k^2} + Q_k \left(\phi(z_k) - (\mathbf{A}(z_k) \mid \dot{z}_k) \right) \right] dt$$

$$+ \frac{1}{8\pi} \int \int (|E|^2 - |H|^2) \, dx \, dt.$$

This is the usual action of charged particles in an electromagnetic field (cf. e.g. [LL70]). Thus, as far as the approximations we have used are accurate, our solitons exibit the same dynamic of charged particles. Notice that our approximations are accurate if the δ-support $K(\psi(t, \cdot))$ of the solitons is so small that on it $\mathbf{A}$ and ϕ can be considered constant.

11 Conclusions

The field equation related to the Lagrangian (2) is probably the simplest Lorentz invariant equation, in three space dimensions, which has static solitons. Nevertheless, its soliton solutions have some interesting properties since they behave just as relativistic particles.

So some facts of Relativity theory (the space- contraction and time- dilation, the increasing of mass with the velocity, the equivalence between mass and energy) can be deduced, without extra assumptions, from a " mechanical" model described by a single equation. This equation might be interpreted as the equation of an "elastic medium" in a Newtonian space-time. Thus this model shows that, from a purely formal point of view, the main features of the special relativity can be deduced from a partial differential equation in a Newtonian space-time.

An algebraic approach to nonstandard analysis

V. Benci

Abstract. In this paper we present the main features of Nonstandard Analysis without any use of Logic or any deep tool from set theory. The hyperreal numbers are introduced by three axioms of algebraic type whose consistency is proved by algebraic means. Probably this way of introducing Nonstandard Analysis, is not the most interesting and elegant. Moreover, part of the richness and the power of Nonstandard Analysis (as the Transfer and the Saturation Principles) are lost. Nevertheless, most of its elementary applications can be carried out without any problem. In order to show this, the last four sections of this paper are devoted to simple applications to Elementary Calculus.

1 Introduction

Most mathematicians do not like Nonstandard Analysis. One of the reasons could be the large use of logic and axiomatic set theory which is employed in its presentation. Actually the logic is not necessary and the field of hyperreal can be introduced only by algebraic tools. The aim of this paper is just to show how this is possible. The hyperreal numbers will be introduced by three axioms of algebraic type (section 2.1). The consistency of these axioms will proved building a model, still using algebraic tools (section 7). Probably this way of introducing the hyperreal numbers is not the most interesting and elegant. Moreover, part of the richness and the power of Nonstandard Analysis is lost. Nevertheless, there are at least three good reasons to do this:

I - In this way the *infinitesimal* numbers are associated with *infinitesimal* sequences. The reader who already thinks of infinitesimals as "infinitesimal sequences" will find the "infinitesimal numbers" closer to familiar objects.

II - It is not necessary to introduce any use of logic or any deep tool from set theory (except than theorem 20 which is a well known theorem of Algebra whose proof relies on Zorn's lemma).

III - Simplifying the language in a suitable way. I think that this way of introducing the infinitesimals could be employed to teach calculus at elementary level. And, I hope, to convince some people about this. For this reason, the last sections of this paper present an elementary introduction to Calculus.

2 The hyperreal numbers

2.1 The axioms for the hyperreal numbers

In this section we will introduce the hyperreal numbers by mean of three Axioms. The first Axiom just state that the hyperreal numbers constitute a numerical field.

Axiom 1 *The set of the hyperreal numbers $\mathbf{R}^*$ is an ordered field which contains $\mathbf{R}$ as subfield.*

In order to introduce the second Axiom, consider the set $\mathcal{F}(\mathbf{N}, \mathbf{R})$ of functions from the natural numbers to the real numbers (i.e. the set of sequences). This set has the structure of commutative ring with respect to the following operations:

$$(\varphi + \psi)(n) = \varphi(n) + \psi(n)$$

$$(\varphi \cdot \psi)(n) = \varphi(n) \cdot \psi(n)$$

Axiom 2 *There exists a surjective ring homomorphism*

$$J : \mathcal{F}(\mathbf{N}, \mathbf{R}) \to \mathbf{R}^*$$

i.e. a map which satisfies the following properties:

- *J is onto;*
- *$J(\varphi + \psi) = J(\varphi) + J(\psi)$,*
- *$J(\varphi \cdot \psi) = J(\varphi) \cdot J(\psi)$.*

Notice that also $\mathbf{R}$ satisfies Axioms 1 and 2; it is sufficient to set

$$J(\varphi) = \varphi(\bar{n}) \tag{1}$$

where $\bar{n}$ is any fixed natural number. The third Axiom prevents J to have the form 1 and implies that $\mathbf{R}^*$ is strictly bigger than $\mathbf{R}$.

Axiom 3 *If there exists $k \in \mathbf{N}^+$ such that*

$$\forall n \in \mathbf{N}^+, \; \varphi(kn) \geq a,$$

with $a \in \mathbf{R}$, then

$$J(\varphi) \geq a.$$

2.2   The number α_0

Before discussing the consequences of the Axioms, we will introduce a number belonging to $\mathbf{R}^* - \mathbf{R}$ which we will call α_0. It is defined by the following formula

$$\alpha_0 = J(i)$$

where $i : \mathbf{N} \to \mathbf{R}$ is the natural imbedding i.e.

$$i(n) = n. \tag{2}$$

The number α_0 plays a very special role in our presentation of Nonstandard Analysis. We have decided to call it α_0 since, as we will see, it is an infinite number and α_0 reminds the symbol used for the infinite number $\aleph_0$.

2.3   The natural preextension of a function

The number α_0 allows to define the notion of natural pre-extension of a function which is a very important tool for getting an elementary presentation of N.S.A. Given any function $\varphi \in \mathcal{F}(\mathbf{N}, \mathbf{R})$, we want to extend it to a new function (for simplicity, denoted by the same symbol)

$$\varphi : \mathbf{N} \cup \{\alpha_0\} \to \mathbf{R}^*$$

Clearly, if φ is a rational function (i.e. the ratio of two polynomial) defined on $\mathbf{R}$, it can be extended without any problem on $\mathbf{R}^*$ just using the field operations. In fact we have the following proposition:

Proposition 1. *(i) If c_a is the function which takes the constant value a, then*

$$J(c_a) = a.$$

(ii) If P is a polynomial, then

$$J(P) = P(\alpha_0).$$

(iii) If ρ is a rational function, then

$$J(\rho) = \rho(\alpha_0). \tag{3}$$

Proof. (i) - Since $c_a(n) \geq a$, by Axiom 3,

$$J(c_a) \geq a. \tag{4}$$

Moreover,

$$-c_a(n) \geq -a,$$

and so

$$J(-c_a) \geq -a;$$

also, by Axiom 2

$$J(-c_a) = J((-1) \cdot (c_a)) = J(-1) \cdot J(c_a) = -J(c_a);$$

thus, we get $J(c_a) \leq a$ and using the (4) we get the conclusion. (ii) - Let

$$P(x) = \sum_m a_m x^m,$$

then P restricted to $\mathbf{N}$ can be written as follows

$$P = \sum_m a_m i^m,$$

where i is defined by (2). Then by the Axiom 2 and (i), we have

$$J(P) = \sum_m J(a_m i^m) = \sum_m J(a_m) J(i^m) =$$

$$\sum_m a_m J(i)^m = \sum_m a_m \alpha_0{}^m = P(\alpha_0).$$

(iii) - Now let $\rho = \frac{P}{Q}$ be a rational function; then, by (ii) and Axiom 2,

$$J(\rho) \cdot Q(\alpha_0) = J\left(\frac{P}{Q}\right) \cdot J(Q) = J\left(\frac{P}{Q}Q\right) = J(P) = P(\alpha_0),$$

thus,

$$J(\rho) = \frac{P(\alpha_0)}{Q(\alpha_0)} = \rho(\alpha_0)$$

□

Remark 1. Notice that by (i) of the above proposition, identifying the constant function c_λ with the real number λ, we have that

$$J(\lambda\varphi) = J(c_\lambda\varphi) = J(c_\lambda)J(\varphi) = \lambda J(\varphi);$$

thus it follows that J is an algebras homomorphism if we regard $\mathbf{R}^*$ as an algebra over $\mathbf{R}$.

The identity (3) suggests the following definition:

Definition 1. *Given any function $\varphi \in \mathcal{F}(\mathbf{N}, \mathbf{R})$, we extend it to $\mathbf{N} \cup \{\alpha_0\}$ by setting*

$$\varphi(\alpha_0) = J(\varphi). \tag{5}$$

This extension is called the natural preextension of φ.

2.4   The canonical representation of a hyperreal number

By Axiom 2, a unique hyperreal number $J(\varphi)$ corresponds to every function $\varphi \in \mathcal{F}(\mathbf{N}, \mathbf{R})$. The equation (5) provides a simple and meaningful way to represent this hyperreal number; this representation will be called the "canonical representation" of an hyperreal number. For example,

$$\sin\frac{2}{\alpha_0}, \ \arctan\alpha_0, \ \frac{3 + 4\alpha_0}{7 + \alpha_0}, \ \log\alpha_0, \ e^{\alpha_0}, \ \alpha_0!, \tag{6}$$

are canonical representations of hyperreal numbers. The following theorem helps in proving relations between hyperreal numbers.

Proposition 2. *Let $\varphi, \psi \in \mathcal{F}(\mathbf{N}, \mathbf{R})$; then*

- *(i) $\exists k \in \mathbf{N}^+, \forall n \in \mathbf{N}^+, \ \varphi(kn) \geq \psi(kn) \Rightarrow \varphi(\alpha_0) \geq \psi(\alpha_0)$;*
- *(ii) $\exists k \in \mathbf{N}^+, \forall n \in \mathbf{N}^+, \ \varphi(kn) = \psi(kn) \Rightarrow \varphi(\alpha_0) = \psi(\alpha_0)$;*
- *(iii) $\exists k \in \mathbf{N}^+, \forall n \in \mathbf{N}^+, \ \varphi(kn) \neq \psi(kn) \Rightarrow \varphi(\alpha_0) \neq \psi(\alpha_0)$.*
- *(iv) $\exists k \in \mathbf{N}^+, \forall n \in \mathbf{N}^+, \ \varphi(kn) > \psi(kn) \Rightarrow \varphi(\alpha_0) > \psi(\alpha_0)$.*

Proof. (i) By Axiom 3,

$$\varphi(kn) - \psi(kn) \geq 0 \Rightarrow J(\varphi - \psi) \geq 0;$$

then using Axiom 2 and equation (5), we have

$$0 \leq J(\varphi - \psi) = J(\varphi) - J(\psi) = \varphi(\alpha_0) - \psi(\alpha_0)$$

and hence the conclusion. (ii) It follows directly from (i). (iii) Set $\sigma(n) = \varphi(n) - \psi(n)$, and

$$\tau(m) = \begin{cases} \frac{1}{\sigma(m)} & if\ m = kn \\ 1 & if\ m \neq kn \end{cases};$$

then,

$$\tau(kn) \cdot \sigma(kn) = 1$$

and hence, by (ii)

$$\tau(\alpha_0) \cdot \sigma(\alpha_0) = 1.$$

Since $\mathbf{R}^*$ is a field, $\sigma(\alpha_0) \neq 0$ and so $\varphi(\alpha_0) \neq \psi(\alpha_0)$ (iv) it follows trivially from (i) and (iii). $\qquad\qquad\square$

Using the canonical representation of hyperreal numbers and the (ii) of Proposition 2, it is possible to extend identities relative to real numbers to hyperreal numbers; for example, from the identity

$$\log(xy) = \log(x) + \log(y)$$

we get

$$\log(3\alpha_0) = \log(3) + \log(\alpha_0) \tag{7}$$

in fact, for every $n \in \mathbf{N}^+$, we have that

$$\log(3n) = \log(3) + \log(n)$$

and so Proposition 2 (ii) gives the (7) Similarly, using Proposition 2 (i) and (iii), we can get inequalities. For example, it is easy to check that all the numbers of (6) are different from each other and are ordered from the smaller to the greater. At this point, it turns out to be useful to have at least an other special symbol: from now on the symbol ε_0 will denote the number $1/\alpha_0$ i.e. the number $J(\psi)$ where $\psi(n) = \frac{1}{n}$. Notice that ε_0 is the inverse of α_0; in fact, for any $n \in \mathbf{N}^+$,

$$\psi(n) \cdot i(n) = 1$$

and hence

$$\psi(\alpha_0) \cdot i(\alpha_0) = \varepsilon_0 \cdot \alpha_0 = 1.$$

Using this symbol, the first three number of (6) can be written in a simpler and more meaningful way as follows:

$$\sin 2\varepsilon_0, \quad \frac{\pi}{2} - \arctan \varepsilon_0, \quad \frac{4 + 3\varepsilon_0}{1 + 7\varepsilon_0}.$$

3   The structure of the hyperreal line

3.1   Infinitesimal and infinite numbers

Now, let us give some definitions relative to the field of the hyperreal numbers.

Definition 2. *A number $\xi \in \mathbf{R}^*$ is called infinitesimal if*

$$\forall k \in \mathbf{N}, |\xi| < \frac{1}{k}$$

We say that a number $\xi \in \mathbf{R}^*$ is bounded (or finite) if

$$\exists k \in \mathbf{N}, |\xi| < k$$

We say that a number $\xi \in \mathbf{R}$ is infinite (or unbounded) if it is not bounded, i.e.

$$\forall k \in \mathbf{N}, |\xi| > k$$

Notice that the above definition make sense not only for the hyperreal numbers but for every non-Archimedean field. By Proposition 2, it is immediate to verify the following theorem:

Theorem 1. α_0 *is infinite, ε_0 is infinitesimal.*

Proof. Choose arbitrarily $k \in \mathbf{N}^+$, then, for every $n \in \mathbf{N}^+$ we have

$$i(kn) \geq k;$$

Thus by Proposition 2 (i),

$$\alpha_0 = i(\alpha_0) \geq k$$

By the arbitrariness of k it follows that α_0 is infinite. Since,

$$\forall k \in \mathbf{N}, \ \alpha_0 > k$$

by the properties of ordered fields, we have that

$$\forall k \in \mathbf{N}, \ \frac{1}{\alpha_0} < \frac{1}{k}$$

hence, $\varepsilon_0 = \frac{1}{\alpha_0}$ is infinitesimal. $\square$

Using Proposition 2, it is easy to check that the numbers

$$\varepsilon_0, \ \sin \varepsilon_0, \ \frac{\alpha_0}{7 + \alpha_0^3}, \ \log(1 - \varepsilon_0^3)$$

are infinitesimal. As examples of infinite numbers consider

$$\alpha_0, \ \frac{5 + \alpha_0^4}{7 + \alpha_0^3}, \ \log(1 + \alpha_0^3), \ \alpha_0!, \ \varepsilon_0 e^{\alpha_0}.$$

All the infinitesimals and all the real numbers are bounded, however there are finite numbers which are not infinitesimal nor real as, for example

$$5 + \varepsilon_0, \ 7 + \sin \alpha_0, \ \frac{5 + \alpha_0}{7 + 2\alpha_0}, \ \log(6 - \varepsilon_0^3) \qquad (8)$$

Two other definitions are useful

Definition 3. *A hyperreal number* $\xi \in \mathbf{R}^* - \mathbf{R}$ *is called ideal; a hyperreal number which is real is called standard.*

Thus a hyperreal number is either standard or ideal; the infinite numbers are all ideal, the infinitesimals are all ideal except 0, the bounded numbers can be standard or ideal.

Definition 4. *Two numbers* $\xi, \eta \in \mathbf{R}^*$ *are called infinitely close if* $\xi - \eta$ *is infinitesimal; in symbol we shall write*

$$\xi \sim \eta$$

For example

$$\cos(\varepsilon_0) \sim \frac{2\alpha_0}{7 + 2\alpha_0}$$

3.2 Monads and Galaxies

Definition 5. *Given a hyperreal number* $\xi \in \mathbf{R}^*$, *the monad of* ξ *is the set*

$$mon(\xi) = \{x \in \mathbf{R}^* \mid x \sim \xi\}$$

The galaxy of ξ *is the set*

$$gal(\xi) = \{x \in \mathbf{R}^* \mid x - \xi \text{ is finite}\}$$

Thus $mon(0)$ is the set of infinitesimal and $gal(0)$ is the set of bounded numbers. All the real numbers are contained in $gal(0)$; the unique real number contained in $mon(0)$ is "0".

Theorem 2. *The set* $gal(0)$ *of bounded elements is a subring of* $\mathbf{R}^*$, *i.e. sum, differences and products of bounded elements are bounded.*

Proof. If x and y are bounded, there exists k and $h \in \mathbf{N}$ such that

$$|x| \le k, \ |y| \le h$$

hence,

$$|x + y| \le k + h,$$
$$|x - y| \le k + h,$$
$$|x \cdot y| \le k \cdot h$$

so $x + y$, $x - y$ and xy are bounded.   □

Corollary 1. *Any two galaxies are equal or disjoint.*

Proof. For each $x \in \mathbf{R}^*$, the galaxy of x is the coset of x modulo $gal(0)$:

$$gal(x) = \{x + t \mid t \in gal(0)\}$$

□

Theorem 3. *The set $mon(0)$ of infinitesimal is a subring of $\mathbf{R}^*$ and an ideal in $gal(0)$. That is:*

- *sum, differences and products of infinitesimals are infinitesimals.*
- *the product of an infinitesimal and a finite element is infinitesimal.*

Proof. For any $k > 0$, since x and y are infinitesimal

$$|x| \le \frac{1}{2k}, \quad |y| \le \frac{1}{2k}$$

hence,

$$|x \pm y| \le \frac{1}{k},$$

and

$$|x \cdot y| \le \frac{1}{k},$$

so $x + y$, $x - y$ and xy are infinitesimal. Let x be bounded and y be infinitesimal; then it exists $M \in \mathbf{N}$ such that $|x| \le M$; since

$$|y| \le \frac{1}{Mk},$$

we have that

$$|x \cdot y| \le |x| \cdot |y| \le \frac{1}{k}.$$

□

Corollary 2. *Any two monads are equal or disjoint.*

Proof. For each $x \in \mathbf{R}^*$, the monad of x is the coset of x modulo $mon(0)$:

$$mon(x) = \{x + t \mid t \in mon(0)\}$$

□

3.3   The Standard Part

In this section we will introduce the notion of *standard part* and the notion of *trace* which play a central role in the definition of the basic notions of Calculus as the notion of derivative, integral, series, etc.

Theorem 4 (Standard Part Theorem). *Any bounded number $\xi \in \mathbf{R}^*$ is infinitely close to a unique real number $x_0 \in \mathbf{R}$; this number is called the standard part of ξ; in symbols we shall write*

$$x = st(\xi).$$

Proof. Set

$$A = \{x \in \mathbf{R} \mid x \leq \xi\}$$

and

$$B = \{x \in \mathbf{R} \mid x \geq \xi\}$$

By the completeness of $\mathbf{R}$, there exists $x_0 \in \mathbf{R}$:

$$\forall a \in A, \; \forall b \in B, \; a \leq x_0 \leq b$$

It is easy to check that $x_0 \sim \xi$.   □

By the above theorem, it follows that any bounded hyperreal number ξ has the following representation, (called normal form):

$$\xi = x_0 + \varepsilon$$

where $x_0 := st(\xi)$ and $\varepsilon := \xi - st(\xi)$. For example the first two numbers of the list (8) is in normal form; the last two have the following normal form:

$$\frac{1}{2} + \frac{3\varepsilon_0}{4 + 14\varepsilon_0}, \; \log(6) + \log(1 - \frac{\varepsilon_0^3}{6})$$

Also the number $7 + \sin \alpha_0$ has a normal form, but it cannot be computed explicitly using only our axioms (cf. the discussion in section 7.2)

Theorem 5. *The function*

$$st : gal(0) \to \mathbf{R}$$

is a ring homomorphism, i.e.

- $st(x + y) = st(x) + st(y)$
- $st(x - y) = st(x) - st(y)$
- $st(x \cdot y) = st(x) \cdot st(y)$

Moreover,

- *if $st(y) \neq 0$, $st\left(\frac{x}{y}\right) = \frac{st(x)}{st(y)}$*

3.4   The function *trace*

The function *st* is not defined for infinite numbers. Thus for the applications it is convenient to introduce the enlarged real line

$$\hat{\mathbf{R}} = \mathbf{R} \cup \{+\infty, -\infty\}$$

and the function trace

$$tr : \mathbf{R}^* \to \hat{\mathbf{R}}$$

defined as follows:

$$tr(\xi) = \begin{cases} st(\xi) & \text{if } \xi \text{ is bounded} \\ +\infty & \text{if } \xi \text{ is unbounded and positive} \\ -\infty & \text{if } \xi \text{ is unbounded and negative} \end{cases}$$

Using the usual algebra of $+\infty$, $-\infty$ and function *trace*, we get the following theorem:

Theorem 6. *Using the conventional algebra of $+\infty$, $-\infty$ we have*

- $tr(x + y) = tr(x) + tr(y)$
- $tr(x - y) = tr(x) - tr(y)$
- $tr(x \cdot y) = tr(x) \cdot tr(y)$
- $tr\left(\frac{x}{y}\right) = \frac{y}{|y|} \cdot \frac{tr(x)}{tr(y)}$

We recall that in the conventional algebra of $\pm\infty$, the following expressions are called indeterminate form:

$$+\infty - \infty, \quad \pm\infty \cdot 0, \quad \frac{\pm\infty}{\pm\infty}, \quad \pm\frac{0}{0}.$$

Clearly, in these cases the theorem 6 is meaningless and, in order to compute the trace, it is necessary to use other tools (the same story of the limits......).

4   The "∗" operator

The "∗" operator plays a fundamental role in nonstandard analysis as we will see in the following sections. It allows to extend subsets of real numbers into subsets of hyperreal numbers; moreover, it extends the functions defined on subsets of real numbers to subsets of hyperreal numbers.

4.1   The natural extension of sets

If $A \subset \mathbf{R}$, we set
$$A^* = \{\varphi(\alpha_0) \mid \varphi \in \mathcal{F}(\mathbf{N}, A)\}$$

The set A^* is called the **-transform* of A. Since $A \subseteq A^*$, A^* is also called the *natural extension* of A. The natural extesions $\mathbf{N}^*$, $\mathbf{Z}^*$, $\mathbf{Q}^*$ and $\mathbf{R}^*$ of the sets of the natural, integer, rational and real numbers are particularly important; they are called the sets of hypernatural, hyperinteger, hyperrational and hyperreal numbers respectively. Since $i \in \mathcal{F}(\mathbf{N}, \mathbf{N})$, α_0 is a hypernatural number. An other set which will be important for us is the natural extension of an interval $[a, b]^*$ with $a, b \in \mathbf{R}$; it is not difficult to prove that

$$[a, b]^* = \{\xi \in \mathbf{R}^* \mid a \le \xi \le b\}. \tag{9}$$

The proof of the following theorem is somewhat more delicate

Theorem 7. . *If A and B are subsets of* $\mathbf{R}$, *then*

$$(A \cup B)^* = A^* \cup B^*$$

and

$$(A \cap B)^* = A^* \cap B^*$$

Proof. It is immediate to verify that $A^* \cup B^* \subseteq (A \cup B)^*$, in fact

$$\mathcal{F}(\mathbf{N}, A) \cup \mathcal{F}(\mathbf{N}, B) \subset \mathcal{F}(\mathbf{N}, A \cup B).$$

So we have to prove that $\xi = \varphi(\alpha_0)$ with $\varphi \in \mathcal{F}(\mathbf{N}, A \cup B)$ implies $\xi \in A^* \cup B^*$; we consider two cases: $\xi \ne 0$ and $\xi = 0$. If $\xi \ne 0$, we set

$$\varphi_1(n) = \begin{cases} \varphi(n) & \text{se } \varphi(n) \in A \\ 0 & \text{se } \varphi(n) \in B - A \end{cases}$$

and

$$\varphi_2(n) = \begin{cases} \varphi(n) & \text{se } \varphi(n) \in B - A \\ 0 & \text{se } \varphi(n) \in A \end{cases} ;$$

thus

$$\begin{cases} \varphi_1(\alpha_0) \cdot \varphi_2(\alpha_0) = 0 \\ \varphi_1(\alpha_0) + \varphi_2(\alpha_0) = \xi \end{cases}$$

From the first equation it follows that $\varphi_1(\alpha_0)$ or $\varphi_2(\alpha_0)$ equals 0, so, using the second equation we get $\xi = \varphi(\alpha_0)$. Now we consider the case $\xi = 0$. In this case there exists $n_0 \in \mathbf{N}$ such that $\varphi(n_0) = 0$ since otherwise the function

$$\frac{1}{\varphi(n)}$$

would be well defined and hence also $\frac{1}{\xi}$ would be defined contradicting the fact that $\xi = 0$. The fact that $\varphi(n_0) = 0$, implies that $0 \in A \cup B$ and hence $\xi = 0 \in A^* \cup B^*$. The equality $A^* \cap B^* = (A \cap B)^*$, follows straightforward from the following fact:

$$\mathcal{F}(\mathbf{N}, A) \cap \mathcal{F}(\mathbf{N}, B) = \mathcal{F}(\mathbf{N}, A \cap B).$$

$\square$

From the above theorem it follows that

$$(A_1 \cup ... \cup A_n)^* = A_1^* \cup ... \cup A_n^* \tag{10}$$

and

$$(A_1 \cap ... \cap A_n)^* = A_1^* \cap ... \cap A_n^*. \tag{11}$$

where $A_1, ..., A_n$ is a finite family of subsets of $\mathbf{R}$. Observe that these equalities, in general, do not hold for infinite families of sets; for example

$$\left(\bigcup_{n \in \mathbf{N}^+} [-n, n] \right)^* = \mathbf{R}^* \neq \bigcup_{n \in \mathbf{N}^+} [-n, n]^*$$

since the last set does not contain unbounded numbers. The following theorem states when the natural extensions are proper extensions.

Theorem 8. . $A = A^*$ if and only if A is a finite subset of $\mathbf{R}$.

Proof. First of all notice that $\{a\}^* = \{a\}$ for every $a \in \mathbf{R}$ since $\mathcal{F}(\mathbf{N}, \{a\})$ contains only the constant function c_a. Suppose that $A = \{a_1, ..., a_m\}$ is a finite set. Then, by the 10,

$$\{a_1, ..., a_m\}^* = \{a_1\}^* \cup ... \cup \{a_m\}^* = \{a_1\} \cup ... \cup \{a_m\} = \{a_1, ..., a_m\}$$

Now suppose that A is infinite. Then there exists a injective function

$$\psi : \mathbf{N} \to A;$$

it is sufficient to prove that $\psi(\alpha_0) \in A^* - A$. For every $x \in A$, $\psi(n) = x$ for one value of n at most and hence, from proposition 2(iii), $\psi(\alpha_0) \neq x$; thus, for every $x \in A$, $\psi(\alpha_0) \neq x$ and so $\psi(\alpha_0) \in A^* - A$. $\square$

An other important set of numbers is the following one

$$\Omega \overset{def}{=} \mathbf{N}^* - \mathbf{N}; \tag{12}$$

it is not difficult to prove that Ω is not the natural extension of any set; moreover we have the following theorem.

Theorem 9. *The set Ω contains only unbounded numbers.*

Proof. Let $\xi \in \mathbf{N}^*$ be a bounded number. Then $\exists M \in \mathbf{R}$ such that

$$|\xi| \leq M.$$

Thus, exploiting the (9) and the theorem 7, we have

$$\xi \in [-M, M]^* \cap \mathbf{N}^* = ([-M, M] \cap \mathbf{N})^*$$

Since by the Th. 8

$$([-M, M] \cap \mathbf{N})^* = [-M, M] \cap \mathbf{N}$$

we have that $\xi \in \mathbf{N}$ and hence Ω contains only unbounded numbers.   □

4.2   The natural extension of functions

Definition 6. *Given any function*

$$f : A \rightarrow B, \quad A, B \subset \mathbf{R},$$

we define a new function

$$f^* : A^* \rightarrow B^*$$

as follows:

$$f^*(\varphi(\alpha_0)) = (f \circ \varphi)(\alpha_0)$$

We need to prove show that the above definition is well posed namely we have to show that the value of $f^*(\xi)$, $\xi = \varphi(\alpha_0) \in A^*$, depends only on ξ and not on the function φ.

Theorem 10. *The 6 is well posed, i.e.* $\varphi(\alpha_0) = \psi(\alpha_0)$, *implies that*

$$(f \circ \varphi)(\alpha_0) = (f \circ \psi)(\alpha_0).$$

Proof. Set

$$I(n) = \begin{cases} 1 \text{ if } \varphi(n) - \psi(n) = 0 \\ \\ 0 \text{ if } \varphi(n) - \psi(n) \neq 0 \end{cases}$$

Since, $\forall n \in \mathbf{N}$,

$$\varphi(n) - \psi(n) + I(n) \neq 0$$

then, by Proposition 2, (iii),

$$\varphi(\alpha_0) - \psi(\alpha_0) + I(\alpha_0) \neq 0$$

and since $\varphi(\alpha_0) - \psi(\alpha_0) = 0$, we have that

$$I(\alpha_0) \neq 0. \tag{13}$$

Now, notice that

$$\forall n \in \mathbf{N}, \ [(f \circ \varphi)(n) - (f \circ \psi)(n)] \cdot I(n) = 0$$

In fact, if $\varphi(n) - \psi(n) = 0$ the first factor vanishes; on the other hand, if $\varphi(n) - \psi(n) \neq 0$ the second factor vanishes. Thus, by Proposition 2, (ii),

$$[(f \circ \varphi)(\alpha_0) - (f \circ \psi)(\alpha_0)] \cdot I(\alpha_0) = 0$$

and so, by the (13) we get the conclusion:

$$(f \circ \varphi)(\alpha_0) - (f \circ \psi)(\alpha_0) = 0$$

□

For every $x \in A$, $f(x) = f^*(x)$; then f^* is an extension of f : it is called the *natural extension* of f. When no ambiguity is possible the "$*$" will be omitted; and this fact will happen most of the times. Clearly, if $A^* = \mathbf{N}^*$, we have that $f^*(\alpha_0)$ coincides with the natural preextension of f as defined by the 5.

4.3 The measure μ

We now introduce a measure μ on $\mathbf{N}$ which plays a crucial role. This measure is defined for every $A \in \mathcal{P}(\mathbf{N})$ as follows:

$$\mu(A) = I_A(\alpha_0) \tag{14}$$

where I_A is the characteristic function of A i.e.

$$I_A(n) = \begin{cases} 1 & if \ n \in A \\ 0 & if \ n \in \mathbf{N}-A \end{cases}.$$

Let us prove that μ is actually a measure.

Theorem 11. *The function μ defined by the (14) is a measure, i.e.*

- *(i) $\mu(\emptyset) = 0$*
- *(ii) $A \cap B = \emptyset \Rightarrow \mu(A \cup B) = \mu(A) + \mu(B)$*

Moreover it satisfies the following properties:

- *(iii) $\mu(A) = \mu(B) = 1 \Rightarrow \mu(A \cap B) = 1$*

Proof. (i) is trivial. (ii) since, $A \cap B = \emptyset$, we have that

$$I_{A \cup B}(\alpha_0) = I_A(\alpha_0) + I_B(\alpha_0)$$

and so

$$\mu(A \cup B) = I_{A \cup B}(n) = I_A(n) + I_B(n) = \mu(A) + \mu(B).$$

(iii) We have

$$\mu(A \cap B) = \mu(A \cap B) + \mu(A) - \mu(B) \geq$$
$$\geq \mu(A \cap B) + \mu(A - B) = \mu(A) = 1$$

thus

$$\mu(A \cap B) = 1.$$

□

Notice that this measure is defined for every subset of N but it is not countably additive. If φ and ψ are two functions defined on N, we say that $\varphi = \psi$ almost everywhere if

$$\mu\left(\{n \in N \,|\, \varphi(n) = \psi(n)\}\right) = 1$$

and we will use the notation

$$\varphi = \psi \;\; \mu\text{-}a.e.$$

or, more simply, the notation

$$\varphi = \psi \;\; a.e.$$

Also, we shall use the notations

$$\varphi > \psi \;\; a.e; \;\; \varphi \neq \psi \;\; a.e; \;\; etc.$$

with obvious meaning. For us, the importance of the measure μ relies on the following theorem:

Theorem 12. *Two hyperreal numbers $\varphi(\alpha_0)$ and $\psi(\alpha_0)$ are equal if and only if*

$$\varphi = \psi \;\; a.e.$$

Proof. We set

$$A = \{n \in N \,|\, \varphi(n) = \psi(n)\}\,.$$

Now suppose that $\varphi = \psi$ *a.e.* We have that

$$[\varphi(n) - \psi(n)] \cdot I_A(n) = 0$$

and hence

$$[\varphi(\alpha_0) - \psi(\alpha_0)] \cdot I_A(\alpha_0) = 0.$$

Since $\mu(A) = 1$, $I_A(\alpha_0) = 1$ and so $\varphi(\alpha_0) - \psi(\alpha_0) = 0$. Now suppose that

$$\varphi(\alpha_0) - \psi(\alpha_0) = 0 \tag{15}$$

and set

$$\chi(n) = \begin{cases} 0 & if \;\; n \in A \\ \frac{1}{\varphi(n) - \psi(n)} & if \;\; n \in N - A \end{cases}.$$

then

$$[\varphi(n) - \psi(n)] \cdot \chi(n) = 1 - I_A(n);$$

hence

$$[\varphi(\alpha_0) - \psi(\alpha_0)] \cdot \chi(\alpha_0) = 1 - I_A(\alpha_0);$$

by the (15), we get $1 - I_A(\alpha_0) = 0$ and thus $I_A(\alpha_0) = 1$. □

From the above theorem, the following corollary is obtained strightforward

Corollary 3. *Given two hyperreal numbers* $\varphi(\alpha_0)$ *and* $\psi(\alpha_0)$

- *(i)* $\varphi(\alpha_0) \geq \psi(\alpha_0)$ *if and only if*

$$\varphi \geq \psi \ \ a.e.$$

- *(ii)* $\varphi(\alpha_0) > \psi(\alpha_0)$ *if and only if*

$$\varphi > \psi \ \ a.e.$$

- *(iii)* $\varphi(\alpha_0) \neq \psi(\alpha_0)$ *if and only if*

$$\varphi \neq \psi \ \ a.e.$$

An other consequence of theorem 12 is the following theorem which turns out to be quite useful.

Theorem 13. *(Principio di Cauchy) As usual, let* $f, g \in \mathcal{F}(\mathbf{N}, \mathbf{R})$ *and set* $\Omega = \mathbf{N}^* - \mathbf{N}$; *then*

- *if* $\exists \bar{n}, \forall n \geq \bar{n}, f(n) = g(n)$, *then*

$$\forall \omega \in \Omega, \ f(\omega) = g(\omega);$$

- *if* $\exists \bar{n}, \forall n \geq \bar{n}, f(n) \geq g(n)$, *then*

$$\forall \omega \in \Omega, \ f(\omega) \geq g(\omega);$$

- *if* $\exists \bar{n}, \forall n \geq \bar{n}, f(n) > g(n)$, *then*

$$\forall \omega \in \Omega, \ f(\omega) > g(\omega);$$

- *if* $\exists \bar{n}, \forall n \geq \bar{n}, f(n) \neq g(n)$, *then*

$$\forall \omega \in \Omega, \ f(\omega) \neq g(\omega).$$

Proof. We shall prove the first relation since the others can be proved in an analogous way. Let $\omega = \varphi(\alpha_0)$; since ω is unbounded, $\omega \geq \bar{n}$ and by corollary 3 (i),

$$\varphi(n) \geq \bar{n} \ \ for \ a.e. \ n;$$

then

$$f(\varphi(n)) = g(\varphi(n)) \ \ for \ a.e. \ n$$

and hence, by theorem 12 $f(\omega) = g(\omega)$. □

5   Hyperfinite sets

The hyperfinite sets are subsets of $\mathbf{R}^*$ which share many properties of the finite subsets of $\mathbf{R}$. In the second part of this book, we will give a more general definition of hyperfinite sets and we will see that this notion applies not only to the subsets of $\mathbf{R}$.

5.1   Definition

Definition 7. *A set of hyperreal numbers $\Gamma \subset \mathbf{R}^*$ is called hyperfinite if there exists a family of finite sets $\{\Gamma_n\}$, $\Gamma_n \subset \mathbf{R}$, such that*

$$\xi \in \Gamma \Leftrightarrow \exists \varphi \in \mathcal{F}(\mathbf{N}, \mathbf{R}), \ \xi = \varphi(\alpha_0) \ \& \ \varphi(n) \in \Gamma_n$$

Thus, a hyperfinite set can be represented as follows:

$$\Gamma = \{\varphi(\alpha_0) \mid \forall n \in \mathbf{N}, \varphi(n) \in \Gamma_n\} \tag{16}$$

It is immediate to check that a finite subset F of $\mathbf{R}$ is hyperfinite (it sufficient to choose $\forall n \in \mathbf{N}, \Gamma_n = F$; in this case, by theorem 8, $\Gamma = F^* = F$). The interesting fact is the existence of hyperfinite sets which are not finite.
Example Let us see an example: the set

$$\{0, 1, ..., \nu\} := \{n \in \mathbf{N}^* \mid n \leq \nu\} \tag{17}$$

is hyperfinite. In fact, if $\nu = \psi(\alpha_0)$, with $\psi \in \mathcal{F}(\mathbf{N}, \mathbf{N})$, take

$$\Gamma_n := \{k \in \mathbf{N}^* \mid k \leq \psi(n)\}$$

$$= \{0, 1, 2, ..., \psi(n) - 1, \psi(n)\}$$

Clearly the above set is finite if and only if $\psi(\alpha_0)$ is a finite number.
   The following theorem shows some elementary properties which the hyperfinite sets share with the finite sets.

Theorem 14. *The hyperfinite sets satisfy the following properties:*

- *(i) if Γ and Δ are hyperfinite sets, then $\Gamma \cup \Delta$ and $\Gamma \cap B$ are hyperfinite;*
- *(ii) if Γ is hyperfinite and $A \subset \mathbf{R}$ then $\Gamma \cap A^*$ is hyperfinite;*
- *(iii) if Γ is hyperfinite and $f \in \mathcal{F}(\mathbf{R}, \mathbf{R})$, then $f(\Gamma)$ is hyperfinite;*
- *(iv) every nonempty hyperfinite set has the maximum and the minimum.*

Proof. (i) - Set
$$\Gamma = \{\varphi(\alpha_0) \mid \forall n \in \mathbf{N}, \varphi(n) \in \Gamma_n\}$$

and
$$\Delta = \{\varphi(\alpha_0) \mid \forall n \in \mathbf{N}, \varphi(n) \in \Delta_n\}$$

with Γ_n and Δ_n finite. Then,

$$(\Gamma \cup \Delta)_n = \Gamma_n \cup \Delta_n$$

and

$$(\Gamma \cap \Delta)_n = \Gamma_n \cap \Delta_n$$

are finite and hence, $\Gamma \cup \Delta$ and $\Gamma \cap \Delta$ are hyperfinite. (ii)since

$$A^* = \{\varphi(\alpha_0)| \ \varphi \in \mathcal{F}(\mathbf{N}, A)\}$$

$$= \{\varphi(\alpha_0)| \ \varphi(n) \in A\},$$

we have that

$$(\Gamma \cap A^*) = \{\varphi(\alpha_0)| \ \varphi(n) \in A \cap \Gamma_n\}$$

the conclusion follows from the fact that the sets $A \cap \Gamma_n$ are finite. (iii) it follows directly from the fact that

$$f(\Gamma) = \{((f \circ \varphi)(\alpha_0))| \ \varphi(n) \in \Gamma_n\}$$

$$= \{\varphi(\alpha_0)| \ \varphi(n) \in f(\Gamma_n)\}$$

and that the family of sets $\{f(\Gamma_n)\}$ is finite. (iv) - Set

$$\psi(n) = \max \Gamma_n$$

Clearly, $\psi(\alpha_0)$ is the maximum of Γ. Analogously for the minimum.   □

Notice that it is not true in general that a subset of an hyperfinite set is hyperfinite: for example, we have that

$$\mathbf{N} \subset \{n \in \mathbf{N}^* \ | \ n \leq \alpha_0\}$$

but $\mathbf{N}$ is not hyperfinite since it does not have the maximum.

5.2   Hyperfinite sums

A hyperfinite sum is an operation which corresponds to the intuitive idea of adding all the elements of a hyperfinite set. A hyperfinite sum is denoted by

$$\sum_{x \in \Gamma} x.$$

The value of this sum, by definition, is given by $S(\alpha_0)$ where

$$S(n) := \sum_{x \in \Gamma_n} x \tag{18}$$

and $\{\Gamma_n\}$ is the sequence of finite sets given by the (16). Notice that the set Γ determines the sequence $\{\Gamma_n\}$ for μ-almost every $n \in \mathbf{N}$; thus also $S(n)$ is uniquely determined for μ-a.e $n \in \mathbf{N}$ and so $S(\alpha_0)$ is well defined. For example, let Γ be the set (17); then

$$S(n) = 0 + 1 + 2 + \ldots + \psi(n) = \frac{1}{2}\psi(n) \left[\psi(n) + 1\right]$$

thus

$$\sum_{x \in \Gamma} x = S(\alpha_0) = \frac{1}{2}\psi(\alpha_0)[\psi(\alpha_0) + 1] = \frac{\nu(\nu + 1)}{2}$$

since $\nu = \psi(\alpha_0)$. Given a function

$$f : \mathbf{N} \times \mathbf{N} \to \mathbf{R}$$

an expression of the type

$$\sum_{k=0}^{\nu} f(k, \alpha_0), \ \nu \in \mathbf{N}^*$$ (19)

will denote a hyperfinite sum with $\Gamma_n = \{f(0, n), ..., f(\psi(n), n)\}$ where $\psi(\alpha_0) = \nu$. In this case Γ can be denoted by the following expressive notation

$$\Gamma = \{f(0, \alpha_0), ..., f(\nu, \alpha_0)\}$$

Clearly, the value of such a sum is given by $S(\alpha_0)$ where

$$S(n) = \sum_{k=0}^{\psi(n)} f(k, n)$$

Finally an expression of the type

$$\sum_{k=\omega}^{\nu} f(k, \alpha_0), \ \omega, \nu \in \mathbf{N}^*$$

will denote the number

$$\sum_{k=0}^{\nu} f(k, \alpha_0) - \sum_{k=0}^{\omega} f(k, \alpha_0)$$

Examples If Γ is the set (17), then we can write

$$\sum_{x \in \Gamma} x = \sum_{k=0}^{\nu} k = \frac{\nu(\nu + 1)}{2}$$

since

$$\Gamma_n = \{0, 1, 2, ..., \psi(n)\}$$

where $\psi(\alpha_0) = \nu$ and

$$\Gamma = \{0, 1, 2, ..., \nu\}$$

Let us compute

$$\sum_{k=1}^{\alpha_0} k^2$$

In this case

$$S(n) = \sum_{k=1}^{n} k^2 = \frac{1}{3}n^3 + \frac{1}{2}n^2 + \frac{1}{6}n$$

hence

$$\sum_{k=1}^{\alpha_0} k^2 = \frac{1}{3}\alpha_0^3 + \frac{1}{2}\alpha_0^2 + \frac{1}{6}\alpha_0 \tag{20}$$

Let us see an other example:

$$\sum_{k=\alpha_0}^{\alpha_0^2} \varepsilon_0^k$$

In this case

$$S(n) = \sum_{k=n}^{n^2} \left(\frac{1}{n}\right)^k = \frac{\left(\frac{1}{n}\right)^n - \left(\frac{1}{n}\right)^{n^2}}{1 - \frac{1}{n}}$$

and so

$$\sum_{k=\alpha_0}^{\alpha_0^2} \varepsilon_0^k = \frac{\varepsilon_0^{\alpha_0} - \varepsilon_0^{\alpha_0^2}}{1 - \varepsilon_0}.$$

5.3   The hyperfinite grid

Now we introduce a very important hyperfinite set:

$$\mathbf{H} = \{\xi \in \mathbf{R}^* \mid |\xi| \leq \alpha_0 \ \& \ \exists \nu \in \mathbf{Z}^*, \ \xi = \nu\varepsilon_0\}$$

This set will be called the hyperfinite grid.

Theorem 15. H *is a hyperfinite set.*

Proof. **H** has the form (16) with

$$\Gamma_n = \left\{x \in \mathbf{R} \mid \forall n \in \mathbf{N}^+, \exists q \in \mathbf{Z}, \ |q| \leq n^2 \ \& \ x = \frac{q}{n}\right\}$$

$\square$

The property of the hyperfinite grid which has more interest for us is the following:

Theorem 16. *For any real number $x \in \mathbf{R}$ there is a number $\sigma \in \mathbf{H}$ such that $\sigma \leq x < \sigma + \varepsilon_0$*

Proof. The set $(-\infty, x]^* \cap \mathbf{H}$ is hyperfinite by Th. 14(ii), and by the same theorem, (iv), it has the maximum which we call σ. It immediate to see that σ has the required properties.

$\square$

6   The order of a hyperreal number

6.1   Definition

The notion of order, which usually is given for functions or sequences, can be defined for hyperreal numbers.

Definition 8. *Let* $\xi, \eta \in (\mathbf{R}^+)^*$:

- *we say that* ξ *and* η *have the same order if* $\frac{\xi}{\eta}$ *and* $\frac{\eta}{\xi}$ *are bounded; in this case we write*

$$\xi \approx \eta$$

- *we say that* ξ *has higher order than* η *if* $\frac{\xi}{\eta}$ *is unbounded; in this case we write*

$$\xi \gg \eta$$

- *we say that* ξ *has lower order than* η *if* $\frac{\xi}{\eta}$ *is an infinitesimal; in this case we write*

$$\xi \ll \eta.$$

The following theorem lists some properties easy to check:

Theorem 17. *Let* ξ, η *e* $\vartheta \in (\mathbf{R}^*)^+$ *be positive hyperreal numbers:*

- *(i) if* $\xi \approx \eta$ *and* $\eta \approx \vartheta$, *then* $\xi \approx \vartheta$ *(so* $\approx$ *is an equivalence relation on* $(\mathbf{R}^*)^+$*);*
- *(ii) if* $\xi \gg \eta$ *e* $\eta \gg \vartheta$, *then* $\xi \gg \vartheta$ *(so* $\gg$ *is a partial order relation on* $(\mathbf{R}^*)^+$*);*
- *(iii) if* $\xi \gg \eta$, *then* $\xi + \eta \approx \xi$;
- *(iv) if* η *is unbounded, then* $\xi\eta \gg \xi$ *for every* $\xi \in (\mathbf{R}^*)^+$;
- *(v) if* η *is bounded but not infinitesimal, then* $\xi\eta \approx \xi$ *for every* $\xi \in (\mathbf{R}^*)^+$;
- *(vi) if* η *is infinitesimal, then* $\xi\eta \ll \xi$, *for every* $\xi \in (\mathbf{R}^*)^+$;
- *(vii) if* $\eta \gg \xi$, *then* $\frac{1}{\eta} \ll \frac{1}{\xi}$;
- *(viii) if* $\xi \approx \eta$ *and* $\xi_1 \approx \eta_1$, *then* $\xi\xi_1 \approx \eta\eta_1$.

The following theorem illustrates a possible application of the notion of order in the computation of the trace of a number:

Theorem 18. *Let* $\xi, \eta, \xi_1, \eta_1 \in \mathbf{R}^* - \{0\}$ *and suppose that* $|\xi_1| \ll |\xi|$ *and* $|\eta_1| \ll |\eta|$. *Then*

$$tr\left(\frac{\xi + \xi_1}{\eta + \eta_1}\right) = tr\left(\frac{\xi}{\eta}\right)$$

Proof. It is sufficient to observe that

$$\frac{\xi + \xi_1}{\eta + \eta_1} = \frac{\xi}{\eta} \cdot \frac{1 + \frac{\xi_1}{\xi}}{1 + \frac{\eta_1}{\eta}}$$

and that

$$tr\left(\frac{1 + \frac{\xi_1}{\xi}}{1 + \frac{\eta_1}{\eta}}\right) = 1.$$

□

In order to apply the theorems 17 and 18 to the computation to the trace of a number, it is necessary to compare the order of some basic numbers. Tis is done in the following theorem:

Theorem 19. *Let ω be an unbounded number, then*

- *(i) if $a < b$ are natural numbers, then $\omega^a \ll \omega^b$*
- *(ii) for every $a, b \in \mathbf{R}^+$, $\omega^a \ll b^\omega$*
- *(iii) for every $b \in \mathbf{R}^+$, $b^\omega \ll \omega!$*
- *(iv) $\omega! \ll \omega^\omega$*

Proof. (i) trivial since ω^{b-a} is an infinite number. (ii) we know that for n sufficiently large

$$\frac{n^a}{b^n} < \frac{1}{n},$$

thus by the Cauchy Principle, (theorem 13), we have that

$$\forall \omega \in \Omega, \frac{\omega^a}{b^\omega} < \frac{1}{\omega}.$$

Thus

$$\forall \omega \in \Omega, \omega^a \ll b^\omega.$$

The other results can be proved in an analogous way once the corresponding results for numerical sequences are known.   □

The theorems 18 and 19 make easier the computation of the trace of a number; for example

$$tr\left(\frac{\alpha_0! + 7\alpha_0^3}{5\alpha_0! + 2\alpha_0^7}\right) = tr\left(\frac{\alpha_0!}{5\alpha_0!}\right) = \frac{1}{5} \tag{21}$$

$$tr\left(\frac{\alpha_0^7 + 8\alpha_0^3}{3\alpha_0! + 2\alpha_0^7}\right) = tr\left(\frac{\alpha_0^7}{3\alpha_0!}\right) = 0 \tag{22}$$

6.2   Relative order

Given two hyperreal numbers ξ and ω it might be useful to compare their order using real numbers; then the following definition turns out to be useful:

Definition 9. *The number $\xi \in (\mathbf{R}^*)^+$ is said to have order "a" ($a \in \mathbf{R}$) relatively to an unbounded number ω if*

$$\xi \approx \omega^a$$

ξ is said to have order " $+\infty$ " relatively to ω if

$$\forall a \in \mathbf{R}, \xi \gg \omega^a$$

Analogously, the number ξ is said to have order "a" $(a \in \mathbf{R})$ relatively to an infinitesimal ε if

$$\xi \approx \varepsilon^a$$

ξ is said to have order " $+\infty$ " relatively to ε if

$$\forall a \in \mathbf{R}^+, \; \xi \ll \varepsilon^a$$

If we say that ξ has order "a" $(a \in \mathbf{R})$ (or "$+\infty$") without specifying ω or ε, we mean the order relative to α_0 i.e.

$$\xi \approx \alpha_0^a \quad (or \; \forall a \in \mathbf{R}, \; \xi \gg \alpha_0^a)$$

Examples The number

$$5\alpha_0^2 + 9\alpha_0 + 2 + 3\varepsilon_0$$

has order 2 relatively to α_0, order 1 relatively to α_0^2 and order -2 relatively to ε_0. The number

$$5\varepsilon_0^3 + 7\varepsilon_0^4$$

has order 3 relatively to ε_0 and order -3 relatively to α_0. The number

$$\sqrt{\varepsilon_0^3 + 7\varepsilon_0^2}$$

has order $3/2$ relatively to ε_0 and order $-3/2$ relatively to α_0 . The bounded numbers which are not infinitesimal have order 0 relatively to every unbounded number and to every infinitesimal.

Given given a hyperreal ξ and an unbounded number (e.g. α_0) it might not exist a real number a such that

$$\xi \approx \alpha_0^a$$

Then it might be useful to have the following definition:

Definition 10. *The number $\xi \in (\mathbf{R}^*)^+$ is said to have order of type "a^+" $(a \in \mathbf{R})$ if*

$$\forall x \in \mathbf{R}, \; x > a \Rightarrow \alpha_0^a \ll \xi \ll \alpha_0^x;$$

similarly, the number $\xi \in \mathbf{R}^{+*}$ is said to have order of type "a^-" $(a \in \mathbf{R})$ if

$$\forall x \in \mathbf{R}, \; x < a \Rightarrow \alpha_0^x \ll \xi \ll \alpha_0^a$$

For example, the number $\log \alpha_0$ has order 0^+, the number

$$\frac{\alpha_0^2}{\log \alpha_0}$$

has order 2^-. Notice that two numbers may have order of the same type but different; for example

$$\frac{\alpha_0^2}{\log \alpha_0} \quad and \quad \frac{\alpha_0^2}{\log \log \alpha_0}$$

have both order of type 2^-, but they have different order:

$$\frac{\alpha_0^2}{\log \alpha_0} \ll \frac{\alpha_0^2}{\log \log \alpha_0}.$$

7 Some remarks on the Axioms

In this section we shall make some remarks on the Axioms 1,..., 3 which we have used to introduce the hyperreal numbers.

7.1 An algebraic construction of the hyperreal numbers

First we recall some known facts of algebra.

Definition 11. *Let R be a commutative ring with identity. A subset $I \subset R$ is called ideal if*

- *(i) if $\varphi, \psi \in I \Rightarrow \varphi + \psi, \ \varphi - \psi, \ \varphi \cdot \psi \in I$;*
- *(ii) $\varphi \in I \ \& \ \psi \in R \Rightarrow \varphi \cdot \psi \in I$.*

Definition 12. *An ideal $I \subset R$ is called maximal if for any ideal J*

$$I \subset J \subseteq R \Rightarrow J = R.$$

As example of maximal ideal in the ring $\mathcal{F}(\mathbf{N}, \mathbf{R})$ consider the following set:

$$I(a) = \{\sigma \in \mathcal{F}(\mathbf{N}, \mathbf{R}) \mid \sigma(a) = 0\}$$

where $a \in \mathbf{N}$. The ideal

$$I(a, b) = \{\sigma \in \mathcal{F}(\mathbf{N}, \mathbf{R}) \mid \sigma(a) = \sigma(b) = 0\}, \ a \neq b$$

is not maximal since $I(a, b) \subset I(a)$. The following proposition provides an example of a nontrivial ideal which will be useful for us:

Proposition 3. *The set*

$$\Im_0 = \{\sigma \in \mathcal{F}(\mathbf{N}, \mathbf{R}) \mid \exists k \in \mathbf{N}^+, \forall n \in \mathbf{N}^+, \sigma(kn) = 0\}$$

is an ideal.

Proof. If φ and $\psi \in \Im_0$, then $\exists k_1, k_2 \in \mathbf{N}^+$ such that

$$\forall n \in \mathbf{N}^+, \varphi(k_1 n) = \psi(k_2 n) = 0.$$

now set $k = k_1 \cdot k_2$; then we have

$$\forall n \in \mathbf{N}^+, \varphi(kn) = \psi(kn) = 0$$

and hence

$$\forall n \in \mathbf{N}^+, \varphi(kn) + \psi(kn) = \varphi(kn) \cdot \psi(kn) = 0$$

Then the (i) of definition 11 holds. (ii) can be checked straightforward.   □

We now need some theorem of algebra; let us we recall them:

Theorem 20. *Let I be an ideal of R. Then there exists a maximal ideal I_m containing I.*

Theorem 21. *Let I be an ideal of R. Then the family of sets*

$$R/I = \{a + I \mid a \in R\}$$

is a ring (with the obvious operation) and it is called the quotient ring.

Theorem 22. *Let*

$$J : R \to R'$$

be a ring homomorphism. Then $\ker(J)$ is an ideal in R. Moreover, if J is onto, R' is isomorphic to $R/\ker(J)$.

Theorem 23. *Let R be a commutative ring with identity and let I be an ideal in R. Then I is maximal if and only if R/I is a field.*

Using the above theorems we can construct a model which satisfies the Axioms 1, 2, 3. Let $\Im$ be a maximal ideal in $\mathcal{F}(\mathbf{N}, \mathbf{R})$ containing $\Im_0$. Such an ideal exists by the theorem 20. Define

$$\mathbf{R}^* := \mathcal{F}(\mathbf{N}, \mathbf{R})/\Im \tag{23}$$

By the theorem 23, $\mathbf{R}^*$ is a field. Now we want to equip $\mathbf{R}^*$ with an order structure. $\mathcal{F}(\mathbf{N}, \mathbf{R})$ has a natural structure of partial ordering defined by the following formulas:

$$\varphi \geq 0 :\Leftrightarrow \forall n \in \mathbf{N}, \; \varphi(n) \geq 0$$

$$\varphi \geq \psi :\Leftrightarrow \varphi - \psi \geq 0.$$

As usual, we define

$$\varphi^+(n) := \max(0, \varphi(n)); \quad \varphi^-(n) := -\min(0, \varphi(n)).$$

Now we define an order relation on the equivalence classes $[\varphi] := \varphi + \Im$ of $\mathbf{R}^*$ as follows:

$$[\varphi] \geq 0 :\Leftrightarrow [\varphi] = [\varphi^+] \tag{24}$$

$$[\varphi] \geq [\psi] :\Leftrightarrow [\varphi - \psi] \geq 0.$$

Now, we can prove the following theorem:

Theorem 24. $\mathbf{R}^*$, *defined by the (23) satisfies the Axioms 1,2,3. provided that* $\mathbf{R}$ *is identified with the subfield of* $\mathcal{F}(\mathbf{N}, \mathbf{R})/\Im$ *of the constant functions.*

Proof. We already know that $\mathbf{R}^*$ is a field. So we have prove that $\mathbf{R}^*$ is an ordered field. First let us prove that the (24) is a liner order relation i.e.

$$[\varphi] \leq [\psi] \ \& \ [\psi] \leq [\varphi] \Leftrightarrow [\varphi] = [\psi]$$

To see this it is sufficient to prove that

$$[\varphi] \geq 0 \ \& \ [\varphi] \leq 0 \Leftrightarrow [\varphi] = 0.$$

Thus let us suppose that $[\varphi] \geq 0 \ \& \ [\varphi] \leq 0$; clearly

$$[\varphi] \geq 0 \Rightarrow [\varphi] = [\varphi^+] \Rightarrow \varphi - \varphi^+ \in \mathfrak{F}$$

Analogously

$$[\varphi] \leq 0 \Rightarrow [\varphi] = [-\varphi^-] \Rightarrow \varphi + \varphi^- \in \mathfrak{F}$$

Thus, by the above formulas

$$(\varphi - \varphi^+) + (\varphi + \varphi^-) = 2\varphi - \varphi^+ + \varphi^- = \varphi \in \mathfrak{F}$$

which implies $[\varphi] = 0$. It is immediate to check that the order relation is consistent with the field operation; thus $\mathbf{R}^*$ is an ordered field and Axiom 1 is satisfied. Now consider the natural projection

$$J : \mathcal{F}(\mathbf{N}, \mathbf{R}) \to \mathcal{F}(\mathbf{N}, \mathbf{R})/\mathfrak{F}$$

defined by

$$J(\varphi) = [\varphi]$$

Since J is a ring homomorphism, it satisfies the Axiom 2. The Axiom 3 is a consequence of the fact that $\mathfrak{F}_0 \subset \mathfrak{F}$. □

7.2   Explicit formulas

Two different expressions may represent the same hyperreal number. For example, by 2 (ii), it is easy ti check that the expressions

$$(-1)^{\alpha_0+1} \cdot \alpha_0 \text{ and } -\alpha_0$$

represent the same number. However, it is not always possible to know when two expressions represent the same number. For example, consider the function:

$$\sigma(n) = \begin{cases} 1 & \text{if } n = m^2 \text{ is a square number} \\ 0 & \text{otherwise} \end{cases}$$

Since

$$\sigma(n) \cdot [1 - \sigma(n)] = 0,$$

by Proposition 2 (ii), $\sigma(\alpha_0)$ satisfies the following equation:

$$x - x^2 = 0,$$

Thus, since $\mathbf{R}^*$ is a field, $\sigma(\alpha_0)$ is 0 or 1. It remains to decide which one is true. This question is equivalent to know if σ or $1 - \sigma$ belong to the ideal $\mathfrak{J}$. Which alternative holds cannot be deduced by our Axioms. Of course, we can make Axiom 3 stronger in order to answer to the above question. For example we can replace Axiom 3 with the following:

Axiom 3' If there exists $k_1, k_2 \in \mathbf{N}^+$ such that, for every $n \in \mathbf{N}^+$

$$\varphi\left((k_1 n)^{k_2}\right) \geq a, \ a \in \mathbf{R}$$

then

$$J(\varphi) \geq a.$$

Using this Axiom, it is easy to see that $\sigma(\alpha_0) = 1$. However, at this point it is not difficult to construct an other proposition which cannot be proved by these new Axioms. This is a peculiarity of N.S.A. which, probably, cannot be avoided. If we construct $\mathbf{R}^*$ as in the section 7.1, this kind of questions are equivalent to an explicit knowledge of the maximal ideal $\mathfrak{J}$, but the existence of $\mathfrak{J}$ is guaranteed by the lemma of Zorn and no explicit construction of $\mathfrak{J}$ exists. Notice that the knowledge of $\mathfrak{J}$ is equivalent to the knowledge of the measure μ, since

$$\mathfrak{J} = \{\sigma \in \mathcal{F}(\mathbf{N}, \mathbf{R}) \mid \sigma(n) = 0 \ for \ \mu\text{-a.e. } n\} \tag{25}$$

Since $\mathfrak{J}$ cannot be given explicitly, there are propositions containing α_0 whose truth cannot be deduced by the axioms. This fact might be psychologically unpleasant for many people, however it does not affect the efficiency of N.S.A. On the other hand it can be proved that, if the continuum hypotheses holds, the ordered field $\mathbf{R}^*$ is unique up to isomorphism.

8   Continuity

8.1   Definition of Continuity

Probably, the definition of continuity is the most delicate notion of Calculus. The definition which we use today has been introduced relatively late; in fact it has been introduced by Weierstrass towards the end of last century. Earlier, the notion of continuity was based on that of infinitesimal. These reasons make the continuity a good topic to test the power of nonstandard analysis.

Definition 13. *Given a set $A \subset \mathbf{R}$, a function*

$$f : A \to \mathbf{R}$$

is called continuous in a point $x_0 \in A$ if for every $\xi \in A^*$,

$$\xi \sim x_0 \Rightarrow f(\xi) \sim f(x_0)$$

The function f is called continuous in A if it is continuous in every point of A

Thus, to check that a function is continuous in x_0 you have just to evaluate the number $f(\xi) - f(x_0)$. For example, in order to prove the continuity of the function $f(x) = x^2$ at the point $x_0 = 2$, you have to evaluate $f(2 + \varepsilon) - f(2)$, where ε is a generic infinitesimal and to check if this difference is infinitesimal:

$$f(2 + \varepsilon) - f(2) = (2 + \varepsilon)^2 - 4 = 4\varepsilon - \varepsilon^2 \sim 0$$

Example Let us consider a nontrivial example. Define the following function:

$$f(x) = \begin{cases} \frac{1}{q} & \text{if } x = \frac{p}{q} \in \mathbf{Q} \text{ and } \frac{p}{q} \text{ is a reduced fraction;} \\ 0 & \text{if } x \in \mathbf{R} - \mathbf{Q} \end{cases}$$

This function is continuous for $x \in \mathbf{R} - \mathbf{Q}$ and discontinuous for $x \in \mathbf{Q}$. In order to prove this fact, first we extend f to the hyperreal numbers:

$$f^*(\xi) = \begin{cases} \frac{1}{q} & \text{if } \xi = \frac{p}{q} \in \mathbf{Q}^* \text{ and } \frac{p}{q} \text{ is a reduced fraction;} \\ 0 & \text{if } \xi \in (\mathbf{R} - \mathbf{Q})^* \end{cases}$$

Now take $x \in \mathbf{R} - \mathbf{Q}$ and $\xi \sim x$. If $\xi \in (\mathbf{R} - \mathbf{Q})^*$, $f(\xi) = 0 = f(x)$ and hence $f(\xi) \sim f(x)$; if $\xi = \frac{p}{q} \in \mathbf{Q}^*$ we have that $q \in \mathbf{Z}^* - \mathbf{Z}$ since

$$q \in \mathbf{Z} \Rightarrow \begin{cases} tr\,(\xi) = \frac{p}{q} \in \mathbf{Q} & \text{if } p \in \mathbf{Z}, \text{ absurd since } x \in \mathbf{R} - \mathbf{Q} \\ tr\,(\xi) = \pm\infty & \text{if } p \in \mathbf{Z}^* - \mathbf{Z}, \text{ absurd since } x \text{ is bounded} \end{cases}$$

this implies that q is unbounded and hence $f(\xi) = \frac{1}{q} \sim 0 = f(x)$. Now take $x = \frac{p}{q} \in \mathbf{Q}$. since there is $\xi \sim x$ with $\xi \in (\mathbf{R} - \mathbf{Q})^*$, $f(\xi) = 0 \not\sim \frac{1}{q} = f(x)$.

The next proposition, whose proof is very simple and left to the reader, gives different characterizations of continuity:

Proposition 4. *Given a function*

$$f : A \to \mathbf{R}$$

and a point $x_0 \in A$ the following statements are equivalent:

- *(i) f is continuous in x_0;*
- *(ii) $\forall \xi \in mon(x_0) \cap A^*$, $tr(f(\xi)) = f(x_0)$;*
- *(iii) $f(mon(x_0) \cap A^*) \subset mon(f(x_0))$;*
- *(iv) $\varepsilon \sim 0$ & $x_0 + \varepsilon \in A^* \Rightarrow f(x_0 + \varepsilon) - f(x_0) \sim 0$.*

8.2   The semicontinuity and the theorem of Weierstrass

Also the semicontinuity can be easily characterized with nonstandard tools.

Definition 14. *Given a set $A \subset \mathbf{R}$, a function*

$$f : A \to \mathbf{R}$$

is called lower semicontinuous in a point $x_0 \in A$ if for every $\xi \in mon(x_0) \cap A^*$,

$$tr(f(\xi)) \geq f(x_0);$$

if the above inequality is reversed, f is called upper semicontinuous.

Using the (ii) of the Prop. 4, we get that a function is continuous at a point x_0 if and only if it is upper and lower semicontinuous at that point. Before proving the Weierstrass theorem we recall the nonstandard definition of supremum and infimum:

Definition 15. *If $A \subset \mathbf{R}$, the supremum of A is a number $l \in \hat{\mathbf{R}}$ such that*

- *(i) $\forall x \in A$, $l \geq x$*

- *(ii) $\exists x \in A^*$, $tr(x) = l$*

The supremum of A is denoted by

$$\sup(A)$$

If the inequality in (i) is reversed, we get the definition of infimum.

Now we shall give a nonstandard proof of Weierstrass theorem

Theorem 25 (Weierstrass). *Let $f : [a, b] \to \mathbf{R}$ be a upper semicontinuous function. Then it has the maximum. Analogous fact for lower semicontinuous functions and the minimum.*

Proof. We argue for the maximum. Set

$$l = \sup(f([a, b])).$$

By the definition of supremum, there exists $\eta \in f([a, b])^*$ with $tr(\eta) = l$. So, there is $\xi \in [a, b]^*$, such that $\eta = f(\xi)$. We claim that

$$x_0 = tr(\xi)$$

is the maximum point. Since the interval $[a, b]$ is bounded and closed, $x_0 \in [a, b]$; hence $f(x_0)$ is defined and by the upper semicontinuity of f,

$$f(x_0) \geq tr(f(\xi)) = tr(\eta) = l.$$

$\square$

8.3   The theorem of Bolzano

Theorem 26 (Bolzano). *Let $f : [a, b] \to \mathbf{R}$ be a continuous function such that*

$$f(a) < 0 \text{ and } f(b) > 0.$$

Then there is a point $x_0 \in [a, b]$ such that

$$f(x_0) = 0$$

Proof. Set

$$A = \{x \in [a, b]^* \cap \mathbf{H} \mid f(x) \leq 0\}$$

Since $\exists \xi \in \mathbf{H}$, $\xi \geq a$, $\xi \sim a$, it results that $A \neq \emptyset$. Then, by theorem (14), you can take

$$\xi = \max A.$$

$\xi \in A$, so $f(\xi) \leq 0$; since $f(b) > 0$, we have that $\xi < b$; so by the continuity of f, it follows that $\xi \not\sim b$; this implies that $\xi + \varepsilon_0 < b$ and that $\xi + \varepsilon_0 \in [a, b]$. So $\xi + \varepsilon_0 \in \mathbf{H} - A$, which implies that $f(\xi + \varepsilon_0) > 0$. Thus, we have

$$f(\xi) \leq 0 < f(\xi + \varepsilon_0).$$

Taking $x_0 = tr(\xi)$, using the continuity of f and taking the trace of the terms of the above inequality we get

$$f(x_0) \leq 0 \leq f(x_0)$$

$\square$

8.4   The uniform continuity and the theorem of Cantor

In the usual Analysis, the distinction between continuity and uniform continuity is very subtle (or at least, it appears subtle to the person who sees it for the first time). On the contrary, in N.S.A. this distinction arises naturally.

Definition 16. *Given a set $A \subset \mathbf{R}$ function*

$$f : A \to \mathbf{R}$$

is called uniformly continuous in A if for every $\xi, \eta \in A^*$,

$$\xi \sim \eta \Rightarrow f(\xi) \sim f(\eta)$$

Thus, in the continuity we compare the values which a function takes on two infinitely close points *one of which is standard;* in the uniform continuity we do not impose this restriction. For example, consider the function $\frac{1}{x}$ defined in the set $A = \mathbf{R} - \{0\}$. Clearly it is continuous, but it is not uniformly continuous. In fact

$$\varepsilon_0, 2\varepsilon_0 \in A^*; \quad \varepsilon_0 \sim 2\varepsilon_0$$

but $f(\varepsilon_0) \not\sim f(2\varepsilon_0)$ since

$$f(\varepsilon_0) - f(2\varepsilon_0) = \frac{1}{\varepsilon_0} - \frac{1}{2\varepsilon_0} = \frac{1}{2}\alpha_0 \not\sim 0.$$

The nonstandard proof of the theorem of Cantor on uniform continuity turns out to be almost a triviality:

Theorem 27 (Cantor). *Any continuous function $f : [a, b] \to \mathbf{R}$ is uniformly continuous.*

Proof. Take $\xi, \eta \in A^*$ with $\xi \sim \eta$. Since $[a, b]$ is bounded and close; $tr(\xi) \in [a, b]$. We set $x = tr(\xi)$ and since $\eta \sim \xi$ we have also that $x = tr(\eta)$. By the continuity of f, $f(\xi) \sim f(x)$ and $f(\eta) \sim f(x)$. Hence

$$f(\xi) \sim f(\eta).$$

$\square$

9 Derivation

9.1 Definition

Let us give the usual definition of derivative:

Definition 17. *We say that a function $f : (a, b) \to \mathbf{R}$ is differentiable in a point $x_0 \in (a, b)$ if there is a number $l \in \mathbf{R}$ such that*

$$\forall \varepsilon \in mon(0) - \{0\}, \ tr\left(\frac{f(x_0 + \varepsilon) - f(x_0)}{\varepsilon}\right) = l.$$

The number l is called the derivative of f at the point x_0 and it is denoted by

$$f'(x_0)$$

By the definition of derivative we have that

$$f(x_0 + \varepsilon) = f(x_0) + \varepsilon f'(x_0) + \varepsilon\delta \tag{26}$$

with $\delta \sim 0$. If f is derivable in every point of (a, b) then the derivative function

$$f' : (a, b) \to \mathbf{R}$$

is defined and hence it is defined its natural extension

$$f' : (a, b)^* \to \mathbf{R}^*$$

9.2   The theorem of Lagrange

The proof of the elementary theorem of the differential calculus done using non-standard analysis are not very different from the conventional ones; so we omit them; we only consider the Lagrange theorem since it has a useful corollary without any analogous in the conventional analysis.

Theorem 28. *of Lagrange Given a continuous function* $f : [a, b] \to \mathbf{R}$, *derivable in* (a, b), *there exists a point* $x_0 \in (a, b)$ *such that*

$$\frac{f(b) - f(a)}{b - a} = f'(x_0)$$

The theorem of Lagrange, as well as the other theorems of analysis can be extended from the world of the real numbers to the world of the hyperreal numbers:

Theorem 29. *Given a derivable function* $f : (a, b) \to \mathbf{R}$, *and two points* $\xi, \eta \in (a, b)^*, \xi < \eta$, *there exists a point* $\theta \in (a, b)^*, \xi < \theta < \eta$ *such that*

$$\frac{f(\eta) - f(\xi)}{\eta - \xi} = f'(\theta)$$

Proof. Let $\varphi(n)$ and $\psi(n)$ be functions such that $\varphi(\alpha_0) = \xi$, $\psi(\alpha_0) = \eta$ and $\forall n \in \mathbf{N}$, $\varphi(n), \psi(n) \in (a, b)$. Then, $\forall n \in \mathbf{N}$, we can apply the Lagrange theorem to the interval $[\varphi(n), \psi(n)]$ and to obtain a point $x_n \in (\varphi(n), \psi(n))$ such that

$$\frac{f(\psi(n)) - f(\varphi(n))}{\psi(n) - \varphi(n)} = f'(x_n)$$

The conclusion follows taking $\theta = x_{\alpha_0}$. □

Now, we expose a corollary of the theorem of Lagrange which is meningless in the Standard Analysis; this corollary will be used to prove Th. 30. Among the corollaries of the above theorem there is the following one.

Corollary 4. *If* $f \in C^1[a, b]$ *then if* $\xi, \eta \in (a, b)^*$, $\xi \sim \eta$, *we have*

$$f(\eta) - f(\xi) = f'(\xi) \cdot (\eta - \xi) + \delta(\eta - \xi)$$

with $\delta \sim 0$.

Proof. By theorem 29,

$$f(\eta) - f(\xi) = f'(\theta) \cdot (\eta - \xi) ; \tag{27}$$

now, setting

$$\delta = f'(\theta) - f'(\xi), \tag{28}$$

we have that $\delta \sim 0$, since by the continuity of f', $f'(\xi) \sim f(st(\xi)) \sim f'(\theta)$. Comparing the (27) and the (28), we get the conclusion. □

10   Integration

10.1   Definition of definite integral

Given a interval $(a, b) \subset \mathbf{R}$, the set $(a, b)^* \cap \mathbf{H}$ is a hyperfinite set. It consists of "*hypermultiple*" of ε_0 which belong to $(a, b)^*$; this set allows to give the following definition of integral:

Definition 18. *Given a function $f : (a, b) \to \mathbf{R}$ the integral of f in (a, b) is the number*

$$\int_a^b f(x)\, dx := tr\left(\varepsilon_0 \sum_{\xi \in (a,b)^* \cap \mathbf{H}} f(\xi) \right) \tag{29}$$

For example, let us compute $\int_0^1 x^2\, dx$. Since

$$(0, 1)^* \cap \mathbf{H} = \{ k\varepsilon_0 \mid k = 1, 2, .., \alpha_0 - 1 \},$$

we have

$$\int_0^1 x^2\, dx = tr\left(\varepsilon_0 \sum_{k=1}^{\alpha_0-1} (k\varepsilon_0)^2 \right) = tr\left(\varepsilon_0^3 \sum_{k=1}^{\alpha_0} k^2 \right)$$

$$= tr\left(\varepsilon_0^3 \left(\frac{1}{3}\alpha_0^3 + \frac{1}{2}\alpha_0^2 + \frac{1}{6}\alpha_0 \right) \right) = tr\left(\frac{1}{3} + \frac{1}{2}\varepsilon_0 + \frac{1}{6}\varepsilon_0^2 \right) = \frac{1}{3}$$

In general, if $[\xi]$ denotes the integer part of the hyperreal number ξ, (i.e. the largest hyperinteger smaller or equal to ξ), the (29) can be written as follows:

$$\int_a^b f(x)\, dx = tr\left(\varepsilon_0 \sum_{k=[a\alpha_0]+1}^{[b\alpha_0]-1} f(k\varepsilon_0) \right) ;$$

this formula can help in computations. Notice that in the definition 18, it is possible to assume that a and/or b are equal to $\pm\infty$. For example, let us consider the following integral:

$$\int_a^{+\infty} e^{-x}\, dx;$$

since

$$\sum_{k=M}^N x^k = \frac{x^{N+1} - x^M}{x - 1}$$

we have

$$\int_a^{+\infty} e^{-x}\, dx = tr\left[\varepsilon_0 \sum_{k=[a\alpha_0]+1}^{\alpha_0^2} (e^{-\varepsilon_0})^k \right]$$

$$= tr\left(\varepsilon_0 \cdot \frac{e^{-\varepsilon_0 \alpha_0^2} - e^{-\varepsilon_0([a\alpha_0]+1)}}{e^{-\varepsilon_0} - 1} \right)$$

$$= tr\left(\frac{\varepsilon_0}{e^{-\varepsilon_0}-1}\right) \cdot tr\left[\left(e^{-\alpha_0}\right) - e\left(-\varepsilon_0[\alpha\alpha_0]-\varepsilon_0\right)\right]$$

$$= (-1) \cdot tr\left[-e\left(-\varepsilon_0[\alpha\alpha_0]-\varepsilon_0\right)\right] = e^{-\alpha}.$$

Remark 2. The integral (29) not only has a very simple definition, but it is defined for every function; actually it could be defined also for every set $E \subset \mathbf{R}$, (it is enough to replace $(a,b)^* \cap \mathbf{H}$ with $E^* \cap \mathbf{H}$); of course it coincides with the usual Riemann integral when f is Riemann integrable. Clearly if f is not Riemann integrable, or worst if it is not even Lebesgue measurable, the integral (29) might have some pathological behavior (for example it is not invariant for translations). Nevertheless, the integral (29) has many nice properties; for example it coincides with the improper integral when this is well defined as the above example has shown.

Now let us define the integral on a oriented interval.

Definition 19. *Let f be a function defined in an interval I and let $a, b \in I$ be two real numbers. If $a = b$ we set*

$$\int_a^b f(x)\,dx = 0.$$

If $b < a$ we set.

$$\int_a^b f(x)\,dx = -\int_b^a f(x)\,dx$$

We conclude this section defining the integral for hyperreal intervals:

Definition 20. *Let f be a function defined in an interval I and let $a, b \in I^*$ be two hyperreal numbers. If $a = \varphi(\alpha_0)$ and $b = \psi(\alpha_0)$ (with $\varphi(n), \psi(n) \in I$), we set*

$$\int_a^b f(x)\,dx = G(\alpha_0).$$

where

$$G(n) = \int_{\varphi(n)}^{\psi(n)} f(x)\,dx$$

10.2   The fundamental theorem of calculus

Now we shall prove the fundamental theorem of calculus. Even if this theorem is an immediate consequence of the Theorem of the Primitive 32, we will give here an independent proof; in fact, using nonstandard analysis, this proof is not too technical and it gives a better understanding of this theorem, particularly when you compare it with its higher dimensional analogous, i.e. the Stokes formula and the Gauss Divergence theorem.

Theorem 30. *Let $F \in C^1[a, b]$, then*

$$\int_a^b F'(x)\, dx = F(b) - F(a). \tag{30}$$

Proof. By the definition of integral we have

$$\int_a^b F'(x)\, dx \sim \varepsilon_0 \sum_{k=[a\alpha_0]+1}^{[b\alpha_0]-1} F'(k\varepsilon_0);$$

recalling the Corollary 4, we have

$$F'(k\varepsilon_0) = \frac{F(k\varepsilon_0 + \varepsilon_0) - F(k\varepsilon_0)}{\varepsilon_0} + \eta_k$$

thus

$$\int_a^b F'(x)\, dx \sim \varepsilon_0 \sum_{k=[a\alpha_0]+1}^{[b\alpha_0]-1} \left[\frac{F(k\varepsilon_0 + \varepsilon_0) - F(k\varepsilon_0)}{\varepsilon_0} + \eta_k \right]$$

$$= \sum_{k=[a\alpha_0]+1}^{[b\alpha_0]-1} F((k+1)\,\varepsilon_0) - \sum_{k=[a\alpha_0]+1}^{[b\alpha_0]-1} F(k\varepsilon_0) + \varepsilon_0 \sum_{k=[a\alpha_0]+1}^{[b\alpha_0]-1} \eta_k. \tag{31}$$

Setting $\eta = \max(|\eta_k|)$ (notice that the maximum exists by theorem 14 (iv))

$$\left| \varepsilon_0 \sum_{k=[a\alpha_0]+1}^{[b\alpha_0]-1} \eta_k \right| \leq \varepsilon_0 \sum_{k=[a\alpha_0]+1}^{[b\alpha_0]-1} \eta \leq$$

$$\varepsilon_0 \eta \left([b\alpha_0] - [a\alpha_0] - 2 \right) \leq \eta(b - a).$$

Thus the last term of the (31) is infinitesimal; so, shifting an index in the first hyperfinite sum we get

$$\int_a^b F'(x)\, dx \sim \sum_{k=[a\alpha_0]+2}^{[b\alpha_0]} F(k\varepsilon_0) - \sum_{k=[a\alpha_0]+1}^{[b\alpha_0]-1} F(k\varepsilon_0)$$

$$= F([b\alpha_0]\,\varepsilon_0) - F(([a\alpha_0]+1)\,\varepsilon_0)$$

$$\sim F(b) - F(a)$$

since F is continuous. In the above formula the first and the last term are standard numbers, hence

$$\int_a^b F'(x)\, dx = F(b) - F(a)$$

$\square$

10.3   Main properties of the definite integral

The following theorem states the main properties of the integral (29) which are shared with the Riemann integral:

Theorem 31. *Let f and g be two functions defined in (a, b) then*

- (i)
$$\int_a^b \lambda f(x)\, dx = \lambda \int_a^b f(x)\, dx$$

- (ii)
$$\int_a^b [f(x) + g(x)]\, dx = \int_a^b f(x)\, dx + \int_a^b g(x)\, dx$$

- (iii) *if* $c \in (a, b)$ *then*
$$\int_a^b f(x)\, dx = \int_a^c f(x)\, dx + \int_c^b f(x)\, dx$$

- (iv) *if* $f(x) \le g(x)$ *in* (a, b), *then*
$$\int_a^b f(x)\, dx \le \int_a^b g(x)\, dx$$

- (v)
$$\left| \int_a^b f(x)\, dx \right| \le \int_a^b |f(x)|\, dx$$

- (vi) *if* $m \le f(x) \le M$ *for every* $x \in [a, b]$, *then*
$$m \cdot (b - a) \le \int_a^b f(x)\, dx \le M \cdot (b - a)$$

- (vii) *if* f *is continuous, then there exists* $x_0 \in (a, b)$ *such that*
$$f(x_0) = \frac{1}{b - a} \int_a^b f(x)\, dx$$

Proof. (i)
$$\int_a^b \lambda f(x)\, dx = tr \left[\varepsilon_0 \sum_{x \in H \cap (a,b)^\bullet} \lambda f(x) \right] =$$

$$= \lambda\, tr \left[\varepsilon_0 \sum_{x \in H \cap (a,b)^\bullet} f(x) \right] = \lambda \int_a^b f(x)\, dx.$$

(ii)
$$\int_a^b [f(x) + g(x)]\, dx = tr \left[\varepsilon_0 \sum_{x \in H \cap (a,b)^\bullet} [f(x) + g(x)] \right] =$$

$$= tr\left[\varepsilon_0 \sum_{z\in \mathbf{H}\cap(a,b)^{\bullet}} f(x)\right] + tr\left[\varepsilon_0 \sum_{z\in \mathbf{H}\cap(a,b)^{\bullet}} g(x)\right] =$$

$$= \int_a^b f(x)\,dx + \int_a^b g(x)\,dx.$$

(iii) Since the set $\mathbf{H}\cap(a,b)^*$ and the set

$$[\mathbf{H}\cap(a,c)^*]\cup[\mathbf{H}\cap(c,b)^*] = \mathbf{H}\cap[(a,c)^*\cup(c,b)^*]$$

differ at most for the point c, we have that

$$\varepsilon_0 \sum_{z\in \mathbf{H}\cap(a,b)^{\bullet}} f(x) \sim \varepsilon_0 \left(\sum_{z\in \mathbf{H}\cap(a,c)^{\bullet}} f(x) + \sum_{z\in \mathbf{H}\cap(c,b)^{\bullet}} f(x)\right);$$

the remaining part of the proof is trivial:

$$\int_a^b f(x)\,dx = tr\left[\varepsilon_0 \sum_{z\in \mathbf{H}\cap(a,b)^{\bullet}} f(x)\right] =$$

$$= tr\left[\varepsilon_0 \left(\sum_{z\in \mathbf{H}\cap(a,c)^{\bullet}} f(x) + \sum_{z\in \mathbf{H}\cap(c,b)^{\bullet}} f(x)\right)\right] =$$

$$= tr\left[\varepsilon_0 \sum_{z\in \mathbf{H}\cap(a,c)^{\bullet}} f(x)\right] + tr\left[\varepsilon_0 \sum_{z\in \mathbf{H}\cap(c,b)^{\bullet}} f(x)\right] =$$

$$= \int_a^c f(x)\,dx + \int_c^b f(x)\,dx.$$

(iv)

$$\int_a^b f(x)\,dx = tr\left[\varepsilon_0 \sum_{z\in \mathbf{H}\cap(a,b)^{\bullet}} f(x)\right] \le$$

$$\le tr\left[\varepsilon_0 \sum_{z\in \mathbf{H}\cap(a,b)^{\bullet}} g(x)\right] = \int_a^b g(x)\,dx.$$

(v) We have that

$$-|f(x)| \le f(x) \le |f(x)|;$$

thus, from (iv) it follows that

$$-\int_a^b |f(x)|\,dx \le \int_a^b f(x)\,dx \le \int_a^b |f(x)|\,dx$$

(vi) Since
$$\forall x \in [a, b], \ m \leq f(x) \leq M$$

by (iv), we have that

$$\int_a^b m \, dx \leq \int_a^b f(x) \, dx \leq \int_a^b M \, dx;$$

and the conclusion follows straightforward (vii) Set

$$m = \min_{x \in [a,b]} f(x) \ ed \ M = \max_{x \in [a,b]} f(x).$$

m and M are well defined by Weierstrass theorem. By (v) it follows that

$$\frac{1}{b-a} \int_a^b f(x) dx \in [m, M];$$

then, by the Intermediate Values Theorem, there exists $c \in [a, b]$ such that f gets the value

$$\frac{1}{b-a} \int_a^b f(x) dx$$

$\square$

By the definitions 19 and 20, the following theorem follows straightforward:

Corollary 5. *Let f be a function defined in an interval I and let $a, b \in I^*$ be two hyperreal numbers. Then, all the properties of Th. 31 hold except that the (iv) and (vi) which hold only if $b \geq a$.*

10.4   The theorem of the primitive

We recall the definition of primitive:

Definition 21. *A function $F : (a, b) \to \mathbf{R}$ is called primitive of $F : (a, b) \to \mathbf{R}$ if*
$$\forall x \in (a, b), \ F'(x) = f(x).$$

Now we can for mulate the Theorem of the Primitive

Theorem 32. *If $f : (a, b) \to \mathbf{R}$ is a continuous function, then, $\forall c \in \mathbf{R}, \forall x_0 \in (a, b)$, the function*

$$F(x) := c + \int_{x_0}^x f(t) \, dt \tag{32}$$

is a primitive of f.

Proof. Use the definitions 19, 20 and the corollary 5; then, for any $\varepsilon \in mon(0) - \{0\}$, we have

$$F'(x) = st \left[\frac{c + \int_{x_0}^{x+\varepsilon} f(t)\, dt - \left(c + \int_{x_0}^{x} f(t)\, dt \right)}{\varepsilon} \right]$$

$$= st \left[\frac{\int_{x_0}^{x+\varepsilon} f(t)\, dt + \int_{x_0}^{x} f(t)\, dt}{\varepsilon} \right]$$

$$= st \left[\frac{\int_{x}^{x+\varepsilon} f(t)\, dt}{\varepsilon} \right] = st \left[\frac{\varepsilon f(\xi)}{\varepsilon} \right] = st \left[f(\xi) \right]$$

where $\xi \sim x$; using the continuity of f we get the conclusion:

$$F'(x) = st \left[f(\xi) \right] = f(x).$$

$\square$

As it is well known, the theorem of the Primitive allows a very simple proof of the Fundamental Theorem of Calculus.

Proof. **Second proof of Th. 30.** Let F be the primitive of f defined by the (32); then

$$\int_a^b f(t)\, dt = \int_a^{x_0} f(t)\, dt + \int_{x_0}^b f(t)\, dt = c + \int_{x_0}^b f(t)\, dt - \left(c + \int_{x_0}^a f(t)\, dt \right)$$

$$= c + \int_{x_0}^b f(t)\, dt - \left(c + \int_{x_0}^a f(t)\, dt \right) = F(b) - F(a).$$

Since

$$\int_a^b f(t)\, dt = \int_a^b F'(t)\, dt$$

we get the (30).

$\square$

11   Series and limits

In this section we will just give the nonstandard definitions of series and limits

11.1   Definition of series

A series is an expression of the type

$$\sum_{k=0}^{\infty} a_k \tag{33}$$

For every $\omega \in \Omega$, we consider the hyperfinite sum

$$\sum_{k=0}^{\omega} a_k \tag{34}$$

where

$$\Omega = \mathbf{N}^* - \mathbf{N} = \{n \in \mathbf{N}^* \mid n \text{ is unbounded}\}$$

(cf. Th. 9). We recall that the value of the hyperfinite sum (34) is given by $S(\omega) = S(\varphi(\alpha_0))$ where

$$\omega = \varphi(\alpha_0) \quad \text{and} \quad S(m) := \sum_{k=0}^{m} a_k$$

The (34) is a particular type of hyperfinite sum. Intuitively it represent the sum of an infinite (actually ω) number of terms. The sum (34) allows to give a meaning to the sum (33) according to the following definition:

Definition 22. *We say that the series (33) converges to $s \in \mathbf{R}$, if*

$$\forall \omega \in \Omega, \ tr\left(\sum_{k=0}^{\omega} a_k\right) = s$$

We say that the series (33) diverges if the above formula holds with $s = \pm\infty$; we say that the series (33) is undefined if $tr(S(\omega))$ depends on the particular choice of $\omega \in \Omega(\mathbf{N})$.

Notice the difference between the sum (34) and (33): the (34) is defined for every value of ω, it depends on ω and it is an hyperreal number; the (33) might not be defined, but if it is defined it is real number.

Examples Consider the series

$$\sum_{k=0}^{\infty} \frac{1}{2^k}$$

Since

$$S(n) = \sum_{k=0}^{n} \frac{1}{2^k} = 2 - \left(\frac{1}{2}\right)^n$$

we have that

$$\forall \omega \in \Omega, \ tr\left(\sum_{k=0}^{\omega} \frac{1}{2^k}\right) = tr\left(2 - \left(\frac{1}{2}\right)^{\omega}\right) = 2$$

and so

$$\sum_{k=0}^{\infty} \frac{1}{2^k} = 2.$$

The series

$$\sum_{k=0}^{\infty} k^2$$

diverges since

$$\forall \omega \in \Omega, \ tr\left(\sum_{k=0}^{\omega} k^2\right) = tr\left(\frac{1}{3}\omega^3 + \frac{1}{2}\omega^2 + \frac{1}{6}\omega\right) = +\infty$$

The series

$$\sum_{k=0}^{\infty} (-1)^k$$

is undefined diverges since

$$\omega \text{ even} \Rightarrow tr\left(\sum_{k=0}^{\omega} (-1)^k\right) = 1; \quad \omega \text{ odd} \Rightarrow tr\left(\sum_{k=0}^{\omega} (-1)^k\right) = 0$$

11.2   Limits

In the usual expositions of calculus, the notion of limit is the tool which allows the definition of continuity, derivative and integral In nonstandard analysis the infinitesimals and the *tr* operator can replace the notion of limit quite well. Thus this notion is not necessary to Nonstandard Analysis. Nevertheless, its relation with Nonstandard analysis is very interesting: we can say that the "limit" is the points on which the Standard Analysis and the Nonstandard Analysis touch each other. Probably the "Standard Mathematicians" consider the hyperreal numbers as sequences of real numbers, the infinitesimals as "infinitesimal sequences" and the operator *tr* as a strange way to take the limit of a subsequence: so they might conclude that Nonstandard Analysis is only a different way of defining limits. On the other hand, the "Nonstandard Mathematicians"-and particularly the Platonic ones- think that infinitesimal numbers do in fact exist and that the notion of limit is just an awkward way to talk about infinitesimals without mentioning them. It is like wanting to say a curse word without pronouncing it; you have to use a lengthy description thus employing ten words instead of one. Anyhow, the Nonstandard definition of limit is quite simple:

Definition 23. *Given a function* $f : A \to \mathbf{R}$ *we say that the limit of* f *for* $x \to x_0 \in \hat{\mathbf{R}}$ *is* $l \in \hat{\mathbf{R}}$ *if*

$$\forall \xi \in \mathbf{R}^*, \ tr(\xi) = x_0 \Rightarrow tr\left[f(\xi)\right] = l$$

and we write

$$\lim_{x \to x_0} f(x) = l$$

Notice that the above definition makes sense not only if $l \in \mathbf{R}$ but also if $tr\,[f(x)] = +\infty$ or $tr\,[f(x)] = -\infty$. Also, it makes sense even if $x_0 = +\infty$ or $-\infty$. We conclude this paper giving the definition of maximum and minimum limit which has a very simple nonstandard definition.

$$\text{max-}\lim_{x \to x_0} f(x) = \sup\,\{tr\,[f(\xi)] \mid tr(\xi) = x_0\}$$

$$\text{min-}\lim_{x \to x_0} f(x) = \inf\,\{tr\,[f(\xi)] \mid tr(\xi) = x_0\}\,.$$

References

Part I

[AA99]   G. Alberti and L. Ambrosio. A geometrical approach to monotone functions in $\mathbf{R}^n$. *Mathematische Zeitschrift*, 230:259–316, 1999.

[AAC92]   G. Alberti, L. Ambrosio, and P. Cannarsa. On singularities of convex functions. *Manuscripta Math.*, 76:421–436, 1992.

[AAG95]   S. Altschuler, S. Angenent, and Y. Giga. Mean curvature flow through singularities for surfaces of rotation. *J. Geom. Anal.*, 5:293–358, 1995.

[AB90]   L. Ambrosio and A. Braides. Functionals defined on partitions of sets of finite perimeter, I: integral representation and Γ-convergence. *J. Math. Pures. Appl.*, 69:285–305, 1990.

[AB97]   L. Ambrosio and A. Braides. Energies in SBV and variational models in fracture mechanics. *in Homogenization and Applications to Material Sciences, (D. Cioranescu, A. Damlamian, P. Donato eds), GAKUTO, Gakkotosho, Tokyo*, pages 1–12, 1997.

[AB98a]   G. Alberti and G. Bellettini. A non-local anisotropic model for phase transitions: asymptotic behaviour of rescaled energies. *Eur. J. Appl. Math.*, 9:261–284, 1998.

[AB98b]   G. Alberti and G. Bellettini. A non-local anisotropic model for phase transitions, I: the optimal profile problem. *Math. Ann.*, 310:527–560, 1998.

[ABCP96]   G. Alberti, G. Bellettini, M. Cassandro, and E. Presutti. Surface tension in Ising systems with Kac potential. *J. Stat. Phys.*, 82:743–796, 1996.

[ABG98]   R. Alicandro, A. Braides, and M.S. Gelli. Free discontinuity problems generated by singular perturbation. *Proc. Roy. Soc. Edinburgh*, 128-A:1115–1129, 1998.

[ABS94]   G. Alberti, G. Bouchitté, and P. Seppecher. Un résultat de perturbations singulières avec la norm $H^{1/2}$. *C. R. Acad. Sci. Paris*, 319-I:333–338, 1994.

[ABS98]   G. Alberti, G. Bouchitté, and P. Seppecher. Phase transitions with line-tension effect. *Arch. Ration. Mech. Anal.*, 144:1–49, 1998.

[AC79]   S.M. Allen and J.W. Cahn. A macroscopic theory for antiphase boundary motion and its application to antiphase domain coarsening. *Acta Metal.*, 27:1085–1095, 1979.

[ACI]   S. Angenent, D.L. Chopp, and T. Ilmanen. A computed example of nonuniqueness of mean curvature flow in $\mathbf{R}^n$. *Comm. PDE*, to appear.

[ACI95]   S. Angenent, D.L. Chopp, and T. Ilmanen. Nonuniqueness of mean curvature flow in $\mathbf{R}^3$. *Comm. Partial Differ. Equations*, 20:1937–1958, 1995.

[ACM97]   L. Ambrosio, A. Coscia, and G. Dal Maso. Fine properties of functions with bounded deformation. *Arch. Ration. Mech. Anal.*, 139:201–238, 1997.

[ACS93]   L. Ambrosio, P. Cannarsa, and H.M. Soner. On the propagation of singularities of semi-convex functions. *Ann. Sc. Norm. Sup.*, 20:597–616, 1993.

[AFF93]   E. Acerbi, I. Fonseca, and N. Fusco. Regularity results for equilibria in a variational model for fracture. *Proc. Roy. Soc. Edinburgh*, A,127:889–902, 1993.

[AFP97]   L. Ambrosio, N. Fusco, and D. Pallara. Partial regularity of free discontinuity sets II. *Ann. Sc. Norm. Sup. di Pisa, ser. IV*, 24:39–62, 1997.

[AG92]   S. Altschuler and M.A. Grayson. Shortening space curves and flow through singularities. *J. Differential Geom.*, 35:283–398, 1992.

[AIVa]   S. Angenent, T. Ilmanen, and J. Velazquez. Nonuniqueness in geometric heat flows. To appear.

[AIVb]   S. Angenent, T. Ilmanen, and J. Velazquez. Nonuniqueness of motion by mean curvature in dimensions four through seven. In preparation.

[AIVc]   S. Angenent, T. Ilmanen, and J. Velazquez. Slow blow-up and nonuniqueness of motion by mean curvature. In preparation.

[Alb93]   G. Alberti. Rank-one property for derivatives of functions with bounded variation. *Proc. Roy. Soc. Edinburgh*, 123-A:239–274, 1993.

[ALCM93]   L. Alvarez, P.L. Lions, M.G. Crandall, and J.M. Morel. Axioms and fundamental equations of image processing. *Arch. Ration. Mech. Anal.*, 123:199–257, 1993.

[Ale93]   A.D. Alexandroff. Almost everywhere existence of the second differential of a convex function and some properties of convex surfaces connected with it. *Len. Gos. Univ. Uchen. Zap. Ser. Mat. Nauk.*, 6:3–35, 1993. (in russian).

[ALM92]   L. Alvarez, P.L. Lions, and J.M. Morel. Imagine selective smoothing and edge detection by nonlinear diffusion I, II. *SIAM J. Num. Anal.*, 29:182–193, 845–866, 1992.

[Alt91]   S. Altschuler. Singularities of the curve shortening flow for space curves. *J. Differential Geom.*, 34:499–514, 1991.

[AM96]   L. Ambrosio and C. Mantegazza. Curvature and distance function from a manifold. Accepted for publication on J. Geometric Analysis, (available on the preprint server of Scuola Normale, http://www.sns.it), 1996.

[AM97]   G. Alberti and C. Mantegazza. A note on the theory of SBV functions. *Boll. Unione Mat. Ital., VII. Ser., B 11*, 2:375–382, 1997.

[Amb89]   L. Ambrosio. A compactness theorem for a new class of functions of bounded variation. *Boll. Unione Mat. Ital., VII. Ser., B 3*, 4:857–881, 1989.

[Amb90]   L. Ambrosio. Existence theory for a new class of variational problems. *Arch. Ration. Mech. Anal.*, 111:291–322, 1990.

[Amb94]   L. Ambrosio. On the lower semicontinuity of quasi-convex integrals defined in $SBV(\Omega; \mathbf{R}^k)$. *Nonlinear Anal.*, 23:405–425, 1994.

[Amb95a]   L. Ambrosio. Movimenti minimizzanti. *Rend. Accad. Naz. Sci. XL Mem. Mat.*, 113:191–246, 1995.

[Amb95b]   L. Ambrosio. A new proof of the SBV compactness theorem. *Calc. Var. Partial Differ. Equ.*, 3:127–137, 1995.

[Amb97]   L. Ambrosio. *Corso introduttivo alla Teoria Geometrica della Misura ed alle Superfici Minime.* Scuola Normale Superiore, Pisa, 1997.

[Amb99]   L. Ambrosio. *Lecture notes on geometric evolution problems, distance function and viscosity solutions.* This volume, 1999.

[And94a]   B. Andrews. Contraction of convex hypersurfaces in euclidean space. *Calc. Var. Partial Differ. Equ.*, 2:151–171, 1994.

[And94b]   B. Andrews. Contraction of convex hypersurfaces in riemannian spaces. *J. Differential Geom.*, 34:407–431, 1994.

[Ang90]   S. Angenent. Parabolic equations for curves and surfaces Part I. Curves with p-integrable curvature. *Ann. of Math.*, 132:451–483, 1990.

[Ang91]   S. Angenent. Parabolic equations for curves and surfaces. Part II. Inter-
           sections, blow-up and generalized solutions. *Ann. of Math.*, 133:171–215,
           1991.
[Ang92]   S. Angenent. Shrinking doughnuts. In *Nonlinear Diffusion Equations and
           Their Equilibrium States*, volume 3, pages 171–215. Birkhäuser, 1992.
[AP97]   L. Ambrosio and D. Pallara. Partial regularity of free discontinuity sets
           I. *Ann. Sc. Norm.*, 24:1–38, 1997.
[AS96]   L. Ambrosio and H.-M. Soner. A level set approach to the evolution of
           surfaces of any codimension. *J. Differential Geom.*, 43:693–737, 1996.
[AT90]   L. Ambrosio and V.M. Tortorelli. Approximation of functionals depend-
           ing on jumps by elliptic functionals via Γ-convergence. *Comm. Pure
           Appl. Math.*, 43:999–1036, 1990.
[AT92]   L. Ambrosio and V.M. Tortorelli. On the approximation of free discon-
           tinuity problems. *Boll. Unione Mat. Ital.*, VII. Ser., B 6, 1:105–123,
           1992.
[ATW93]   F. Almgren, J.E. Taylor, and L. Wang. Curvature-driven flows: a varia-
           tional approach. *SIAM J. Control Optim.*, 31:387–437, 1993.
[AW75]   D.G. Aronson and H.F. Weinberger. Nonlinear diffusion in population
           genetics, combustion and nerve pulse propagation. In *PDE and related
           topics*, volume 446 of *Lect. Notes Math.* Springer-Verlag, Berlin, 1975.
[Bal89]   J.M. Ball. A version of the fundamental theorem for Young measures. In
           M. Rascle et al., editor, *PDE's and Continuum Models of Phase Tran-
           sitions*, volume 344 of *Lect. Notes in Physics*, pages 207–215. Springer-
           Verlag, Berlin, 1989.
[Bal90]   S. Baldo. Minimal interface criterion for phase transitions in mixtures
           of Cahn-Hilliard fluids. *Ann. Inst. H. Poincaré Anal. Non Linéaire*,
           7:289–314, 1990.
[BBH94]   F. Bethuel, H. Brezis, and F. Hélein. *Ginzburg-Landau vortices*, vol-
           ume 13 of *Progress in Nonlinear Diff. Eq. and Appl.* Birkhäuser, Boston,
           1994.
[BC93]   A. Braides and A. Coscia. A singular perturbation approach to problems
           in fracture mechanics. *Math. Mod. Meth. Appl. Sci.*, 3:302–340, 1993.
[BD98a]   M. Bardi and I. Capuzzo Dolcetta. *Optimal control and viscosity solu-
           tions of Hamilton-Jacobi-Bellmann equations.* Birkhäuser, Boston, 1998.
[BD98b]   A. Braides and A. Defranceschi. *Homogenization of Multiple Integrals*,
           volume 12 of *Oxford Lecture Series in Math. and its Appl.* Oxford Uni-
           versity Press, 1998.
[BDV96]   A. Braides, A. Defranceschi, and E. Vitali. Homogenization of free dis-
           continuity problems. *Arch. Ration. Mech. Anal.*, 135:297–356, 1996.
[Bel97]   G. Bellettini. Alcuni risultati sulle minime barriere per movimenti geo-
           metrici di insiemi. *Boll. Unione Mat. Ital.*, VII. Ser., A 11, 2:485–512,
           1997.
[BF94]   A.C. Barroso and I. Fonseca. Anisotropic singular perturbations - The
           vectorial case. *Proc. Roy. Soc. Edinburgh*, 124-A:527–571, 1994.
[BFP97]   G. Bellettini, P. Colli Franzone, and M. Paolini. Convergence of front
           propagation for anisotropic bistable reaction-diffusion equations. *Asymp-
           totic Anal.*, 15:325–358, 1997.
[BG98]   A. Braides and A. Garroni. On the nonlocal approximation of free dis-
           continuity problems. *Comm. Partial Differ. Equations*, 23:817–829, 1998.
[BK91]   L. Bronsard and R. Kohn. Motion by mean curvature as the singular
           limit of Ginzburg-Landau models. *J. Diff. Equations*, 90:211–237, 1991.

[BM97]   A. Braides and G. Dal Maso. Nonlocal approximation of the Mumford-Shah functional. *Calc. Var. Partial Differ. Equ.*, 5:293–322, 1997.

[BMO92]   J. Bence, B. Merriman, and S. Osher. Diffusion generated motion by mean curvature. *CAM Report, Dept. of Math., Univ. of California*, 92-18, 1992.

[BN]   G. Bellettini and M. Novaga. A result on motion by mean curvature in arbitrary codimension. *Diff. Geom. and its Appl.*, to appear.

[BN97a]   G. Bellettini and M. Novaga. Barriers for a class of geometric evolution problems. *Atti Accad. Naz. Lincei Cl. Sci. Fis. Mat. Natur. Rend (9), Mat. Appl.*, 8:119–128, 1997.

[BN97b]   G. Bellettini and M. Novaga. Minimal barriers for geometric evolutions. *J. Differential Equations*, 139:76–103, 1997.

[BN98]   G. Bellettini and M. Novaga. Comparison results between minimal barriers and viscosity solutions for geometric evolutions. *Ann. Sc. Norm. Sup., PISA, Cl. Sci.*, XXVI:97–131, 1998.

[BNP98]   G. Bellettini, M. Novaga, and M. Paolini. An example of three dimensional fattening for linked space curves evolving by curvature. *Comm. Partial Differ. Equations*, 23:1475–1492, 1998.

[Bon96]   A. Bonnet. On the regularity of edges in image segmentation. *Ann. Inst. H. Poincaré Anal. Non Linéaire*, 13:485–528, 1996.

[Bou90]   G. Bouchitté. Singular perturbations of variational problems arising from a two-phase transition model. *Appl. Math. Optimization*, 21:289–315, 1990.

[BP87]   G. Barles and B. Perthame. Discontinuous solutions of deterministic optimal stopping problems. *Math. Modelling Numerical Analysis*, 21:557–579, 1987.

[BP94]   G. Bellettini and M. Paolini. Two examples of fattening for the mean curvature flow with a driving force. *Atti Accad. Naz. Lincei Cl. Sci. Fis. Mat. Natur. Mat. App.*, 5:229–236, 1994.

[BP95a]   G. Bellettini and M. Paolini. Quasi-optimal error estimates for the mean curvature flow with a forcing term. *Differ. Integral Equ.*, 8:735–752, 1995.

[BP95b]   G. Bellettini and M. Paolini. Some results on minimal barriers in the sense of De Giorgi applied to motion by mean curvature. *Rend. Atti Accad. Naz. XL, XIX*, pages 43–67, 1995.

[BP95c]   G. Bellettini and M. Paolini. Teoremi di confronto tra diverse nozioni di movimento secondo la curvatura media. *Atti Accad. Naz. Lincei Cl. Sci. Fis. Mat. Natur. Mat. App.*, 6:45–54, 1995.

[BP96a]   G. Bellettini and M. Paolini. Anisotropic motion by mean curvature in the context of Finsler geometry. *Hokkaido Math. J.*, 25:537–566, 1996.

[BP96b]   A. Braides and V. Chiadò Piat. Integral representation results for functionals defined on $SBV(\Omega; \mathbf{R}^m)$. *J. Math. Pures Appl.*, 75:595–626, 1996.

[Bra78]   K.A. Brakke. *The motion of a surface by its mean curvature*. Princeton University Press, Princeton N.J., 1978.

[Bra98]   A. Braides. *Approximation of free-Discontinuity Problems*, volume 1694 of *Lecture Notes in Math.* Springer-Verlag, Berlin, 1998.

[Bre73]   H. Brezis. *Opérateurs maximaux monotones et semigroups de contraction dand les espaces de Hilbert*. North Holland Mathematics Studies, 5. North Holland, Amsterdam, 1973.

[Bre92]   H. Brezis. *Analisi Funzionale, Teoria e Applicazioni*. Liguori Napoli, 1992.

[BS98]   G. Barles and P.E. Souganidis. A new approach to front propagation: theory and applications. *Arch. Ration. Mech. Anal.*, 141:237–296, 1998.

[BSS93]   G. Barles, H.M. Soner, and P.E. Souganidis. Front propagation and phase-field theory. *SIAM. J. Cont. Opt.*, 31:439–469, 1993.

[BZ87]   A. Blake and A. Zisserman. *Visual Reconstruction.* MIT Press, Cambridge, 1987.

[Car76]   M.P. Do Carmo. *Differential geometry of curves and surfaces.* Prentice-Hall, New Jersey, 1976.

[Car95]   M. Carriero. Variational models in image segmentation: first and second order energy approaches. *Preprint del Dip. Matem. Univ. Lecce*, 6, 1995.

[CEL84]   M.G. Crandall, L.C. Evans, and P.L. Lions. Some properties of viscosity solutions to Hamilton–Jacobi equations. *Trans. Amer. Math. Soc.*, 282:487–502, 1984.

[CGG91]   Y.-G. Chen, Y. Giga, and S. Goto. Uniqueness and existence of viscosity solutions of generalized mean curvature flow equations. *J. Differential Geom.*, 33:749–786, 1991.

[CH58]   J.W. Cahn and J.E. Hilliard. Free energy of a non-uniform system I: interfacial free energy. *J. Chem. Phys.*, 28:258–267, 1958.

[Che92]   X. Chen. Generation and propagation of interfaces by reaction diffusion equation. *J. Diff. Equations*, 96:116–141, 1992.

[Cho94]   D. Chopp. Numerical computation of self-similar solutions for mean curvature flow. *Exper. Math.*, 3:1–16, 1994.

[CI90]   M.G. Crandall and H. Ishii. The maximum principle for semicontinuous functions. *Differ. Integral Equ.*, 3:1001–1014, 1990.

[CIL92]   M.G. Crandall, H. Ishii, and P.L. Lions. User's guide to viscosity solutions of second order partial differential equations. *Bull. Am. Math. Soc.*, 27:1–67, 1992.

[CL83]   M.G. Crandall and P.L. Lions. Viscosity solutions of Hamilton–Jacobi equations. *Trans. AMS*, 277:1–43, 1983.

[CL91]   M. Carriero and A. Leaci. S^k-valued maps minimizing the L^p-norm of the gradient with free discontinuities. *Ann. Sc. Norm. Super. Pisa, Cl. Sci. Ser. IV, 18*, 3:321–352, 1991.

[CLT92]   M. Carriero, A. Leaci, and F. Tomarelli. Special bounded Hessian and elastic-plastic plate. *Rend. Accad. Naz. delle Scienze (dei XL)*, 109:223–258, 1992.

[CLT94]   M. Carriero, A. Leaci, and F. Tomarelli. Strong solution for an elastic plastic plate. *Calc. Var. Partial Differ. Equ.*, 2:219–240, 1994.

[CLT96]   M. Carriero, A. Leaci, and F. Tomarelli. A second order model in image segmentation: Blake & Zisserman functional. In F. Tomarelli R. Serapioni, editor, *Variational Methods for Discontinuous Structures*, pages 57–72. Birkäuser, 1996.

[Cor98]   G. Cortesani. Sequences of non-local functionals which approximate free-discontinuity problems. *Arch. Ration. Mech. Anal.*, 144:357–402, 1998.

[Cra86]   M.G. Crandall. Nonlinear semigroups and evolution governed by accretive operators. *Proc. Symp. Pure Math.*, 45:305–337, 1986.

[CT91]   G. Congedo and I. Tamanini. On the existence of solutions to a problem in multidimensional segmentation. *Ann. Inst. H. Poincaré Anal. Non Linéaire*, 2:175–195, 1991.

[CT94]   J.W. Cahn and J.E. Taylor. Surface motion by surface diffusion. *Acta Metall. Mater.*, 42:1045–1063, 1994.

[Dib94]   F. Dibos. Uniform rectifiability of image segmentations obtained by a variational method. *J. Math. Pures Appl.*, 73:389–412, 1994.

[DS96]   G. David and S. Semmes. On the singular set of minimizers of the Mumford-Shah functional. *J. Math. Pures Appl.*, 75:299–342, 1996.

[DZ94]   M. Delfour and J.P. Zolésio. Shape analysis via oriented distance function. *J. Functional Analysis*, 123:129–201, 1994.

[EG92]   L.C. Evans and R.F. Gariepy. *Measure theory and fine properties of functions.* CRC Press, Boca Raton, 1992.

[EH89]   K. Ecker and G. Huisken. Mean curvature evolution of entire graphs. *Ann. Math.*, 130:453–471, 1989.

[EH91]   K. Ecker and G. Huisken. Interior estimates for hypersurfaces moving by mean curvature. *Invent. Math.*, 105:547–569, 1991.

[ES91]   L.C. Evans and J. Spruck. Motion of level sets by mean curvature I. *J. Differential Geom.*, 33:635–681, 1991.

[ES92a]   L.C. Evans and J. Spruck. Motion of level sets by mean curvature II. *Trans. AMS*, 330:635–681, 1992.

[ES92b]   L.C. Evans and J. Spruck. Motion of level sets by mean curvature III. *J. Geom. Analysis*, 2:121–150, 1992.

[ES95]   L.C. Evans and J. Spruck. Motion of level sets by mean curvature IV. *J. Geom. Analysis*, 5:77–114, 1995.

[ESS92]   L.C. Evans, H.M. Soner, and P.E. Souganidis. Phase transitions and generalized motion by mean curvature. *Comm. Pure Appl. Math.*, 45:1097–1123, 1992.

[Eva93]   L.C. Evans. Convergence of an algorithm for mean curvature motion. *Indiana Univ. Math. J*, 42:533–557, 1993.

[Fed59]   H. Federer. Curvature measures. *Trans. Am. Math. Soc.*, 93:418–491, 1959.

[Fed69]   H. Federer. *Geometric Measure Theory.* Springer-Verlag, Berlin, 1969.

[Fed70]   H. Federer. The singular sets of area minimizing rectifiable currents with codimension one and of area minimizing flat chains modulo two with arbitrary codimension. *Bull. Am. Math. Soc.*, 76:767–771, 1970.

[FF95]   I. Fonseca and G. Francfort. A model for the interaction between fracture and damage. *Calc. Var. Partial Differ. Equ.*, 3:407–446, 1995.

[FH88]   P.C. Fife and L. Hsiao. The generation and propagation of internal layers. *Nonlinear Anal.*, 12:19–41, 1988.

[FK95]   I. Fonseca and M.A. Katsoulakis. Γ-convergence, minimizing movements and generalized motion by mean curvature. *Differ. Integral Equ.*, 8:1619–1656, 1995.

[FM92]   I. Fonseca and S. Müller. Quasiconvex integrands and lower semicontinuity in L^1. *SIAM J. Math. Anal.*, 23:1081–1098, 1992.

[Foo84]   R.L. Foote. Regularity of the distance function. *Proc. Am. Math. Soc.*, 92:153–155, 1984.

[FP96]   F. Fierro and M. Paolini. Numerical evidence of fattening for the mean curvature flow. *Math. Models Methods Appl. Sci.*, 6:1103–1118, 1996.

[Fri64]   A. Friedman. *Partial differential equations of parabolic type.* Prentice-Hall, Englewood Cliffs N.J., 1964.

[FS93]   W.H. Fleming and H.M. Soner. *Controlled Markov Processes and Viscosity Solutions.* Springer-Verlag, New York, 1993.

[Fu85]   J.H.G. Fu. Tubular neighbourhoods in euclidean spaces. *Duke Math. J.*, 52:1025–1046, 1985.

[GA88]   E. De Giorgi and L. Ambrosio. Un nuovo funzionale del calcolo delle variazioni. *Atti Accad. Naz. Lincei Rend. Cl. Sci. Fis. Mat. Natur.*, 82:199–210, 1988.

[Gag83]   M.E. Gage. An isoperimetric inequality with applications to curve shortening. *Duke Math. J.*, 50:1225–1229, 1983.

[Gag84]   M.E. Gage. Curves shortening makes convex curves circular. *Invent. Math.*, 76:357–364, 1984.

[GCL89]   E. De Giorgi, M. Carriero, and A. Leaci. Existence theorem for a minimum problem with free discontinuity set. *Arch. Ration. Mech. Anal.*, 108:195–218, 1989.

[GF75]   E. De Giorgi and T. Franzoni. Su un tipo di convergenza variazionale. *Atti Accad. Naz. Lincei Rend. Cl. Sci. Fis. Mat. Natur.*, 58:842–850, 1975.

[GG92a]   Y. Giga and S. Goto. *Geometric evolution of phase-boundaries, IMA VMA 43, M.E. Gurtin and G.B. MacFadden, editors.* Springer-Verlag, Berlin, 1992.

[GG92b]   Y. Giga and S. Goto. Motion of hypersurfaces and geometric equations. *J. Math. Soc. Japan*, 44:99–111, 1992.

[GGIS91]   Y. Giga, S. Goto, H. Ishii, and M.H. Sato. Comparison principle and convexity preserving properties for singular degenerate parabolic equations on unbounded domains. *Indiana Math. J.*, 40:443–470, 1991.

[GH86]   M. Gage and R.S. Hamilton. The heat equation shrinking convex plane curves. *J. Differential Geom.*, 23:69–95, 1986.

[Gio90a]   E. De Giorgi. New conjectures on flow by mean curvature. In M.L. Benevento, T. Bruno, and C. Sbordone, editors, *Methods of Real Analysis and Partial Differential Equations*, pages 120–127. Liguori, Napoli, 1990.

[Gio90b]   E. De Giorgi. Conjectures on limits of some quasilinear parabolic equations and flow by mean curvature, September 3-7, 1990. Lecture delivered at the meeting on "Partial Differential Equations and Related Topics" in honour of L. Nirenberg, Trento.

[Gio91a]   E. De Giorgi. Congetture sui limiti delle soluzioni di alcune equazioni paraboliche quasi lineari. In *Nonlinear Analysis. A Tribute in Honour of G. Prodi*, pages 173–187. Scuola Normale Superiore, Pisa, 1991.

[Gio91b]   E. De Giorgi. Free discontinuity problems in calculus of variations. In R. Dautray, editor, *Frontiers in Pure and Applied Mathematics*, pages 55–62. North-Holland, Amsterdam, 1991.

[Gio92]   E. De Giorgi. New ideas in calculus of variations and geometric measure theory. In *Motion by Mean Curvature and Related Topics*, pages 63–69. Walter de Gruyter, Berlin, 1992.

[Gio93a]   E. De Giorgi. New problems on minimizing movements. In *Boundary value problems for PDE and applications, Vol. 29*, pages 81–98. Masson, 1993.

[Gio93b]   E. De Giorgi. Congetture riguardanti barriere, superfici minime, movimenti secondo la curvatura media. *Manuscript*, Lecce, November 4, 1993.

[Gio94a]   E. De Giorgi. Barriere, frontiere, e movimenti di varietà, 1994. Manuscript, proseguimento della prima bozza di Pavia, Pisa.

[Gio94b]   E. De Giorgi. Barriers, boundaries, motion of manifolds. *Lectures held in Pavia*, March 18, 1994.

[Gio96]   E. De Giorgi. Congetture riguardanti alcuni problemi di evoluzione. A paper in honour of J. Nash. *Duke Math. J.*, 81:255–26, 1996.

[Giu84]   E. Giusti. *Minimal Surfaces and Functions of Bounded Variation*, volume 80 of Monographs in Math. Birkhäuser, Boston, 1984.

[GMT80]   E. De Giorgi, A. Marino, and M. Tosques. Problemi di evoluzione in spazi metrici e curve di massima pendenza. *Atti Accad. Naz. Lincei, Rend. Cl. Sci. Fis. Mat. Natur.*, 68:180–187, 1980.

[GMT82]   E. De Giorgi, A. Marino, and M. Tosques. Funzioni (p,q)–convesse. *Atti Accad. Naz. Lincei, Rend. Cl. Sci. Fis. Mat. Natur.*, 73:6–14, 1982.

[GMT85]   M. De Giovanni, A. Marino, and M. Tosques. Evolution equations with lack of convexity. *Nonlinear Analysis TMA*, 12:1401–1443, 1985.

[Gob]   M. Gobbino. Minimizing movements and evolution problems in euclidean spaces. *Ann. Mat. Pura Appl.*, to appear.

[Gob98]   M. Gobbino. Finite difference approximation of Mumford-Shah functional. *Comm. Pure Appl. Math.*, 51:197–228, 1998.

[Gra87]   M.A. Grayson. The heat equation shrinks embedded plane curves to round points. *J. Differential Geom.*, 26:285–314, 1987.

[Gra89a]   M.A. Grayson. A short note on the evolution of surfaces via mean curvature. *Duke Math. J.*, 58:555–558, 1989.

[Gra89b]   M.A. Grayson. Shortening embedded curves. *Ann. Math.*, 129:71–111, 1989.

[Ham75]   R.S. Hamilton. *Harmonic Maps of Manifolds with Boundary*. Lect. Notes in Math., Springer-Verlag, Berlin, Vol. 471, 1975.

[Ham89]   R. Hamilton. CBMS Conference Notes, Hawaii, 1989.

[HI]   G. Huisken and T. Ilmanen. The inverse mean curvature flow and the Penrose inequality. In preparation.

[Hui84]   G. Huisken. Flow by mean curvature of convex surfaces into spheres. *J. Differential Geom.*, 20:237–266, 1984.

[Hui90]   G. Huisken. Asymptotic behavior for singularities of the mean curvature flow. *J. Differential Geom.*, 31:285–299, 1990.

[Hui98]   G. Huisken. A distance comparison principle for evolving curves. *Asian J. Math.*, 2:127–133, 1998.

[Ilm92]   T. Ilmanen. Generalized flow of sets by mean curvature on a manifold. *Indiana Univ. Math. J.*, 41:671–705, 1992.

[Ilm93a]   T. Ilmanen. Convergence of the Allen–Cahn equation to Brakke's motion by mean curvature. *J. Differential Geom.*, 38:417–461, 1993.

[Ilm93b]   T. Ilmanen. The level set flow on a manifold. *Proc. Symp. Pure Math.*, 54:193–204, 1993.

[Ilm94]   T. Ilmanen. Elliptic regularization and partial regularity for motion by mean curvature. *Mem. Am. Math. Soc.*, 108, No. 520:1–90, 1994.

[Ilm95]   T. Ilmanen. Some techniques in geometric heat flows. In *Lectures delivered at the Conference on Partial Differential Equations and Applications to Geometry*. International Centre for Theoretical Physics, Trieste, 1995.

[IS95]   H. Ishii and P.E. Souganidis. Generalized motion of noncompact hypersurfaces with velocity having arbitrary growth on the curvature tensor. *Tohoku Math. J.*, 47:227–250, 1995.

[Ish89]   H. Ishii. On uniqueness and existence of viscosity solutions of fully nonlinear second order elliptic PDE's. *Comm. Pure Appl. Math.*, 42:15–45, 1989.

[Jen88]   R. Jensen. The maximum principle for viscosity solutions of fully nonlinear second order partial differential equations. *Arch. Ration. Mech. Anal.*, 101:1–27, 1988.

[JKO82]   D. Jasnow, K. Kawasaki, and T. Ohta. Universal scaling in the motion of random interfaces. *Phys. Rev. Lett.*, 49:1223–1226, 1982.

[JLS88]   R. Jensen, P.L. Lions, and P.E. Souganidis. A uniqueness result for viscosity solutions of second order fully nonlinear PDE's. *Proc. AMS*, 102, 1988.

[JSa]   R. Jerrard and H.-M. Soner. Dynamics of vortices. *Arch. Ration. Mech. Anal.*, to appear.

[JSb]   R. Jerrard and H.-M. Soner. Scaling limits and regularity for a class of Ginzburg-Landau systems. *Ann. Inst. H. Poincaré Anal. Non Linéaire*, to appear.

[Kaw85]   B. Kawohl. *Rearrangement and Convexity of Level Sets in PDE.* Lect. Notes Math. 1150. Springer-Verlag, Berlin, 1985.

[KRS89]   J.B. Keller, J. Rubinstein, and P. Sternberg. Fast reaction, slow diffusion and curve shortening. *SIAM J. Appl. Math.*, 49:116–133, 1989.

[KS94]   M. Katsoulakis and P.E. Souganidis. Interacting particle systems and generalized motion by mean curvature. *Arch. Ration. Mech. Anal.*, 127:133–157, 1994.

[KS95]   M. Katsoulakis and P.E. Souganidis. Generalized motion by mean curvature as a macroscopic limit of stochastic Ising models with long range interactions and Glauber dynamics. *Comm. Math. Phys.*, 169:61–97, 1995.

[KS97]   M.A. Katsoulakis and P.E. Souganidis. Stochastic Ising models and front propagation. *J. Stat. Physics*, 87:63–89, 1997.

[Lea92]   A. Leaci. Free discontinuity problems with unbounded data: the two dimensional case. *Manuscripta Math.*, 75:429–441, 1992.

[Lea94]   A. Leaci. Free discontinuity problems with unbounded data. *Math. Models and Meth. in Appl. Sci.*, 4:843–855, 1994.

[Lew83]   J.L. Lewis. Regularity of the derivatives of solutions to certain degenerate elliptic equations. *Indiana Univ. Math. J.*, 32:849–858, 1983.

[Lio82]   P.L. Lions. *Generalized solutions of Hamilton–Jacobi equations.* Research Notes in Math., **69**, Pitman, Boston, 1982.

[Lio83]   P.-L. Lions. Optimal control of diffusion processes and Hamilton-Jacobi-Bellman equations, I. *Comm. Partial Differ. Equations*, 8:1101–1134, 1983.

[LS95]   S. Luckhaus and T. Sturzenhecker. Implicit time discretization for the mean curvature flow equation. *Calc. Var. Partial Differ. Equ.*, 3:253–271, 1995.

[Lun94]   A. Lunardi. *Analytic Semigroups and Optimal Regularity in Parabolic Problems.* Birkhäuser, Boston, 1994.

[Mas93]   G. Dal Maso. *An Introduction to Γ-convergence.* Birkhäuser, Boston, 1993.

[MM77]   L. Modica and S. Mortola. Un esempio di Γ–convergenza. *Boll. Unione Mat. Ital.*, V. Ser., B *4*, 14:285–299, 1977.

[MMS92]   G. Dal Maso, J.M. Morel, and S. Solimini. A variational method in image segmentation: existence and approximation results. *Acta Math.*, 168:89–151, 1992.

[Mod87]   L. Modica. The gradient theory of phase transitions and the minimal interface criterion. *Arch. Ration. Mech. Anal.*, 98:123–142, 1987.

[MOPT93]   A. De Masi, E. Orlandi, E. Presutti, and L. Triolo. Motion by curvature by scaling nonlocal evolution equations. *J. Stat. Phys*, 73:543–570, 1993.

[MOPT94a]   A. De Masi, E. Orlandi, E. Presutti, and L. Triolo. Glauber evolution with Kac potentials. 1. Mesoscopic and macroscopic limits, interface dynamics. *Nonlinearity*, 7:663–696, 1994.

[MOPT94b]   A. De Masi, E. Orlandi, E. Presutti, and L. Triolo. Uniqueness of the instanton profile and global stability in non local evolution equations. *Rendiconti di Matematica*, 14:693–723, 1994.

[MOPT96a]   A. De Masi, E. Orlandi, E. Presutti, and L. Triolo. Glauber evolution with Kac potentials. 2. Fluctuations. *Nonlinearity*, 9:27–51, 1996.

[MOPT96b]   A. De Masi, E. Orlandi, E. Presutti, and L. Triolo. Glauber evolution with Kac potentials. 3. Spinodal decomposition. *Nonlinearity*, 9:53–114, 1996.

[MS89a]   P. De Mottoni and M. Schatzman. Évolution géométriques d'interfaces. *C.R. Acad. Sci. Paris Sér. I*, 309:453–458, 1989.

[MS89b]   D. Mumford and J. Shah. Optimal approximation by piecewise smooth functions and associated variational problems. *Comm. Pure Appl.Math.*, 42:577–685, 1989.

[MS90]   P. De Mottoni and M. Schatzman. Developments of interfaces in $\mathbf{R}^n$. *Proc. Roy. Soc. Edinburgh*, 116-A:207–220, 1990.

[MS94]   J.M. Morel and S. Solimini. *Variational Models in Image Segmentation*. Birkhäuser, Basel, 1994.

[MS95]   P. De Mottoni and M. Schatzman. Geometrical evolution of developed interfaces. *Trans. Amer. Math. Soc.*, 347:1533–1589, 1995.

[Mül]   S. Müller. Mathematical models of microstructure and the calculus of variations. In preparation.

[Mul56]   W.W. Mullins. Two-dimensional motion of idealised grain boundaries. *J. Appl. Phys.*, 27:900–904, 1956.

[Nov96]   M. Novaga. *Evoluzioni di insiemi secondo la curvatura*. Master Thesis, *Pisa University*, 1996.

[NPV94]   R.H. Nochetto, M. Paolini, and C. Verdi. Optimal interface error estimates for the mean curvature flow. *Ann. Sc. Norm. Pisa*, 21:193–212, 1994.

[NPV96]   R.H. Nochetto, M. Paolini, and C. Verdi. A dynamic mesh algorithm for curvature dependent evolving interfaces. *J. Comput. Phys.*, 123:296–310, 1996.

[OS88]   S. Osher and J. Sethian. Fronts propagating with curvature dependent speed. *J. Comp. Phys.*, 79:12–49, 1988.

[OS91]   N. Owen and P. Sternberg. Nonconvex variational problems with anisotropic perturbations. *Nonlinear Anal. TMA*, 16:705–719, 1991.

[Pao97]   M. Paolini. Fattening in two dimensions obtained with a nonsymmetric anisotropy: numerical simulations. In *Proc. ALGORITMY '97, Conference on scientific computing (West Tatra Mountains)*, 1997.

[PV92]   M. Paolini and C. Verdi. Asymptotic and numerical analyses of the mean curvature flow with a space-dependent relaxation parameter. *Asymptotic Anal.*, 5:553–574, 1992.

[PW67]   M.H. Protter and H.F. Weinberger. *Maximum Principles in Differential Equations*. Prentice-Hall, Englewood Cliffs N.J., 1967.

[Set85]   J. Sethian. Curvature and evolution of fronts. *Comm. Math. Physics*, 101:487–495, 1985.

[Sim84]   L. Simon. *Lectures on Geometric Measure Theory*. Centre for Mathematical Analysis, Australian National University, Canberra, 1984.

[Son93a]   H.M. Soner. Ginzburg-Landau equation and motion by mean curvature, I: Convergence; II: Development of the initial interface. *J. Geometric Analysis, to appear*, 1993.

[Son93b]   H.M. Soner. Motion of a set by the curvature of its boundary. *J. Diff. Equations*, 101:313–372, 1993.

[Spo93]   H. Spohn. Interface motion in models with stochastic dynamics. *J. Stat. Physics*, 71:1081–113, 1993.

[SS93]   H.M. Soner and P.E. Souganidis. Singularities and uniqueness of cylindrically symmetric surfaces moving by mean curvature. *Commun. Partial Differ. Equations*, 18:859–894, 1993.

[Str96]   M. Struwe. Geometric evolution problems. *IAS/Park City Mathematical Series*, 2:259–339, 1996.

[Tay92]   J.E. Taylor. II-mean curvature and weighted mean curvature. *Acta Metall. Mater.*, 40:1475–1485, 1992.

[TCH93]   J.E. Taylor, J.W. Cahn, and A.C. Handwerker. Geometric models of crystal growth. *Acta Metal.*, 40:1443–1474, 1993.

[Val94]   M. Valadier. A course on Young measures. *Rend. Ist. Mat. Univ. Trieste*, 26:349–394, 1994.

[Vir89]   E.G. Virga. Drops of nematic liquid crystals. *Arch. Ration. Mech. Anal.*, 107:371–390, 1989.

[Whi]   B. White. Stratification of minimal surfaces, mean curvature flows, and harmonic maps. *J. Reine und Angewandte Math, to appear*.

[Whi89]   B. White. A new proof of the compactness theorem for integral currents. *Comm. Math. Helv.*, 64:207–220, 1989.

[Whi92]   B. White. Some questions of De Giorgi about mean curvature flow of triply-periodic surfaces. In *Motion by Mean Curvature and Related Topics*, pages 210–213. Walter de Gruyter, Berlin, 1992.

[Whi94]   B. White. Partial regularity of mean-convex hypersurfaces flowing by mean curvature. *International Mathematics Research Notes*, 4:185–192, 1994.

[Whi95]   B. White. The topology of hypersurfaces moving by mean curvature. *Comm. Anal. Geom.*, 3:317–333, 1995.

[Zie88]   W.P. Ziemer. *Weakly Differentiable Functions*. Springer-Verlag, New York, 1988.

Part II

[AD70]   D. Anderson and G. Derrick. Stability of time dependent particle like solutions in nonlinear field theories. *J. Math. Phys.*, 11:1336–1346, 1970.

[AM91]   Adimurthi and G. Mancini. The Neumann problem for elliptic equations with critical nonlinearity. *Nonlinear Analysis -A tribute in honour of G. Prodi-Quaderno S.N.S.- Pisa*, pages 9–25, 1991.

[AM94]   Adimurthi and G. Mancini. Geometry and topology of the boundary in the critical Neumann problem. *J. Reine Angew. Math.*, 456:1–18, 1994.

[Ama76]   H. Amann. Fixed point equations and nonlinear eigenvalue problems in ordered spaces. *SIAM Review*, 18:620–709, 1976.

[APY93]   Adimurthi, F. Pacella, and S.L. Yadava. Interaction between the geometry of the boundary and positive solutions of a semilinear Neumann problem with critical nonlinearity. *J. Funct. Anal.*, 113:318–350, 1993.

[Aub76]   T. Aubin. Equations différentielles nonlinéaires et problème de Yamabe concernant la courbure scalaire. *J. Math. Pures et Appl.*, 55:269–293, 1976.

[BC81]   H. Berestycki and T. Cazenave. Instabilitè del etats stationnaires dans des equations de Schrödinger et de Klein-Gordon non linéaires. *C.R.A.S. Paris, série I*, 293:488–492, 1981.

[BC87]   V. Benci and G. Cerami. Positive solutions of some nonlinear elliptic problems in exterior domains. *Arc. Rational Mech. Anal.*, 99:283–300, 1987.

[BC88]   A. Bahri and J.M. Coron. On a nonlinear elliptic equation involving the Sobolev exponent: the effect of the topology of the domain. *Comm. Pure Appl. Math.*, 41:253–294, 1988.

[BC89]   V. Benci and G. Cerami. Existence of positive solutions of the equation $-\Delta u + a(x)u = u^{\frac{N+2}{N-2}}$ in $\mathbb{R}^N$. *J. Funct. Anal.*, 88, 1:90–117, 1989.

[BC91]   V. Benci and G. Cerami. The effect of the domain topology on the number of positive solutions of nonlinear elliptic problems. *Arch. Ration. Mech. Anal.*, 114:79–93, 1991.

[BCP91]   V. Benci, G. Cerami, and D. Passaseo. On the number of the positive solutions of some nonlinear elliptic problems. *Nonlinear Analysis- A tribute in honour of G. Prodi, Quaderno S.N.S., Pisa*, pages 93–107, 1991.

[Ber72]   M.S. Berger. On the existence and structure of stationary states for a nonlinear Klein-Gordon equation. *Jour. Funct. Analysis*, 9:249–261, 1972.

[BF]   V. Benci and D. Fortunato. Solitons and particles. Proc. "Int. Conf. on Nonlinear Diff. Equat. and Appl." Tata Inst.Fund.. Res., Bangalore. To appear.

[BFMP99] V. Benci, D. Fortunato, A. Masiello, and L. Pisani. Solitons and electromagnetic field. *Math. Z., to appear*, 1999.

[BFP96]   V. Benci, D. Fortunato, and L. Pisani. Remarks on topological solitons. *Topological Methods in nonlinear Analysis*, 7:349–367, 1996.

[BFP98]   V. Benci, D. Fortunato, and L. Pisani. Soliton like solutions of a Lorentz invariant equation in dimension 3. *Rev. Math. Phys.*, 10:315–344, 1998.

[BL78]   H. Berestycki and P.L. Lions. Existence d'ondes solitaires dans des problems non linéaires du type Klein-Gordon. *C.R.A.S. Paris, série A*, 287:503–506, 1978.

[BL83a]   H. Berestycki and P.L. Lions. Nonlinear scalar field equations I. Existence of a ground state. *Arch. Ration. Mech. Anal.*, 82:313–346, 1983.

[BL83b]   H. Brezis and E. Lieb. A relation between pointwise convergence of functions and convergence of functionals. *Proc. Amer. Math. Soc.*, 88:486–490, 1983.

[BN83]   H. Brezis and L. Nirenberg. Positive solutions of nonlinear elliptic equations involving critical Sobolev exponents. *Comm. Pure Appl. Math.*, 36:437–477, 1983.

[Böh72]   R. Böhme. Die lösung der verzweigungsgleichungen für nichtlineare eigenwertprobleme. *Math. Z.*, 127:105–126, 1972.

[Bre86]   H. Brezis. Elliptic equations with limiting Sobolev exponents. The impact of topology. *Proceedings 50th Anniv. Courant Inst., Comm. Pure Appl. Math.*, 39:517–539, 1986.

[CFS84]   G. Cerami, D. Fortunato, and M. Struwe. Bifurcation and multiplicity results for nonlinear elliptic problems involving critical Sobolev exponents. *Ann. Inst. H. Poincaré - Anal. Non Linéaire*, 1:341–350, 1984.

[CH62]   R. Courant and D. Hilbert. *Methods of mathematical physics.* Intersciences, 1962.

[CL95]   A. Candela and M. Lazzo. Positive solutions for a mixed boundary value problem. *J. Nonlinear Analysis*, 24:1109–1117, 1995.

[Cor84]   J.M. Coron. Topologie et cas limite des injections de Sobolev. *C.R. Acad. Sci. Paris*, Sér I, 299:209–212, 1984.

[CP96]   G. Cerami and D. Passaseo. High energy positive solutions for mixed and Neumann elliptic problems with critical nonlinearity. *J. Analyse Math.*, 70:1–39, 1996.

[CP97]   G. Cerami and D. Passaseo. Nonminimizing positive solutions for equations with critical exponent in the half space. *SIAM J. Math. Anal.*, 28:867–885, 1997.

[CSS86]   G. Cerami, S. Solimini, and M. Struwe. Some existence results for superlinear elliptic boundary value problems involving critical exponents. *J. Funct. Anal.*, 69:289–306, 1986.

[Dan79]   E.N. Dancer. On radially symmetric bifurcation. *J.London Math. Soc.*, 20:287–292, 1979.

[Dan83]   E.N. Dancer. On the indices of fixed points in cones and applications. *J. Math. Anal. and Applications*, 91:131–151, 1983.

[Dan84]   E.N. Dancer. On positive solutions of some pairs of differential equations. *Trans. Amer. Math. Soc.*, 284:729–743, 1984.

[Dan85]   E.N. Dancer. On positive solutions of some pairs of differential equations ii. *J.Diff. Equations*, 59:236–258, 1985.

[Dan86]   E.N. Dancer. Multiple fixed points of positive maps. *J. Reine Ang. Math.*, 371:46–66, 1986.

[Dan88]   E.N. Dancer. A note on an equation with critical exponent. *Bull. London Math. Soc.*, 20:600–602, 1988.

[Dan91]   E.N. Dancer. On the existence and uniqness of positive solutions for competing species models with diffusion. *Trans. Amer. Math. Soc.*, 326:829–859, 1991.

[Dan92]   E.N. Dancer. Global breaking of symmetry of positive solutions on two dimensional annuli. *Diff. and Int. Equations*, 5:903–914, 1992.

[DD94a]   E.N. Dancer and Y. Du. Competing species equations with diffusion, large interactions and jumping nonlinearities. *Journal of Differential Equations*, 114-2:434–75, 1994.

[DD94b]   E.N. Dancer and Y. Du. Existing of changing sign solutions for some semilinear problems with jumping nonlinearities at zero. *Proceedings Royal Soc. Ed.*, 124-A:1165–1176, 1994.

[DD95]   E.N. Dancer and Y. Du. Existence of positive solution for a 3 species competitive system with diffusion I. *Nonlinear Analysis*, 24:337–357, 1995.

[DEGM82]   R.K. Dodd, J.C. Eilbeck, J.D. Gibbon, and H.C. Morris. *Solitons and Nonlinear Wave Equations*. Academic Press, London, New York, 1982.

[Dei85]   K. Deimling. *Nonlinear functional analysis*. Springer-Verlag, Berlin, 1985.

[DG94]   E.N. Dancer and Z. Guo. Uniqueness and stability for solutions of competing species with large interactions. *Comm. Applied Nonlinear Anal.*, 1:19–45, 1994.

[DG95]   E.N. Dancer and Z. Guo. Some remarks on stable sign changing solutions. *Tohoku Math.*, 47:199–225, 1995.

[DH91]   E.N. Dancer and P. Hess. Stability of fixed points for ordered-preserving discrete-time dynamical systems. *J. Reine Ang. Math.*, 419:125–139, 1991.

[Din89]   W. Ding. Positive solutions of $\Delta u + u^{(n+2)/(n-2)} = 0$ on contractible domains. *J. Partial Differential Equations*, 2:83–88, 1989.

[DNF85]   B. Doubrovine, S. Novikov, and A. Fomenko. *Geometrie et Topologie des Variétés*, volume 2 partie. Editions Mir , Moscou, 1985.

[EL88]   M.J. Esteban and P.L. Lions. Skyrmions and symmetry. *Asymptotic Anal.*, 1:187–192, 1988.

[EPT89]   H. Egnell, F. Pacella, and M.R. Tricarico. Some remarks on Sobolev inequalities. *Nonlinear Analysis T.M.A.*, 13:671–681, 1989.

[Est86]   M.J. Esteban. A direct variational approach to skyrme's model for meson fields. *Comm. Math. Phys.*, 105:571–591, 1986.

[Est91]   M.J. Esteban. Nonsymmetric ground states of symmetric variational problems. *Comm. Pure Appl. Math.*, 44:259–274, 1991.

[Fad85]   E.N. Fadell. Lectures in cohomological index theories of g–spaces, with applications to critical points theory. *Rend. Sem. Dip. Mat. Univ. Calabria*, 6, 1985.

[FN73]   S. Fučik and J. Nečas. *Spectral analysis of nonlinear operators*. Springer-Verlag, Berlin, 1973.

[GF63]   I.M. Gelfand and S.V. Fomin. *Calculus of Variations*. Prentice-Hall, Englewood Cliffs, N.J., 1963.

[Gro95]   M. Grossi. A class of solutions for the Neumann problem $-\Delta u + \lambda u = u^{\frac{N+2}{N-2}}$. *Duke Math. J.*, 79:309–334, 1995.

[GT83]   D. Gilbarg and N.S. Trudinger. *Elliptic partial differential equations of second order*. Springer-Verlag, Berlin, 1983.

[Kic96]   S. Kichenassamy. *Non linear wave equations*. Marcel Dekker Inc., New York, Bael, Hong Kong, 1996.

[KS70]   E.F. Keller and L.A. Segel. Initiation of slime model aggregation viewed as an instability. *J. Theor. Biol.*, 26:399–415, 1970.

[KW75]   J. Kazdan and F. Warner. Remarks on some quasilinear elliptic equations. *Comm. Pure Appl. Math.*, 28:567–597, 1975.

[KZ84]   M.A. Krasnoselškii and P.P. Zabrieko. *Geometric Methods of Nonlinear Analysis*. Springer-Verlag, Berlin, 1984.

[Laz92]   M. Lazzo. Multiple positive solutions of nonlinear elliptic equations involving critical Sobolev exponents. *C. R. Acad. Sci. Paris*, Sér. I Math.:61–64, 1992.

[Lin89]   S.S. Lin. On non radially symmetric bifurcation in the annulus. *J. Diff. Equations*, 80:348–367, 1989.

[Lio85]   P.L. Lions. The concentration-compactness principle in the calculus of variations: The Limit case I - II. *Revista Mat. Ibero-Americana*, 11, 12:145–200; 45–121, 1985.

[LL70]   L. Landau and E. Lifchitz. *Théorie du Champ*. Éditions MIR, Moscou, 1970.

[Llo78]   N. Lloyd. *Degree Theory*. Cambridge University Press, Cambridge, 1978.

[LN88]   C.S. Lin and W.M. Ni. *On the diffusion coefficient of a semilinear Neumann problem*, volume 72 of *Lecture Notes in Math.* Springer-Verlag, 1988.

[LNT88]   C.S. Lin, W.M. Ni, and I. Takagi. Large amplitude stationary solutions to a chemotaxis system. *J. Diff. Equations*, 72:1–27, 1988.

[Mar73]   A. Marino. La biforcazione nel caso variazionale. *Confer. Sem. Mat. Univ. Bari*, 132, 1973.

[MM92]   G. Mancini and R. Musina. The role of the boundary in some semilinear Neumann problems. *Rend. Sem. Mat. Univ. Padova*, 88:127–138, 1992.

[Nar68]   R. Narasimhan. *Analysis on real and complex manifolds*. North Holland, Amsterdam, 1968.

[Ni83]   W.M. Ni. On the positive radial solutions of some semilinear elliptic equations on $I\!R^N$. *Appl. Math. Optim.*, 9:373–380, 1983.

[Nir74]   L. Nirenberg. *Topics in nonlinear functional analysis*. Courant Institute, Lecture notes New York University, 1974.

[NO]   W.M. Ni and Y.G. Oh. Construction of single boundary-peak solutions to a semilinear Neumann problem. In preparation.

[NPT92]   W.M. Ni, X.B. Pan, and I. Takagi. Singular behaviour of least energy solutions of a semilinear Neumann problem involving critical Sobolev exponents. *Duke Math. J.*, 67:1–20, 1992.

[NT91]   W.M. Ni and I. Takagi. On the shape of least energy solutions to a semilinear Neumann problem. *Comm. Pure Appl. Math.*, 45:819–851, 1991.

[NT92]   W.M. Ni and I. Takagi. *On the existence and shape of solutions to a semilinear Neumann problem. Progress in Nonlinear Diff. Equa.* Lloyd, Ni, Peletier, Serrin eds, 1992.

[Nus71]   R. Nussbaum. The fixed point index for locally condensing maps. *Ann. Mat. Pura Appl.*, 89:217–258, 1971.

[Pas]   D. Passaseo. Multiple solutions of elliptic problems with supercritical growth in contractible domains. To appear.

[Pas89]   D. Passaseo. Multiplicity of positive solutions of nonlinear elliptic equations with critical Sobolev exponent in some contractible domains. *Manuscripta Math.*, 65:147–166, 1989.

[Pas90]   D. Passaseo. Esistenza e molteplicità di soluzioni positive per l'equazione $-\Delta u + (\alpha(x) + \lambda)u = u^{\frac{n+2}{n-2}}$ in $\mathbf{R}^n$. *Pubbl. Dip. Mat. Univ. Pisa*, 1990.

[Pas92a]   D. Passaseo. Esistenza e molteplicità di soluzioni positive per equazioni ellittiche con nonlinearità supercritica in aperti contrattili. *Rend. Accad. Naz. Sci. detta dei XL, Memorie di Mat. 110*, XVI, n. 6:77–98, 1992.

[Pas92b]   D. Passaseo. Su alcune successioni di soluzioni positive di problemi ellittici con esponente critico. *Rend. Mat. Acc. Lincei s. 9*, 3:15–21, 1992.

[Pas93a]   D. Passaseo. Multiplicity of positive solutions for the equation $\Delta u + \lambda u + u^{2^*-1} = 0$ in non contractible domains. *Top. Meth. in Nonlin. Anal.*, 2:343–366, 1993.

[Pas93b]   D. Passaseo. Nonexistence results for elliptic problems with supercritical nonlinearity in nontrivial domains. *J. Funct. Anal.*, 1:97–105, 1993.

[Pas94]   D. Passaseo. The effect of the domain shape on the existence of positive solutions of the equation $\Delta u + u^{2^*-1} = 0$. *Top. Meth. in Nonlin. Anal.*, 3:27–54, 1994.

[Pas95]   D. Passaseo. New nonexistence results for elliptic equations with supercritical nonlinearity. *Diff. Int. Equations*, 8, n. 3:577–586, 1995.

[Pas96a]   D. Passaseo. Multiplicity of nodal solutions for elliptic equations with supercritical exponent in contractible domains. *Top. Meth. in Nonlin. Anal.*, 8:245–262, 1996.

[Pas96b]   D. Passaseo. Some sufficient conditions for the existence of positive solutions to the equation $-\Delta u + a(x)u = u^{2^*-1}$ in bounded domains. *Ann. Inst. H. Poincaré*, 13, no 2:185–227, 1996.

[Pas98a]   D. Passaseo. Nontrivial solutions of elliptic equations with supercritical exponent in contractible domains. *Duke Math. J.*, 92, no.2:429–457, 1998.

[Pas98b]   D. Passaseo. Relative category and multiplicity of positive solutions for the equation $\Delta u + u^{2^*-1} = 0$. *J. Nonlinear Anal. T.M.A.*, 33:509–517, 1998.

[Poh65]   S. Pohozaev. Eigenfunctions of the equation $\Delta u + \lambda f(u) = 0$. *Soviet Math. Dokl.*, 6:1408–1411, 1965.

[Rab71]   P.H. Rabinowitz. Some global results for nonlinear eigenvalue problems. *J. Functional Anal.*, 7:487–513, 1971.

[Rab73]   P. Rabinowitz. Some aspects of nonlinear eigenvalues problems. *Rocky Mount. J. Math.*, 3:161–202, 1973.

[Raj88]   R. Rajaraman. *Solitons and instantons*. North Holland, Amsterdam, Oxford, New York, Tokio, 1988.

[RCL92]   A. Carpio Rodriguez, M. Comte, and R. Levandowski. A nonexistence result for a nonlinear equation involving critical Sobolev exponent. *Ann. Inst. H. Poincaré, Analyse Non Linéaire*, 9 no. 3:243–261, 1992.

[Rey89a]   O. Rey. A multiplicity result for a variational problem with lack of compactness. *J. Nonlinear Anal. T.M.A.*, 133:1241–1249, 1989.

[Rey89b]   O. Rey. Sur un problème variationnel non compact: l'effet de petits trous dans le domaine. *C.R. Acad. Sci. Paris, t. 308*, Série I:349–352, 1989.

[Rey90]   O. Rey. The role of the Green's function in a nonlinear elliptic equation involving the critical Sobolev exponent. *J. Funct. Anal.*, 89:1–52, 1990.

[Sch69]   J.T. Schwartz. *Nonlinear functional analysis*. Gordon & Breach, New York, 1969.

[Sch71]   M. Schechter. *Principles of Functional Analysis*. Academic Press, New York, 1971.

[Sch84]   R. Schoen. Conformal deformation of a riemannian metric to a constant scalar curvature. *J. Differential Geom.*, 20:479–495, 1984.

[Sch85]   R. Schaaf. Stationary solutions of chemotaxis systems. *Trans. Amer. Math. Soc.*, 292:531–566, 1985.

[Sky61]   T.H.R. Skyrme. A non-linear field theory. *Proc. Roy. Soc.*, A260:127–138, 1961.

[Spr88]   J. Spruck. *The elliptic sinh Gordon equation and the construction of toroidal soap bubbles*, volume Lectures Notes in Mathematics, Vol. 1340. Springer-Verlag, Berlin, 1988.

[Str]   S. Strocchi. Symmetry breking in classical dynamical systems. Preprint.

[Str77]   W.A. Strauss. Existence of solitary waves in higher dimensions. *Comm. Math. Phys.*, 55:149–162, 1977.

[Str83]   W.A. Strauss. Stable and unstable states of nonlinear wave equations. *Contemporary Math.*, 17:429–441, 1983.

[Str84]   M. Struwe. A global compactness result for elliptic boundary value problems involving limiting nonlinearities. *Math. Z.*, 187:511–517, 1984.

[Str90]   M. Struwe. *Variational methods: Application to nonlinear PDE and Hamiltonian systems*. Springer-Verlag, Berlin, 1990.

[Tal76]   G. Talenti. Best constants in Sobolev inequality. *Ann. Mat. Pura Appl.*, 110:353–372, 1976.

[Tru68]   N. Trudinger. Remarks concerning the conformal deformations of riemannian structures on compact manifolds. *Ann. Sc. Norm. Sup., Pisa*, 22:265–274, 1968.

[Wana]   Z.Q. Wang. The effect of the domain geometry on the number of positive solutions of Neumann problems with critical exponents. Preprint.

[Wanb]   Z.Q. Wang. On the existence and qualitative properties of solutions for a nonlinear Neumann problem with critical exponent. Preprint.

[Wanc]   Z.Q. Wang. Remarks on a nonlinear Neumann problem with critical exponent. Preprint.

[Wan91]   X.J. Wang. Neumann problems of semilinear elliptic equations involving critical Sobolev exponents. *J. Diff. Equations*, 93:283–310, 1991.

[Wan92a]   Z.Q. Wang. On the existence of multiple, single-peaked solutions of a semilinear Neumann problem. *Arch. Ration. Mech. Anal.*, 120:375–399, 1992.

[Wan92b]   Z.Q. Wang. On the existence of positive solutions for semilinear Neumann problems in exterior domains. *Comm. in P.D.E.*, 17:1309–1325, 1992.

[Wit74]   G.B. Witham. *Linear and nonlinear waves*. John Wiley and Sons, New York, 1974.

[Yam60]   H. Yamabe. On a deformation of riemannian structures on compact manifolds. *Osaka Math. J.*, 12:21–37, 1960.

Index

Printing: Weihert-Druck GmbH, Darmstadt
Binding: Buchbinderei Schäffer, Grünstadt